How Is
Quantum Field Theory
Possible?

D1251275

How Is
Quantum Field Theory
Possible?

Sunny Y. Auyang

New York   Oxford
OXFORD UNIVERSITY PRESS
1995

Oxford University Press

Oxford   New York
Athens   Auckland   Bangkok   Bombay
Calcutta   Cape Town   Dar es Salaam   Delhi
Florence   Hong Kong   Istanbul   Karachi
Kuala Lumpar   Madras   Madrid   Melbourne
Maxico City   Nairobi   Paris   Singapore
Taipei   Tokyo   Toronto

and associated companies in
Berlin   Ibadan

Copyright © 1995 by Oxford University Press, Inc.

Published by Oxford University Press, Inc.,
198 Madison Avenue, New York, New York 10016-4314

Oxford is a registered trademark of Oxford University Press

All rights reserved. No part of this publication may be reproduced,
stored in a retrieval system, or transmitted in any form or by any means,
electronic, mechanical, photocopying, recording, or otherwise,
without the prior permission of Oxford University Press.

Library of Congress Cataloging-in-Publication Data
Auyang, Sunny Y.
How is quantum field theory possible? / Sunny Y. Auyang.
p.   cm.   Includes bibliographical references and index.
ISBN (invalid) 0-19-509344-5 — ISBN 0-19-509345-3 (pbk.)
1. Quantum field theory.   2. Physics—Philosophy.
3. Space and time. I. Title.
QC174.45.A88   1995
530.1'43—dc20   95-8562

5 7 9 8 6 4

Printed in the United States of America
on acid-free paper

上獻雙親

To my parents

CONTENTS

How Is
Quantum Field Theory
Possible?

1

Introduction

> The whole of science is nothing more than a refinement of everyday thinking. It is for this reason that the critical thinking of the physicist cannot possibly be restricted to the examination of the concept of his own specific field. He cannot proceed without considering critically a much more difficult problem, the problem of analyzing the nature of everyday thinking.
>
> ALBERT EINSTEIN
> "Physics and Reality"

§ 1. A Human View of the World in Quantum Field Theory

Quantum field theory is the union of quantum mechanics and special relativity. It is our most fundamental physical theory and provides the conceptual framework for the twentieth century's answers to questions about the basic structure of the physical world. However, its philosophical significance has been largely ignored, although relativity and quantum mechanics have separately stimulated much reflection.[1] Consequently, the intimate relations among the meanings of object, experience, space-time, and interaction, which are revealed only in the unified structure of quantum field theory, remain in obscurity. The present work fills the gap in philosophy.

In 1900, Max Planck postulated that the energy spectrum of blackbody radiation is quantized. Five years later, Albert Einstein introduced the notion of light quanta. Physicists realized that quantum ideas are essential to the understanding of atomic structures; classical physics is inadequate. It took two decades of intensive research and the fresh minds of a new generation that included Werner Heisenberg, Erwin Schrödinger, and Paul Dirac to develop nonrelativistic quantum mechanics. Quantum mechanics quickly became the basis of a large part of physics including atomic, molecular, and solid-state phenomena. However, in nuclear and high-energy physics, it is unsatisfactory because it is incompatible with the principle of special relativity advanced by Einstein in 1905.

Dirac founded quantum field theory by uniting quantum mechanics and special relativity. Quantum field theory met with many difficulties that were not solved until after the Second World War when a third generation of physicists from around the world established a satisfactory quantum field theory of the electromagnetic interaction. The extension of the theory to cover the nuclear interactions took another 25 years. During this process, gauge fields or fields with local symmetries, the idea of which first appeared in the general theory of relativity, became dominant. That research effort has yielded a bountiful harvest. Quantum field theory underlies the current standard model in elementary particle physics, through which our intellectual eyes behold both the microscopic structure of matter and the cosmic events occurring within split seconds of the Big Bang.

Despite the success of quantum theories, their philosophical interpretations have always been difficult and controversial. Richard Feynman said: "We always have had a great deal of difficulty in understanding the world view that quantum mechanics represents."[2] Physicists do understand the quantum realm to a significant extent. Their understanding is manifested in the successful application of quantum theories to real-world problems. However, it is poorly articulated. The actions and writings of physicists tacitly uphold a world view rooted in practice and robust common sense. Physicists use subatomic particles as physical tools in laboratory experiments and components in commercial devices; the electronic industry is mainly based on the manipulation of the quantum world. In both technical papers and research proposals, physicists unambiguously assert that they are studying the microscopic structure of matter and the origin of the universe. Thus they uphold a commonsensically realistic view of the quantum world. The philosophical difficulty lies in articulating this world view and defending it against phenomenalist challenges. Such an articulation and defense are the aim of the present work.

The world-view difficulty has several sources. The world described by quantum theories is remote to sensory experiences and is totally different from the familiar classical world. There are scientific puzzles, such as what happens during the process of measurement. Quantum theories disallow certain questions that we habitually ask about physical things, for example, the moment when a radioactive atom decays. The logical structures of quantum theories are more complicated and contain more elements than those of classical mechanics. The working interpretation of quantum theories, which physicists use in practice, invokes the concept of observed results as distinct from physical states. The concept with explicit experiential connotation is not found in classical mechanics and causes interpretive problems. Finally, logical positivism, which dismisses the notion of reality as meaningless metaphysics, casts a long shadow. These factors prompt many interpreters to adopt the phenomenalist position asserting that quantum objects have no definite property, the reality of the microscopic world evaporates, the observer creates what he observes, and what the theories say about the quantum realm are not to be believed for they are fictions that help to predict the behaviors of classical experimental equipments.[3]

The phenomenalist doctrines are espoused by the Copenhagen school that dominates quantum interpretation and are academically influential. However, they conflict with the day-to-day operation of physicists. They would make a scandal out of physicists' requests for public funding of fundamental research. If the microscopic world were fictitious, then billion-dollar accelerators would be monstrous toys that served no purpose except to have their own predicted behaviors verified.

Many physicists find that the chief philosophical issues concern the reality of the microscopic world, the objectivity of quantum theories, and the relation of the world to our experiences. Einstein said: "In the center of the problematic situation I see not so much the question of causality but the question of reality (in a physical sense)." Roger Penrose said: "Like Einstein and his hidden-variable followers, I believe strongly that it is the purpose of physics to provide an *objective* description of reality." Eugene Wigner said: "The principal difficulty is, rather, that it [quantum mechanics] elevates the measurement, that is the observation of a quantity, to the basic concept of the theory."[4]

Wigner's remark shows that the philosophical issue remains if we simply agree that the terms physicists usually deem physically significant are descriptive of quantum properties. Quantum theories include irreducible terms about observed or measured results. These experiential terms threaten objectivity because experiences are intrinsically subjective. Everyone agrees that physics is an empirical science, the validity of whose theories is based on observations. The questions are: Should the empirical aspect be stated explicitly within physical theories, or should it be understood implicitly? Would its incorporation into the theories undermine their objectivity? What does objectivity mean? If we maintain that in physics objectivity should take the strong sense of pertaining to objects, as distinct from mere intersubjective agreement, what do objects mean? What does reality mean?

Ready answers such as objects are given, self-evident, observable, or capable of stimulating our senses will not do, for none is satisfied by the topics of most branches of modern natural science, including physics. Some philosophers say that microscopic entities are not real but can be treated "as if" they are real. The distinction between real and as-if real is based on the direct observability of the entities. However, it is presumptuous to make the limitation of human sense organs into the criterion of reality. Some people try to depict and explain the strange in terms of the familiar. Such analogies lead only to paradoxes because quantum phenomena are radically different. Other people reject quantum objects because they are different, but all their arguement shows is that there is nothing like classical objects in the quantum realm, not that there is no quantum object.

Are there quantum objects? What are they if they are neither given nor resemble anything familiar? To answer these questions we have to abstract from the substantive features of familiar things, delineate the pure logical forms by which we acknowledge objects, and show how the forms are fulfilled in quantum theories. We have to explicate, in general terms and without resorting to the given, what we mean by *objects*. The clear criteria will enable us to affirm the objectivity of quantum theories.

Unfortunately, the explication is not easy. Immanuel Kant had shown that the general concept of objects is a complicated conceptual framework that encompasses the concepts of space and time, the categories ranging from possibility to causality, and most fundamentally, the relation between things and experiences. Its analysis takes up the bulk of the constructive part of the *Critique of Pure Reason*. Kant argued that the concept of objects is the most basic presupposition of empirical knowledge and that it is easily abused. When applied illegitimately, it leads to metaphysical illusions such as God and the soul.

A proper interpretation of quantum theories calls for a general concept of the objective world that adequately accommodates both classical and quantum objects. Thus it demands a critical reexamination of the presuppositions of empirical knowledge and a clarification of the philosophical categories in view of modern physics. It has to delimit the applicability of the categorical framework and criticize illusions such as the observer or the container space-time, which are common in existing interpretations of physical theories. This is a considerable philosophical inquiry. Being a physicist, I would not dare to embark on it if I did not anticipate help from quantum field theory itself.

Presuppositions of knowledge are not stipulations issued from philosophers' armchairs. They are already embedded in all our experiences, even the most common ones, and especially the common ones, for it is in our mundane activities that we first recognize objects. General concepts such as that of objects and properties are what we tacitly understand and automatically use in our daily coping in the world. They make the world intelligible in the first place. Everyone knows them, but to use the words of Gottfried Wilhelm Leibniz, our knowledge is not distinct but confused. Saint Augustine's remark on time, "if no one asks me, I know what it is. If I wish to explain it to someone who asks, I do not know," applies to most general concepts.[5] They lie deeply in the logical structure of our everyday thinking and ordinary language. Philosophers have tried to delineate them, but the task is difficult due to the richness, vagueness, and fluidity of everyday thinking.

Scientific theories are more restrictive in scope and more clearly formulated. However, many do not explicitly incorporate the basic presuppositions of knowledge, for they usually treat their specific topics assuming the common-sense background. They have clear substantive concepts for their topics but rely on everyday understanding for general concepts. Hence their conceptual structures are incomplete. This is the case with Newtonian mechanics. Kant argued from the reflexivity of common experiences that objective knowledge must have a minimum conceptual structure that includes the distinction between the object and the experiences of it. The distinction detaches the object and lets experience know itself as the experience of the object. It is one of Kant's two "highest principles" of empirical knowledge. However, when Kant turned to investigate the structure of Newtonian mechanics, he did not invoke the object–experience distinction.[6] Classical objects are close to the things we handle every day, so that Newtonian mechanics can call upon common sense, perhaps with the help of a little idealization, to distinguish its topics. Mechanics, up to a century after Kant, had taken the notion of objects for granted; it embodies the specific categories but not the deep conceptual structure presupposed by them.

People had a fairly clear notion of planets before the advent of Newtonian mechanics. The same cannot be said of quarks; no one had dreamed of them in the absence of quantum mechanics. Microscopic objects are present to us only through elaborate theories and controlled experiments designed with the help of the theories. They are not obvious at all. We can no longer count on intuitive cognition. Confused knowledge falters in strenuous situations, leading to the doctrines that abandon the reality of the microscopic world.

We can regard the situation in a more positive way: The quantum world goads us to make distinct what we confusedly know. When the topic of a science is so counterintuitive that some tacit presuppositions of knowledge become problematic, we expect scientific theories to address them expressly, to clarify them or reject them. The rationalization of what was instinctive, the explicit assertion of what was implicitly assumed, and the internalization of what was externally granted are no less the advancement of science than the exploration of uncharted territories.

When common sense fails to provide the service it renders to Newtonian mechanics, modern physical theories must take the job upon themselves. They must assume some burden in delineating their topics, and to achieve this they need to state the criterion of objectivity. Furthermore, the quantum world rudely reminds us that observability depends on human conditions that nature need not honor. Thus the theories have to make explicit their intrinsic empirical nature. In short, if quantum theories have conveyed objective knowledge, then they will have distilled the core of the conceptual structure of common sense, formulated it rigorously, incorporated it, and put it to work to overcome the intangibility of their topics. Consequently their conceptual structure would be richer than that of Newtonian mechanics. The enriched structure, which includes elements that explicitly account for the presuppositions of knowledge, would contain the answer to the question about the general meaning of the objective world. Since physical theories are mathematical, their conceptual structures are more clearly exhibited. This greatly helps the philosophical task of uncovering presuppositions.

Until now almost all philosophical investigations of quantum theories have either taken the concept of objectivity for granted or prescribed it as some external criterion, according to which the theories are judged. The judgments often deny the objectivity or even the possibility of microscopic knowledge. I adopt the opposite approach. I start with the premise that quantum field theory conveys knowledge of the microscopic world and regard the general meaning of objects as a question whose answer lies within the theory. This work asks quantum field theory to demonstrate its own objectivity by extracting and articulating the general concept of objects it embodies. We try to learn from it, not only the specifics of elementary particles, but also the general nature of the world and our status in it. What general conditions hold for us and the world we are in so that objects, classical and quantum, which are knowable through observations and experiments, constitute reality? How is knowledge of the quantum world possible? These are part of what Kant asked: How is empirical knowledge in general possible? The title question means: What are the presuppositions and preconditions of quantum field theory, through which we come to know the microscopic world?

The Kantian question suggests that the answer should be sought not externally but internally to the theory by surveying its general conceptual structures and bounds. Quantum field theory still needs the help of common sense to function; we use common sense to apply the theory and recognize experimental data. However, the supplement occurs at the periphery, which is essentially classical. The gist of the theory, whose topics are microscopic and quantum objects, is relatively self-consistent and self-sufficient in its conceptual structure to warrant an internal philosophical investigation.

When the philosophical question is thus turned around, the complicated conceptual structure of quantum field theory appears in a new light. The extra elements absent in classical mechanics are no longer nuisances to be purged but become valuable guides to a world view much more human and reasonable than the popular view promoted in scientism. Scientific theories do not merely depict the world; they depict it in ways comprehensible to us. The exacting condition of the quantum world brings out the power and limitation of human capabilities. Like a man who summons his strength to meet tough challenges and comes to know himself, quantum field theory illuminates not only elementary particles but also ourselves and our being in the world. Its explicit empirical terms indicate a situated view from within the world instead of an absolute view from without. Its conceptual framework is complicated because it takes account of the basic human conditions neglected in the absolute view: the finitude and reflexivity of our knowledge and observation. It reminds us that being objective is not being absolutely unconditional. Science does not and cannot put us in the position of God.

Consider two descriptions of the same phenomenon: (1) Space is homogeneous and isotropic. (2) Space is invariant under translations and rotations of coordinates. Statements in the logical form of (1) were exclusive in pre-twentieth century physics. Statements in the form of (2) dominate twentieth century physics; quantum mechanics contains various representations of the same physical state and rules for transforming among them. Description (1) appears clean; it describes nature without explicit conventional and experiential notions. Description (2) is all contaminated; it invokes conventional coordinates and intellectual transformations of the coordinates. The coordinates are usually interpreted as perspectives of observations; so they are somehow related to human subjects. However, physicists agree that (2) is more objective, for it uncovers the hidden presuppositions of (1) and neutralizes their undesirable effects. They retrofit the conceptual structure embodied in (2) into classical mechanics to make it more satisfactory.

The statement (1) is often interpreted in a framework of *things*; (2) can be interpreted in the framework of *objects.* The object framework includes the thing framework as a substructure and further conveys the epistemological idea that the things are knowable through observations and yet independent of observations. The two frameworks exemplify two different views of the world.

The framework of things is austere and clean cut, and claims to take a completely external and unbiased view of the world. It is widely used in the philosophy of science. It contains only notions of things, their properties and relations; it excludes notions of experiences, viewpoints, and conventions, and it does not allow the possibility of errors and doubts. Thus it conveys the idea that things are absolute, given, and self-evident. The givenness marks certainty and reality.

The logical structure of the framework of things is not different from that of a phenomenal framework of sense impressions. This point was argued by George Berkeley and contributes to the phenomenalist interpretations of quantum theories. "The rose is red" and "the sense impression is red" are formally the same; the two sentences are indistinguishable if written in symbolic forms. Neither framework permits the idea that things and appearances are categorically different. Theories in the two frameworks differ solely in their vocabularies and substantive terms. What counts as a thing or a sense datum, and thus what is admissible to the thing or sense datum vocabulary, is a decision that cannot be made within the theories in either framework. Some people say theories are mirrors of nature; others say the mind is a blank slate upon which sense impressions write. The metaphors illustrate the similarities of the two frameworks: the passivity of the theories or the mind and the givenness of what is reflected or written.

The most common criterion for acknowledging a thing is its empirical evidence. However, just as the notion of things is forbidden in the phenomenal framework, the notion of empirical evidence is extrinsic to the framework of things. Experiences are not mere things; they are intelligible. The framework of things forces interpreters to regard the empirical terms in quantum theories as representatives of things that are not quite physical. The interpretation makes the observer into a peculiar thing and divides the world into the observer and the observed. The result is the illusion of a substantive consciousness.

The framework of things has no notion of conventions. This does not mean that thing theories contain no conventions; it means that even if conventions are present, a thing theory has no means to diagnose and acknowledge them as such. "The sun rises, the sun sets" is perfectly legitimate within a thing theory, which has no means to discern its possible anthropocentric connotation. Consequently, thing theories tend to say too much; unable to disengage descriptions from our idiosyncrasies in making the descriptions, they attribute the whole bundle to things. A thing theory is a convention that is not aware of its own conventionality.

We do not appreciate the restrictiveness and deficiency of the framework of things because we automatically supplement thing theories by common sense, which has a much broader conceptual structure. For instance, when asked about the empirical aspect of things, we say, of course, they are observable. The saying issues from common sense, not from thing theories. With common sense we assess empirical data, discern conventions, and criticize theories. The framework of things is successful in many sciences with the help of common sense.

Some philosophies scorn common sense, throw away the supplement, codify the framework of things or its logical equivalent, the phenomenal framework, and prescribe it as the only "scientific" framework for our thinking about the world. Such philosophies are a source of the intellectual milieu of scientism. They also lead to the backlash of conventionism. A thing theory has no ground to give in the face of a theoretical change, for its intrinsic lack of the notion of conventions means it is absolute. It is all or nothing. As new theories bring new vocabularies of things, entire theories and vocabularies are pronounced to be conventions. Consequently it is now fashionable to say that the supposedly unbiased view of science is merely a deeply prejudiced view from a familiar standpoint, scientific theories are social fabrications, and physicists construct quarks.

Conventionism has a point; the God's-eye view of the philosophies that enshrine the thing framework is illusory. Human beings are radically in the world, so that human views are situated and finite. We cannot view from nowhere but can only view from a particular physical and intellectual standpoint and then try to abstract from it. However, conventionism has overshot the mark in claiming that various scientific theories are incommensurate; not even our everyday theories about the world are so fixed and disjoint. We are finite beings who are aware of their own finitude, active beings who are concerned with the consequences of their activity. This self-awareness induces efforts to abstract from the specific conventions we have adopted. Scientism is blind to the finitude and activity; conventionism rightly points out the finitude but wrongly overlooks the intrinsic self-awareness of the activity. We know, although often vaguely, that our specific viewpoints are conventions and are irrelevant to the objective states of affairs, which somehow unify all conventions. That is how we can talk to aliens. The part of common sense that includes the self-awareness that separates objective states of affairs from their conventional representations embodies the object framework.

An object theory incorporates appropriate thing theories as conventional representations. It regards the conventions as conditions of knowledge, including its empirical aspect, and pronounces objective statements by abstracting from these conditions. "Objects are independent of the mind" is different from "things are"; "to be is distinct from to be perceived" is different from "being." "Being" and "things are" say nothing about us, but they have to appeal to the self-evidence of things to answer how we come to know them. Mind independence says something about knowable objects; to articulate objects we must delineate, in general terms, both the conditions of cognition and the independence thereof. To assert the detachment of objects rationally instead of accepting it on faith, we need some notion of the intellect. The basic notion is the object-experience distinction that Kant expounded but did not apply to Newtonian mechanics. It has been incorporated into modern physical theories, which consequently acquire an elaborate conceptual structure. When interpreters try to squeeze the theories into the simplistic framework of things or phenomena, controversy and illusion result. A proper interpretation of modern physics requires the framework of object. As Kant argued, it is most important that the intellectual notions in the object framework are embodied only in the logical form and not in the substantive content of physical theories. Coordinates are not things, and translation is not a force; they merely express our ways of thinking and describing objective characteristics. They assert the observability and independence of things through the expanded conceptual framework of objects.

Going from "space is homogeneous" to "space is invariant under translations of coordinates," modern physical theories make explicit the meaning of "objective." They simultaneously state the conditions of knowledge and circumscribe the objects of knowledge by literally articulating the common-sense notion that objects are independent of how we look at them or characterize them. Consequently, whereas classical mechanics tells only the structure of things, modern physical theories express the structure of theoretical understanding of the world, which includes the characterization of the objective world and what it generally means to be empirical objects. If classical mechanics has been used by some philosophies to buttress an omniscient view of the world to which we are alien, modern physical theories bring to relief our finitude and offer an understanding of the world in which we are appropriately integrated.

§ 2. The Categorical Framework of Objective Knowledge

How come things, which are not part of us and not created by us, become intelligible objects of experiences? The question easily arises when one confronts the strange quantum world and wonders how we come to know it. However, thoughtful physicists saw that the real marvel lies in common experiences. Schrödinger said: "It is precisely the *common* features of all experiences, such as characterize everything we encounter, which are the primary and most profound occasion for astonishment; indeed, one might almost say that it is the *fact that anything is experienced and encountered at all*." Einstein said: "One may say 'the eternal mystery of the world is its comprehensibility.' It is one of the great realizations of Immanuel Kant that the postulation of a real external world would be senseless without this comprehensibility." Comprehensibility, Einstein explained, is used in the most modest sense. It applies to objects of ordinary experiences and includes the ascription of "real existence" to them. He said that the critical thinking of the physicist cannot proceed without considering "the problem of analyzing the nature of everyday thinking."[7]

Although our chief concerns are quantum fields and elementary particles, we cannot forget that the paradigms of objects are the things we handle everyday, such as tables and chairs, which provide an anchor to the meaning of "object" and "real." It is as unsatisfactory to say that the world of our everyday life is less than real as to say that the microscopic world is as-if real, for the meaning of "real" is obscure in either case. The crux of the philosophical problem lies in the relation between our theories and the world. The relation is articulated and grasped in ordinary language; interpreters must connect the structures of physical theories to general common concepts such as that of objects and proper-ties. Common notions are vague, which implies that they need to be clarified, not that they can be slighted. This clarification is the job of philosophy. However, in the philosophy of physics, it is easy to be engrossed in technical-ities to the neglect of philosophy and bewitched by glamorous notions to the neglect of common sense. Interpreters rehearse mathematics but carelessly invoke general common concepts to explain their meaning. The uncritical usage of common general terms is a major source of quantum mystification. Einstein rightly stressed the indispensability of analyzing the nature of every-day thinking in the interpretation of physical theories.

Some people may object to the emphasis on everyday thinking, arguing that many common notions are discarded by modern physics. The objection is based on the confusion of two types of concepts. We must mark the logical distinction between *substantive* and *general* concepts, or the *substantive content* and the *categorical framework* of a theory. Electron, electrically charged, a dozen, and in between are substantive concepts, which characterize the subject matter of the empirical sciences. Object, property, quantity, and relation are general concepts that constitute the categorical framework within which the substantive contents are acknowledged as descriptive of the world. The eluci-dation of the categorical framework belongs to philosophy, which is concerned not with the specific features but with the general nature of the world and the modes of our knowledge of it. Modern physical theories introduce radically new substantive concepts but maintain the continuity of the categorical frame-work. They do not overthrow general common concepts but rethink them and make them their own, effectively clarifying and reinforcing them.

An electron and an apple bear no resemblance to each other, but both are objects; quantum phases and material textures bear no resemblance to each other, but both are physical properties. What has changed in field theory is not the general concept of individual objects but the specific kind of objects; what has changed in quantum mechanics is not the general concept of properties but the specific type of properties. Since the general concepts must accommodate diverse contents, their articulation requires intensive abstraction. Some people are stuck with certain familiar substantive concepts and are frustrated in the futile attempt to extrapolate them into the world of modern physics. Consequently they conclude that modern physics rejects the concepts of prop-erty and causality, space and time. The conclusion is wrong; it stems from the failure to abstract and the confusion of substantive and general concepts.

Consider the quantum postulate that says the state vector representing an isolated physical system contains a complete description of its characteristics. Despite the technical jargon, the logic is plain: Something has certain proper-ties, which are mathematically represented by a state vector. The logic is immediately grasped and becomes the backbone of our understanding. The understanding is achieved not by extrapolating certain substantive classical characteristics but by extending to the quantum realm certain logical forms of thoughts, such as the subject–predicate form of propositions. Due to the common logical form, a layperson overhearing a conversation among physicists can easily surmise that they are talking about physical objects, although he fails to get the faintest idea what the objects are like. The categorical framework maintains some understanding despite missing details, just as the steel frame of a building keeps it standing despite gaping holes in the walls. It enables us to function with imperfect knowledge, to see the deficiencies, and to learn. It enables physicists to have a working understanding of the quantum world despite the lack of a substantive theory of quantum measurements. As long as the categorical framework remains, the quantum world is strange but not mysterious. If the framework is destroyed, say, by the demise of the concepts of properties and predication, then the quantum world becomes mystical. When "it is such and so" becomes logically illegitimate, we do not know what to think.

In empirical theories, substantive concepts are often represented by definite terms and are prominent. General concepts are not represented by terms; they constitute the complex framework of logical relations among terms. Empirical theories have terms for apples and electrons but no term for objects in general. They include statements such as "electrons are charged" but not "electrons are objects" or "being charged is a property." The general concepts are no more than the logic by which we acknowledge the objectivity of electrons. However, general terms are indispensable for interpretations of empirical theories, for in interpretations we assert: "Electrons are objects."

This work presents a parallel analysis of the conceptual structures of quantum field theory and our everyday thinking. I do not try to describe quantum phenomena in substantive classical terms or vice versa. I try to articulate the categorical framework of objective knowledge, of which quantum field theory is one instance and common sense another. The categorical framework enables us to match logically the formal structures of quantum theories and everyday thinking element by element. The structural fit illuminates their philosophical significance.

The following three chapters contain some introductory material on quantum mechanics, relativity, and quantum field theory. I try to present an overall view and make it intelligible to someone who is neither a stranger nor an expert in the area. I focus on the major concepts, leaving out details not necessary for subsequent reflection. Some technical qualification and philosophical background are given in the notes.

Chapters 5 to 7 contain the central philosophical investigation. The minimal conceptual structure underlying the intelligibility of the objective world is analyzed in three stages. The first primitive structure contains the notion of *an object* as distinct from the *experiences* of it, the second articulates the *plurality* of objects with the help of the concept of *space–time*. In the third stage, it is shown that the reconciliation of these two general concepts requires the introduction of *causal relations*, which completes the idea of the *objective world*. I try to extract the first structure mainly from quantum mechanics, the second from the concept of free fields in field theories with local symmetries, and the third from interacting field theories.

Chapter 5 examines how quantum mechanics embodies the notion that objects are independent of our observations and yet are somehow empirically accessible. It draws two distinctions. First, a purely conceptual distinction is drawn between a physical object and its various representations that include conventional contributions. It reappears in linguistic terms in § 21, which considers how we refer to entities in field theories. Second, physical properties and their empirically assessable symptoms are separated by making explicit the form of our observations. Neither distinction is embodied in pre-twentieth-century physical theories. Relativistic nonquantum mechanics incorporates the first distinction but not the second. Quantum theories incorporate both distinctions; the first is embodied in the observables that provide various representations of the same quantum state, the second in the meanings of quantum states and eigenvalues. Finally, the notion of the observer in quantum mechanics is shown to be an illusion arising from an impoverished categorical framework that forces us to divide the world into the observer and the observed.

Chapter 6 investigates the concept of individual objects in a field and shows that individuals are radically situated and inseparable from the concept of space–time. Space–time is the objective structure of the world by which things are individuated; hence the concept of entities presupposes the concept of space–time. Conversely, space–time is the integral structure that we abstract from the objective world by systematically neglecting the qualitative features and concentrating only on the disposition of objects. Since space–time is an integral part of the concept of individual objects, it is neither substantival nor relational in the prevailing sense. Finally, the problem of individuation in quantum many-body systems is considered. It is shown that the interchangeability *salva veritate* principle of identity is incorporated in the definition of symmetrized quantum states with great consequences.

Chapter 7 explores the category of relations and shows that the relations of similarity and difference, which are taken for granted in many philosophies, are not basic but presuppose causal relations. The discrete objects discussed in Chapter 6 are so individualistic they share no common standard for their properties and provide no base for the relation of similarity. Hence we are led to the philosophical problem of explaining why predicates, for example, "red" or "protonlike," are applicable to different objects. Relational properties, intrinsic to each entity, reconcile the conventions of predication for various entities and effectively cement the universe without sacrificing the individuality of the entities. The replacement of universal conventions of predication by physically significant relations is the major logical step in the development of interacting field theories.

Appendix A examines the categories of quantity, quality, and modality as they are embodied in the theories of probability and statistical inference. The probability calculus provides the means to quantify abstract spaces, which can represent abstract notions such as risk or possibility. "Probability" in the calculus is by definition a relative magnitude. This is the meaning that functions in most physical applications, including quantum theories. However, quantum theories are peculiar for also invoking the ordinary meaning of probability with its connotation of uncertainty. The two senses of probability are used in the main text to interpret quantum theories. They show that the interpretation need not be embroiled in the controversy between determinism and indeterminism.

The structure of quantum field theory is an integral whole. The procedure of structural analysis that I follow is contrary to the tinker-toy-type constructions familiar in many foundational philosophies. We do not start from a given set of primitive elements and build more and more connections among them. Instead, we start with a whole and draw more and more categorical distinctions that accommodate more details; the distinctive contours are the backbone conceptual structures. Perhaps the reader can imagine that he or she is watching the development of a Polaroid film, where details gradually emerge to fill an outline. I hope he or she will recognize the outline as common sense.

2

Nonrelativistic Quantum Mechanics

§ 3. The Structure of Quantum Mechanics

Quantum mechanics was developed to understand atomic structures. In 1925, Heisenberg introduced observables, which embodies the concept of quantum-dynamical variables. Unlike classical variables, observables can be incompatible with each other. With Max Born, Paul Dirac, Pascual Jordan, and Wolfgang Pauli, they developed matrix mechanics. Six months later Schrödinger, working independently, launched wave mechanics. It introduced the wavefunction, which embodies the concept of quantum-mechanical states and the principle of superposition. Born interpreted the absolute square of the wavefunction as the probability of experimental outcomes. The approaches of Heisenberg and Schrödinger were soon shown to be equivalent. Dirac unified them in his transformation theory, which is of great generality and practicality. In 1932, John von Neumann put the theory on a mathematically more rigorous foundation and clinched the Hilbert space formulation of nonrelativistic quantum mechanics.

Since then, several variants formulations and axiomatizations were developed.[8] Despite their mathematical sophistication, all rivals have to make contact with the Hilbert space formulation, which yields almost all experimentally verifiable results. Hence it is fair to say that the Hilbert space formulation has a special status. In this section, I mainly follow Dirac's version. It is mathematically less rigorous than some axiomatizations, but it is unsurpassed in physical insight. It is the version most frequently taught in physics courses and used in physical research. We can say that it represents the working understanding of quantum mechanics. Detailed discussions can be found in the first five chapters of Dirac's classic treatise or many standard quantum mechanics textbooks, for example, the first chapter of Sakurai (1985).

Before presenting the structure of quantum mechanics, let me introduce several primitive concepts that are common in physical theories. The topic of a theory is usually called a *physical system*, which takes the place of what we ordinarily call a thing but avoids the idea of being spatially "well packaged." A system may or may not consist of parts; a partless system is called *single*. The *state* of a physical system gives the abstract summary of all its characteristics at a specific time without committing to any specific way of characterization. Definite characterizations are made by *dynamical variables*. There can be many variables that describe the same state. The state and the dynamical variables are equally important in physical theories. A system can assume

16

different states at various times. All *possible* states of the system are encompassed in its *state space*, or *phase space*, as it is often called in physics. The state spaces of all systems of the same kind have the same structure. The states or dynamical variables of a system evolve according to certain *equations of motion*. The changing states are represented by a curve in the state space.

For example, the state space of a single classical particle is a six-dimensional differentiable manifold. When specific dynamical variables are chosen, the state of the particle can be definitely described, say, in terms of its position and momentum. The particle is governed by Newton's equation of motion, and the temporal variation of its state traces a curve in the state space.

The structure of quantum mechanics has more elements. More specifically, its dynamical variables, called *observables*, have a more complicated conceptual structure. In addition to describing the state, a role it shares with classical-dynamical variables, an observable also gives the possible outcomes of measurements. Quantum mechanics predicts the probability that each outcome obtains. In this section I consider only pure states of single physical systems and observables with discrete nondegenerate spectra.

Q1. *A state of a quantum-mechanical system is represented by a unit vector* $|\phi\rangle$ *in a complex Hilbert space $\mathcal{H}$.*

The topic of quantum mechanics is an *isolated* system, whose interactions with the rest of the world, including instruments for its measurement, are neglected. Often a single system is considered. Single systems are not merely ideal; recent experiments have produced single-photon states and trapped single atoms to record their transitions.[9] Sometimes a system can also be an ensemble of items treated as a unit, such as a collection of electrons prepared in some definite way.

Quantum systems are usually microscopic, but size alone is not a sufficient criterion. Superconducting phenomena can be quite extensive, and the Weber bar in gravitational wave experiments weighs almost a ton, yet both can be represented by a quantum state vector. In quantum cosmology, the entire universe is represented by a quantum state. We can regard the quantum world as where classical mechanics fails. The qualification puts a tentative limit on the validity of quantum mechanics; there is yet no evidence that it applies directly to classical entities. The limit is fuzzy; we have no satisfactory criterion that separates quantum from classical systems.

A *quantum state* is the complete and maximal summary of the characteristics of the quantum system at a moment of time. The qualifications have the same meaning as in "in Newtonian mechanics, the equation of motion with an initial condition gives a complete and maximal description of the trajectory of a classical particle." They apply *within* quantum mechanics, and do not imply any metaphysical completeness or exhaustiveness, which has no place in a physical theory. The state description consists of constant characteristics such as the mass and charge of the system, and variable characteristics that change in time. The collection of all states permissible for a quantum system is theoretically represented by its *state space*, which is a complex Hilbert space.

A *Hilbert space* $\mathscr{H}$ is a generalization of the familiar Euclidean space. It is a linear space whose elements $|\phi\rangle$ are called *vectors*. It is equipped with an *inner product*, which maps a pair of vectors $|\phi_i\rangle$ and $|\phi_j\rangle$ into a complex number $\langle\phi_i|\phi_j\rangle$. Two vectors $|\phi_i\rangle$ and $|\phi_j\rangle$ are *orthogonal* if $\langle\phi_i|\phi_j\rangle = 0$, *orthonormal* if $\langle\phi_i|\phi_j\rangle = \delta_{ij}$, where $\delta_{ij} = 0$ if $i \neq j$, 1 if $i = j$. An orthonormal set of vectors $\{|\alpha_i\rangle\}$ forms a *basis* of $\mathscr{H}$ if every vector in $\mathscr{H}$ can be written as a linear combination of its members. A basis of $\mathscr{H}$ is analogous to a coordinate system in Cartesian geometry. The number of vectors in the basis is the *dimension* of $\mathscr{H}$. For infinite-dimensional cases, we are only interested in separable spaces, which have countable bases.[10]

A quantum state is represented by a unit vector $|\phi\rangle$ up to a phase factor. A *unit vector* $|\phi\rangle$ is one whose norm or magnitude is unity, $\langle\phi|\phi\rangle = 1$. If $|\phi\rangle$ represents a state, so does $e^{i\theta}|\phi\rangle$, where θ is an arbitrary number called the *phase factor*. The characteristics of a state are more concretely revealed in its relation to other states. A vector can be represented as the linear combination of other vectors; similarly a state can be expanded into a linear superposition of other states. This is the quantum-mechanical *superposition principle*. The principle can be understood as follows. The addition or subtraction of quantities generally yields a quantity of the same type; the sum of two lengths is a length, the superposition of the two quantum states is a quantum state. Quantum mechanics is peculiar in that its state space itself possesses the structure required to support the operations of addition and subtraction.

More specifically, $|\phi\rangle$ can be written in terms of a set of basis states $\{|\alpha_i\rangle\}$

$$|\phi\rangle = \sum_i \langle\alpha_i|\phi\rangle|\alpha_i\rangle = \sum_i c_i|\alpha_i\rangle; \qquad \sum_i |c_i|^2 = 1 \qquad (2.1)$$

The sum of $|c_i|^2$ is unity as $|\phi\rangle$ is a unit vector. The complex numbers $c_i \equiv \langle\alpha_i|\phi\rangle$ are called *amplitudes*, or probability amplitudes as distinct from other amplitudes familiar in classical mechanics. The qualification should not confuse us: the absolute square $|c_i|^2$ is a probability; c_i is not. The square root of a quantity is generally not a quantity of the same type; the square root of an area is a length, the square root of a length is I know not what. Thus the meaning of c_i is independently conferred. Generalizing from familiar vector spaces, $c_i|\alpha_i\rangle$ can be regarded as the "component" of $|\phi\rangle$ along the axis $|\alpha_i\rangle$ and c_i the "coordinate" of $|\phi\rangle$. A state $|\phi\rangle$ is completely determined by its set of amplitudes $\{c_i\}$ in the basis $\{|\alpha_i\rangle\}$. (See Fig. 2.1a.)

Q2. *The temporal evolution of the state is governed by the Schrödinger equation.*

The *Schrödinger equation* for the state $|\phi\rangle$ is

$$i\hbar\partial|\phi\rangle/\partial t = H|\phi\rangle \qquad (2.2)$$

where $i = \sqrt{-1}$, $\hbar$ is the Planck constant divided by 2π, and H is the Hamiltonian operator representing the total energy of the system. It is a differential equation like the classical equations of motion. If the state of a system is

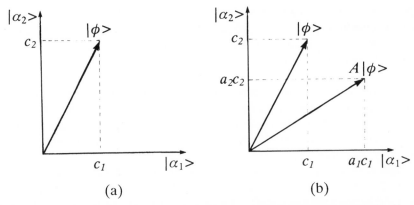

Figure 2.1 (a) A vector $|\phi\rangle$ is decomposed in the orthonormal basis $\{|\alpha_1\rangle, |\alpha_2\rangle\}$; c_i is the "coordinate" of $|\phi\rangle$ in the base $|\alpha_i\rangle$. (b) If $|\alpha_1\rangle$ and $|\alpha_2\rangle$ are eigenvectors of the operator A, then the application of A to a vector $|\phi\rangle$ extracts c_i and multiplies it by the eigenvalue a_i.

given at $t = t_0$, then it is determined for all $t > t_0$. Thus the quantum state evolves deterministically in the same sense that the motion of a classical particle is deterministic. In fact, quantum states are even better behaved. Quantum mechanics does not admit irregular and chaotic motions as classical mechanics does, because the Schrödinger equation is linear and its solutions are periodic or quasiperiodic.

Q3. *An observable associated with a quantum system is represented by a unique self-adjoint operator A on its Hilbert space. The spectrum of the operator comprises all possible values that can be found if the observable is measured.*

An *observable* is a dynamical variable. Familiar examples of observables are energy, position, and momentum. They should not be confused with their classical namesakes. The quantum momentum has a richer structure than classical momentum. Some observables, for example spin, have no classical counterparts.

An *operator* A is a linear transformation of the Hilbert space $\mathscr{H}$ into itself. Many operators are employed in quantum theories, but only a certain class, the self-adjoint operators, represent observables.[11] *Self-adjoint operators* distinguish themselves by having spectra that consist only of real numbers. For an observable A, the *spectrum* $\Lambda(A)$ of its representing operator comprises the set of all possible values obtainable in measurements of A. A spectrum can be discrete, continuous, or a combination of both. Suppose the observable A has a pure point spectrum so that $\Lambda(A) = \{a_i\}$, where the real numbers a_i are called the *eigenvalues* of A. The eigenvalues can be the direct results of experiments. The explicit stipulation of some quantities that can be measured justifies the name *observable*, although eigenvalues or spectral values are only parts of the structure of observables.

As a transformation, an operator usually maps one state into another. However, there are states $|\alpha_i\rangle$ such that for an observable A

$$A|\alpha_i\rangle = a_i|\alpha_i\rangle \qquad (2.3)$$

The states $|\alpha_i\rangle$ are called the *eigenstates* of A. They are invariant under the operation of A, since all A does is to multiply the state $|\alpha_i\rangle$ by a numerical factor a_i. The set of eigenstates $\{|\alpha_i\rangle\}$ constitutes a *basis* of $\mathscr{H}$: in terms of the basis, any state $|\phi\rangle$ can be expressed as in Eq. (2.1). The eigenvalues are all distinct. For observables with *nondegenerate spectra*, each eigenvalue is associated with one eigenstate. From Eqs. (2.1) and (2.3), we can readily derive the *expectation value* $\langle A\rangle_\phi = \langle\phi|A|\phi\rangle$ of the observable A in the state $|\phi\rangle$.

A state $|\phi\rangle$ is not stuck to the particular observable or basis. Suppose B is another observable with eigenstates $\{|\beta_i\rangle\}$. Then, knowing $\{c_i\}$, we can readily write $|\phi\rangle$ in terms of $\{|\beta_i\rangle\}$

$$|\phi\rangle = \sum_i d_i|\beta_i\rangle; \qquad d_i = \langle\beta_i|\phi\rangle = \sum_j c_j\langle\beta_i|\alpha_j\rangle \qquad (2.4)$$

The complex number $\langle\beta_i|\alpha_j\rangle$ is called the *transformation function* between the base states $|\alpha_i\rangle$ and $|\beta_i\rangle$.[12] The different bases associated with different observables form different *representations* of the state, for example, the position or momentum representation. The amplitude in the position representation is called the *wavefunction*.

Two operators A and B *commute* if $AB = BA$. Operators do not generally commute. The order of operations is important because the final state of one operation becomes the initial state of the next. For instance, rotation is an operation or transformation. Rotations about a single axis commute, but rotations about different axes generally do not. Take a book, rotate it by 90 degrees first about the x axis and then about the z axis. The resultant state is different if the two rotations are performed in the reverse order.

For mutually commuting operators with pure point spectrum $A_1, A_2, \ldots$, we can find a representation whose basis is the simultaneous eigenvectors of $A_1, A_2, \ldots$. Two observables are *compatible* if their representative operators commute, *incompatible* if they do not. Incompatible observables admit no simultaneous eigenstates and are peculiar to quantum mechanics. The prominent example of incompatible observables is the position observable Q and momentum observable P. Their commutation relation $[Q, P] = QP - PQ = i\hbar$, is called the *fundamental quantum condition* by Dirac.

An observable introduces a set of basis states in which the characteristics of a state can be revealed. This is part of the Dirac view of quantum physics. With its transformation rules (2.4) among the bases, it is simple, physically intuitive, and exceedingly powerful.

Q4. *The quantity $|c_i|^2$ is the probability that the eigenvalue a_i is observed in a measurement of A on the state $|\phi\rangle$.*

This is the *Born postulate* of quantum mechanics. It says nothing about the probability that the state or the observable has or possesses a certain eigenvalue; the probability and the eigenvalue both pertain to observed data or results of experiments. The concept of *probability* is appropriate here because we are talking about the observed result of a single pure state, which can be just a single click of a Geiger counter. (For more discussion of the notion of probability, see § A3.)

As discussed in § A3, probabilistic statements are not verifiable. A single detected event tells little about the quantum system under study. Thus the Born postulate is always immediately supplemented by explanations that convert it into a statistical assertion. We are told that experiments involve repeated observations or observations on an ensemble of single systems. Thus, Q4 is converted into the following statistical statement.

Q4'. *In a large number of measurements of the observable A, each measurement made on an ensemble of N systems all in the state* $|\phi\rangle$*, if N is large, then in the results of almost all measurements, a fraction* $|c_i|^2$ *of the data points has value* a_i.

This is a phenomenological statement that can be empirically verified. It tells us nothing about the outcome of a single system. Each data point in a measurement pertains to a single system in an ensemble, but we are not told which corresponds to which. The result of each measurement on the ensemble constitutes one sample, which is a distribution of N data points among the values a_i. Q4' describes neither the outcome of a single system nor the outcome of a single sample, but the statistics of a large number of samples. In practice, we often make only a few measurements or as statisticians say, take a few samples. In each sample, the fraction of data points with a_i can deviate from $|c_i|^2$. However, Q4' assures us that if the sample size or the size of the ensemble is large enough, then even if we make only one measurement on one ensemble, the probability is almost 1 that it yields a distribution where a fraction $|c_i|^2$ of the data points has the value a_i. Note that the notion of probability shifts from the outcome of a measurement of a single system to the outcome of a measurement on an ensemble, and the value of the probability shifts from $|c_i|^2$ to approximately 1. This shift is essentially the result of the law of large numbers in probability theory.

The term *expectation value* $\langle A \rangle_\phi$ is best understood in the context of a single system, where it occurs with probability. When we convert to statistical statements, $\langle A \rangle_\phi$ is more appropriately interpreted as the *mean value* of the distribution of data points. If "expectation" occurs in association with the statistical statement, it means prediction, which is a general feature of physical laws. Some people like to put the statistical statement in the subjunctive mood. This is not necessary, for the statement is part of a physical law, and universal laws generally support counterfactual assertions.

A quantity calculated in many practical situations is the *transition probability* from an initial state $|\alpha_i\rangle$ to a final state $|\alpha_f\rangle$ induced by the observable V, $|\langle \alpha_f|V|\alpha_i\rangle|^2$. It should be distinguished from the *transition amplitude* of V between the two states, $\langle \alpha_f|V|\alpha_i\rangle$. The transition amplitude has its own clear and non-probabilistic physical significance of coupling strength. The quantity actually measured in many experiments is the *scattering cross section*, which involves averaging over the initial states and summing over the final states of many transition probabilities.

§ 4. The Quantum Measurement Problem

The structure of quantum mechanics presented in the preceding section shows a peculiarity: The concepts in Q1–Q3 are markedly different from that in Q4, and little explanation is given to mesh the two. Quantum mechanics gives *two descriptions* that differ in nature, subject matter, and treatment. The characteristics described by state vectors are nonclassical; they are irreducibly complex and strangely entangled when expressed in classical terms. Besides its nonclassicality, the state-vector description is decorous; it is essentially of single systems evolving according to an equation of motion. The description offered by the statistics of eigenvalues is just the opposite. The characteristics it describes are classical and familiar. But besides its classicality, it has the appearance of a bastard; it applies not to single systems but only to ensembles, and it is abruptly "thrown out" without any word on its laws of motion; it is merely the result of the "collapse" of nonclassical characteristics. The crux of the problem is that quantum mechanics provides no substantive correlation between the two descriptions. The only relation between them is formal and abstract. It is provided by the observable, whose eigenstates contribute to the state description and whose eigenvalues to the statistical description.

The stepchild treatment accorded the classical description by a theory that many interpreters claim to be universal and fundamental is regrettable, for from a broad perspective that includes but is not limited to quantum mechanics, these characteristics are objective. Classical characteristics are realized in the physical instruments used in quantum experiments, and they are subjected to the laws of classical physics. In the usual understanding, the classical and quantum descriptions are said to be connected in measurements. However, we do not have even a marginally satisfactory account of the measurement process. This is known as the quantum measurement problem.

To see the measurement problem more concretely, consider a quantum object S in the initial state $|\phi_0\rangle = \Sigma_i \, c_i|\alpha_i\rangle$ and an instrument I in the initial state $|\psi_0\rangle$. S and I form a composite system represented by the direct product of their states. S and I interact, and the composite system settles on some final state. Now, if the state of I successfully represents the behavior of a measuring instrument, it should finally realize an eigenvalue of the object system. However, no such process occurs in quantum mechanics, according to which the coupling causes I to develop a phase relation with S

$$|\phi_0\rangle \otimes |\psi_0\rangle = \left(\sum_i c_i|\alpha_i\rangle \right) \otimes |\psi_0\rangle \rightarrow \sum_i c_i(|\alpha_i\rangle \otimes |\psi_i\rangle) \qquad (2.5)$$

The last term follows from the linear nature of temporal evolution according to the Schrödinger equation. The final state of *I* is anything but an eigenstate; it is nonclassical and all entangled with the state of *S*. Von Neumann's analysis shows that adding a second instrument to measure the composite system does not help. Worse, we can arbitrarily bundle up the system with the instrument, and the result does not change.

Intuitively, something physical must have happened during measurements. It is conjectured that the quantum system "jumps" into one of its eigenstates, or the wavefunction "collapses" from a superposition to a single eigenstate. Many efforts were made to substantiate the conjecture. One of the first attempts was by von Neumann. He said there are two kinds of changes, a deterministic change governed by the equation of motion, and an uncontrollable change that occurs only in measurements. His projection postulate, which was modified by G. Lüders, aims to give more substantive accounts of the uncontrollable change. The projection postulate remains controversial. E. G. Beltrametti and G. Cassinelli remarked that it does "not have the status of postulates of quantum theory, necessary to its internal coherence."[13]

L. E. Ballentine argued that the projection postulate leads to wrong results. Even when the quantum system somehow triggers its environment to produce a measurable eigenvalue, its state does not collapse. Consider the track left by a charged particle in a cloud chamber. The incoming particle is usually represented by a momentum amplitude. It encounters the first cloud-chamber atom and ionizes it, leaving the tiny droplet that we observe. This process is sometimes construed as a position measurement that collapses the particle's amplitude into a position eigenstate. The interpretation is untenable. A position eigenstate is a spherical wave that spreads out in all directions. Hence it would be impossible for the particle to ionize subsequent atoms to form a track that indicates the direction of the original momentum, which is allegedly destroyed in the first ionization.[14]

Recently, P. Busch, P. J. Lahti, and P. Mittelstaedt developed a rigorous theory of quantum measurements. They analyzed quantum measurements into two fundamental parts. The first, called *premeasurement*, defines the concept of an instrument that reproduces the probability measure of an object system. The second, called *objectification*, demands that the instrument realizes a definite eigenvalue. They found that the heart of the problem lies in objectification. More disastrously, they proved that "the very concept of a premeasurement, as characterized by the probability reproducibility condition, precludes its realization as a measurement in the sense of objectification." Premeasurement puts the instrument in an explicit relation to the object. The preclusion of objectification shows that whatever stands in explicit relation with quantum objects cannot be classical. Instruments, defined as objects with classical properties, are excluded from quantum mechanics. They concluded: "One would expect, and most researchers on the foundations of quantum mechanics have done so, that the problem of measurement should be solvable *within* quantum mechanics. The long history of this problem shows that, in spite of many important partial results, there seems to be no straightforward route toward its solution. This general impression is confirmed in the present work by means of a number of no-go theorems."[15]

Hidden Variables

Some physicists, dissatisfied with the failure of quantum mechanics to account for the wavefunction collapse, rolled up their sleeves and tackled the problem according to usual procedures in physical research. They make alternative hypotheses, frame theories, and test them with experiments. Some physicists tried to modify quantum mechanics by conjecturing that microscopic systems found with different eigenvalues are actually in different quantum states characterized by *hidden variables* that are averaged over in measurements. Others advanced theorems proving that hidden variables are impossible. The debate clarified the thinking on both sides. John Bell advanced a sharply formulated theory that can be experimentally tested. Experiments have refuted a large class of theories with local hidden variables.[16]

Successful or not, hidden variables are candidates for scientific theories. It is characteristic of scientific theories that their contents are specific and their predictions pinpoint as to be easily verified or falsified by experiments. The specificity of physical theories is their strength; it also limits their liability. The philosophical motivation for developing a physical theory is distinct from and irrelevant to the content of the theory. General relativity and hidden variables are examples of theories driven by a realistic outlook, yet their topics are certain physical properties, not realism. Physical theories are to philosophers as experimental data are to theoretical physicists. The success or failure of certain theories does compel certain ways of formulating philosophical concepts. The experimental results on hidden variables force us to reject certain particular conceptions of reality, such as the identification of reality with things of exclusively classical characteristics.[17] This is the limit of "experimental metaphysics." Experiments can refute certain conceptions of reality but not the concept of reality in general. Philosophical fallacies seldom arise from ignoring experiences; despite impressions disseminated by empiricists about their opponents, few have advocated banging one's head against stone walls. Errors more often stem from wanton extrapolations of limited experiences. This is the source of many mystical doctrines in quantum interpretations. To echo a question of Einstein: Does anyone really believe that the Bell experiments have demonstrated that the moon is not there when no one looks? Has anyone given any reason for this belief?

Hidden variables offer one possible modification of quantum mechanics. There is no proof that it is the only viable alternative. Consider an analogous case. Eighteenth-century physicists, working within the Newtonian framework, had represented gravity by a numerical field and obtained the same result as Newton's action at a distance. The result does not imply that fields cannot possibly make a difference. We now know that fields can be the continuous agent for gravity, but it would require the ten-component tensor field of general relativity. There is no way to arrive at general relativity by tinkering with

Newtonian mechanics. Why should we assume a better theory can be obtained by tinkering with quantum mechanics? Einstein once remarked that the method is "too cheap." It is dangerous to draw metaphysical conclusions hastily, now as then.

Perhaps our present situation regarding the microscopic and macroscopic world is similar to those of Galileo Galilei and Isaac Newton regarding gravity. They have suggested a good answer to these problems. Galileo refused to accept "gravity" as an explanation for the cause of planetary movements, arguing that it is a mere "name." He offered "those wise, clever, and modest words, 'I do not know.'" Newton, having tried but failed to explain the cause of gravity, said: "I scruple not to propose the principles of motion above mentioned, they being of very general extent, and leave their cause to be found out."[18]

3

Relativity and Symmetry

§ 5. Geometry and Space–Time

What is geometry, as distinct from other mathematics? Nowadays the branches of mathematics are so intermingled a clear delineation is hard to make. Loosely speaking, algebraic structures are characterized by a set of operations or laws of composition such as addition or multiplication. Topological structures incorporate the concepts of neighborhood, limit, and continuity, which give rise to the idea of abstract spaces. The fundamental relations of topology, such as that of inclusion or "part of," are qualitative. Algebraic topology and algebraic geometry make a topological structure quantifiable by coordinating it to an algebraic structure, for instance, the real number system. Quantitative concepts such as the metric can also be introduced in geometry, but a metric is not essential to the definition of geometry. There are many geometric structures wherein the concept of distance is not defined.

The rise of non-Euclidean geometry was followed by much philosophizing. However, for physical theories, the crucial difference is between *finite* and *infinitesimal* geometries, not between Euclidean and non-Euclidean geometries, which are both finite. Hermann Weyl reflected on the genesis and essential nature of the new geometry: "The principle of gaining knowledge of the external world from the behavior of its infinitesimal parts is the mainspring of the theory of knowledge in infinitesimal physics as in Riemann's geometry . . . . The question of the validity of the 'fifth postulate,' on which historical development started its attack on Euclid, seems to us nowadays to be a somewhat accidental point of departure."[19]

Georg Bernhard Riemann said in his 1854 inaugural lecture that the problem regarding space persisted because the *general concept of continuous multidimensional manifolds* was lacking.[20] He amended the lack by developing differential geometry. The emphasis is on generality. The familiar Cartesian spaces are multi-dimensional manifolds. However, they are of a very special kind with many built-in peculiarities, which may be obtrusive or erroneous for the representation of space-time. Einstein recalled that he had all the basic ideas for general relativity in 1908. "Why were another seven years required for the construction of the general theory of relativity? The main reason lies in the fact that it is not so easy to free oneself from the idea that co-ordinates must have an immediate metrical meaning."[21] Cartesian geometry rolls together metrical and other spatio-temporal concepts. The conflation hampers the development of physics. The difficulty is solved by differential geometry, which clearly distinguishes various concepts so that physical theories can employ only the necessary ones. The following is a brief comparison of the two geometries.

Cartesian Geometry

Two layers of structure can be distinguished in Cartesian geometry:

C1. *Differentiable and affine structure.* It uniquely represents points in an n-dimensional abstract space by ordered sets of n real numbers and lines by linear equations. The representation constitutes a coordinate system, and a single coordinate system covers the entire Cartesian space. The parallel grid of the coordinate system encompasses an equivalence relation expressing the notion of same direction. The affine structure allows us to compare the direction of lines and lengths of parallel line segments, but it does not provide a general notion of length or distance.

C2. *Metric structure.* It introduces the concepts of lengths and angles through the inner product and leads to a distance function.

Cartesian geometry forces us to invoke points and lines as a package. Also, it applies only to a special kind of spaces that are coverable by a single coordinate system. These requirements are too stringent for the theory of general relativity.

Differential Geometry[22]

The first breakthrough in geometry occurred with Carl Friedrich Gauss's theory of surfaces, where the grid of parallel lines is replaced by an arbitrary dense grid of ordered curves. The curves are not equally spaced; Gauss's important idea is that the spacing between them is not a consideration at all. The Gaussian coordinates consistently assign to each point on a surface a unique pair of ordered numbers, and refrain from going further. *The Gaussian coordinates individuate, but neither relate nor measure.* This is the idea utilized in differential geometry, where the bare differentiable manifold is just a system of identifiable points. If explicit relations or measures are required, extra constructions are introduced. In this way all concepts are kept distinct. Differential geometry has the following layers of structure.

D1. *Differentiable structure.* Like Cartesian geometry, each point in an abstract space is uniquely coordinated by an n-tuple of real numbers, but generally a single coordinate system is incapable of covering a space. The solution is to use a system of local coordinate patches U_α and local coordinate systems $f(U_\alpha)$. Whenever two coordinate patches U_α and U_β overlap, the points in the overlapping region $U_\alpha \cap U_\beta$ are identified by the coordinate transformation function $f_\beta \cdot f_\alpha^{-1}$, which effectively "sews" the patches together. The result is an n-dimensional *differentiable manifold* M^n (Fig. 3.1). The transformation functions

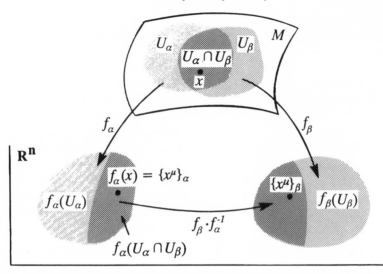

Figure 3.1 A *differentiable manifold M*. The coordinate function f_α establishes a local coordinate system $f_\alpha(U_\alpha)$ for the coordinate patch U_α in M. To each point x in U_α, f_α assigns a set of coordinates $\{x^\mu\}_\alpha$. Within the overlapping region $U_\alpha \cap U_\beta$ of two coordinate patches, the compatibility of the local coordinate systems $f_\alpha(U_\alpha)$ and $f_\beta(U_\beta)$ is secured by the coordinate transformation $f_\beta \cdot f_\alpha^{-1}$. The coordinate transformation maps $\{x^\mu\}_\alpha$ into $\{x^\mu\}_\beta$ and identifies them as the coordinates of the same point x.

ensure that each point in the manifold is uniquely defined, the local coordinate systems are compatible and pass smoothly from one to the other, whatever way we move about in the manifold.

An example is a circle, which is a one-dimensional manifold. A straight line can be continuously bent into an arc, which is the meaning of arc's being "topologically equivalent" to the line; both are Euclidean. The arc is not a circle because it leaves at least one point where the circle is snipped to form a line, or where the ends of a line are glued together to form a circle. Cutting and gluing are not continuous, and the discontinuity makes the line and the circle topologically different. Thus the circle is topologically non-Euclidean; it needs more than one local patch to coordinatize. A single coordinate patch can cover the circle except for a point. To patch up the point we use another finite patch overlapping with the first so there is room for "sewing" by general coordinate transformations. The resultant manifold is the full circle.

Unlike the differentiable structure of Cartesian geometry, the differentiable manifold is a mathematical structure independent of the affine structure. Its definition is the mathematical equivalent of general coordinate transformations in general relativity. The transformations ensure that numerical coordinates can be completely suppressed, so that the points in the manifold are coordinate free.

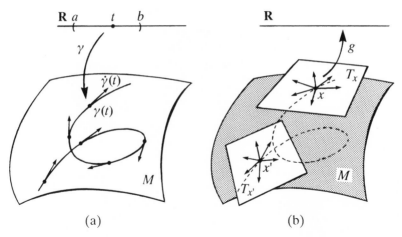

Figure 3.2 (a) A *parameterized curve* $\gamma : \mathbb{R} \to M; t \mapsto \gamma(t)$ is a map that maps a segment of the real line into the differentiable manifold M. The *tangent vectors* $\dot{\gamma}(t)$ are the directional derivatives at each point $\gamma(t)$ of the curve. (b) The collection of all tangent vectors at a point x in M constitutes the *tangent space* T_x over x. The *metric tensor* $g : T_x \to \mathbb{R}$ defines the inner product of two vectors in a single tangent space T_x. Vectors in different tangent spaces generally cannot be compared. However, they become comparable with the construction of a curve and a *connection*, which is not shown.

D2. *Parameterized curves.* The manifold has no intrinsic lines. All *curves* are independently introduced by mapping segments of the real line into the manifold (Fig. 3.2a). Physically, many curves represent world lines of particles.

D3. *Tangent spaces.* An infinitesimal displacement at a point in a manifold gives a *tangent vector* over the point. There are infinitely many tangent vectors over a point corresponding to various directions of displacement. The collection of all these tangent vectors constitutes the *tangent space* over the point. The tangent spaces of an n-dimensional manifold are genuine n-dimensional linear vector spaces (Fig. 3.2b).

D4. *Connection or affine structure.* The tangent spaces, defined over each point, are disjoint. Vectors in different tangent spaces are generally not comparable. To compare tangent spaces at two points requires two extra elements: a *curve* joining the points and a *connection*. Together they generalize the Cartesian affine structure and the concept of parallelism. They enable us to *parallel transport* vectors along the curve. The parallel transport is unique, and it enables us to compare the orientations of distant vectors. All comparisons are path dependent; given a connection, different curves joining two points generally yield different results. Physically, connections represent potentials of interaction fields, for example, the gravitational or electromagnetic potential.

D5. *Metric tensor.* The inner product, called a metric tensor, is defined upon the tangent space over each point in a manifold. The metric tensor gives infinitesimal length elements but not finite distances. It is thus different from the metric structure of Cartesian geometry, which is defined for the space as a whole and yields finite lengths. A metric tensor field assigns a metric tensor to each point in the manifold. A manifold endowed with a metric tensor field is pseudo-Riemannian; it is Riemannian if the metric tensor is positive definite at all points.

D6. *Length of curve.* Finite lengths are obtained by integrating over infinitesimal line elements and are curve dependent.

D7. *Distance function.* It defines the distance between any two points in a Riemannian manifold as the greatest lower bound of the lengths of all curves joining them.

In differential geometry, the Cartesian metric structure is split into the metric tensor for infinitesimal length elements and the distance function for finite distances. The jobs of Cartesian lines are divided between the local coordinate systems fixing positions and curves connecting various points. Cartesian parallelism is taken over by the connection and the curves, which are specified independently in each case. Thus all the basic ideas of Cartesian geometry reappear in some form in differential geometry. If a manifold can be covered by a single coordinate patch, and if a reference frame defined on a tangent space can go from point to point without being rotated, then all tangent spaces are essentially identical and are identified with the space itself, and the manifold becomes a Cartesian space. However, this Cartesian space is a special case of a general structure and its concepts are distinctly defined.

Spatio-temporal Structures[23]

Physical theories parameterized by space–time variables are considered more basic than those that are not. Thus mechanics is more basic than thermodynamics. The characteristics of the parameter spaces we call the spatio-temporal structures. Different spatio-temporal structures parameterize different theories, but they all share a common characteristic. They are all four-dimensional locally Euclidean continua comprising discrete points. They are all mathematically representable by a four-dimensional differential manifold M^4, in which each point can be uniquely designated by an ordered set of four real numbers, at least in a sufficiently small coordinate patch.

Differential geometry makes distinct three roles of geometric concepts, *individuating* or *identifying* points (D1), *relating* points and vectors (D2–D4), and *measuring* quantities (D5–D7). Only the structure supporting the first role is common to all physical theories. The concept of identifiable points in a continuum is most important, but it is often taken for granted because it seems so obvious. Actually it is far from trivial; its precise articulation requires the complicated structure of the differentiable manifold. The construction is needed because the idea of points is so general that in ordinary usages it is invariably invoked with some extraneous ideas, for instance, specific labels of identification. The superfluity has to be erased to bring out the pure idea of identifiable points. The erasure is achieved by coordinate transformations in the differentiable structure.

Beyond the continuum with its individuated points, the spatio-temporal structures differ (Fig. 3.3). In pre-relativistic physics, the four-dimensional continuum can be analyzed into three-dimensional "simultaneous spaces" indexed by the time parameter. There is a metrical structure that includes a *spatial distance* $\Delta x = \sqrt{\Delta x_1^2 + \Delta x_2^2 + \Delta x_3^2}$ and an independent *time interval* Δt.

In Newtonian physics, the structures of space and time are posited independently of the concept of velocity, which is a derivative concept. Special relativity makes the concept of velocity fundamental to the spatio-temporal structure. We should distinguish between a concept and its specific values. Velocity is the parameter of the Lorentz transformations. The values of the parameter are irrelevant and are suppressed by transformations. However, the abstract concept of velocity remains to relate space and time intrinsically and prevents the dissolution of the spatio-temporal structure into independent spatial and temporal components.[24] Consequently, the spatial distance and time interval lose significance. In their place is the *proper time interval* $\Delta\tau$ between two spatio-temporal points

$$\Delta\tau^2 = \Delta t^2 - (\Delta x_1^2 + \Delta x_2^2 + \Delta x_3^2)c^{-2} \tag{3.1}$$

where c is the speed of light. The separation between the points is called *timelike, spacelike,* or *null* depending on whether $\Delta\tau^2$ is positive, negative, or zero. The set of points that are null separated from a point x is called its *light cone*.

In general relativity, the metrical or light cone structure of special relativity is *localized* to the tangent space above a single point. This is in line with the definition of the metric tensor in differential geometry. The localization of the light cone structure is an example of *local symmetries*, which have become prevalent in physics. Note that "local" here means a point and its tangent

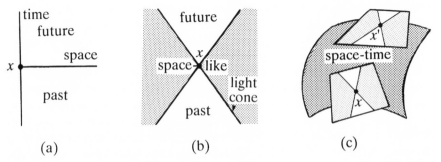

Figure 3.3 The spatio-temporal structures in (a) prerelativistic theories, (b) special relativity, and (c) general relativity.

space, which contains infinitesimal displacements. It should not be confused with "local" in D1, where "local coordinate patch" refers to a finite neighborhood around each point. The result of localization is that in general relativity, the orientations of the light cones on various points are all different from each other. This is in contrast to special relativity, where the orientations are all the same. The difference in the orientations is reconciled by *gravity,* mathematically represented by the connection on a differentiable manifold.

§ 6. Symmetries in Physics

The theory of special relativity rests on two postulates: the principle of special relativity and the constancy of the speed of light. The theory of general relativity also has two basic postulates: the principle of general relativity and the equivalence principle.

The constancy of the speed of light is a feature of electromagnetism. It asserts that the speed of light is the same in all coordinate systems and is independent of the motion of the light source.[25] Of the four postulates, it alone is not controversial; the other three have spawned a great deal of analysis, and we will consider them in the next section. Here we explore one aspect of the relativity principles, the idea of symmetry.

As Einstein stated them, the two principles of relativity have similar forms: Natural laws are the same in a certain class of coordinate systems. The principles say something about both the mathematical form and the physical content of natural laws. Each principle specifies an equivalence class of coordinate systems, which constrains the content of the physical theory. The constraint can be severe, as in special relativity, or it can be unobtrusive, as in general relativity. Whatever the specific constraints, the general idea of the principles is that certain physical quantities are invariant under certain groups of coordinate transformations. Stated in this way, the relativity principles are symmetry principles. The revolutionary concepts they encapsulate, *symmetry* and *invariance*, quickly go beyond spatio-temporal characteristics and find application in many areas. The symmetry structure of physical theories unites many broad principles, of which the conservation law is one example, and the coordinate-free expression of the laws another.

Symmetry: Coordination, Transformation, and Invariance

One common meaning of "symmetry" is balance and proportion. This meaning originated from the Greeks and is closely associated with the notion of beauty.[26] This is not the sense used in physics, where we are concerned with symmetry as in bilateral or radial symmetry. A symmetry pertains to a system as whole. When we say that something has bilateral symmetry, we refer not to its left or right side but to the whole figure. The features characterized by symmetries are meaningful only within the context of the whole, and they are recognizable without reference to external factors.

Technically, the *symmetry* of an object is defined in terms of the transformations that bring the object back into itself, or that leave the object unchanged or invariant. Usually, the larger the group of transformations, the more symmetric the object is. Thus a circle is more symmetric than a triangle. There are infinitely many ways to transform a circle into a position indistinguishable from the original: rotating about the center through any angle or flipping about any axis. Only six transformations can bring an equilateral triangle back into itself: rotations through 120, 240, and 360 degrees about the center, and rotations through 180 degrees about the three bisectors. If no transformation besides the identity can bring a figure back into itself, then the figure is asymmetric.

The set of symmetry transformations forms a group. A group is an algebraic structure; algebraic structures are defined by abstract operations. Abstractly, a *group* is a set of elements with a single rule of composition.[27] An example of a group is the set of integers under the rule of addition. We will consider mostly groups whose elements are operations or transformations. An example is the rotation group, whose elements are various rotations and whose compositional rule is the succession of rotations.

Under a symmetry transformation, the initial and final configurations of the system are identical. Thus two configurations are differentiated and identified in the same breath. *The concept of symmetry contains a concept for difference, another for identity, and a third relating the two.* The difference is marked by some kind of labels or names, usually called *coordinates* in physics. The identity is achieved by the invariance of the system under transformations. *An equivalence class of coordinates, a group of transformations, and a system with certain invariant features form the triad of the concept of symmetry.*

The structure of a symmetry is schematically depicted in Fig. 3.1, which represents a differentiable manifold. For the moment, ignore the patches and look only at the cross-hatched region $U_\alpha \cap U_\beta \equiv U$ in M^n. The *coordinate functions* f_α and f_β map U into the n-dimensional Cartesian space $\mathbb{R}^n$ and define *coordinate systems* $f_\alpha(U)$ and $f_\beta(U)$. To each point x in U, the coordinate function f_α assigns a unique ordered set of n real numbers $\{x^\mu\}_\alpha = \{x^1, x^2, \ldots x^n\}$, which is called the *coordinates* or *coordinate representation* of x. Similarly for f_β. The composite map $f_\beta \cdot f_\alpha^{-1}$ is the *coordinate transformation*, which maps the coordinate system $f_\alpha(U)$ into $f_\beta(U)$ and the coordinates $\{x^\mu\}_\alpha$ into $\{x^\mu\}_\beta$. The point x is *invariant* under the transformations; it is *coordinate-free*. The coordinate transformations form a group.

The concepts of coordinate functions, coordinate transformations, the coordinate-free invariant, and its coordinate representations constitute an integral conceptual structure underlying all symmetries. Coordinates, although conventional, are retained in physical theories employing symmetry structures, for they are required for definite descriptions and conditions of observations. The conditions are irrelevant to physical states of affairs, but are necessary if the physical states are to be empirically verified. "Relativity" stresses the empirical aspect of physical theories and emphasizes that some concepts are defined relative to coordinate systems and observational conditions.

"Invariance" stresses the physical aspect, emphasizing that the pure physical state can be abstracted from the coordinates by symmetry transformations.

The *invariant* features of a symmetry are usually the center of interest. For instance, given the inertial coordinate systems, the invariants under the Galilean group of transformations are the spatial distance and temporal intervals, and the invariant under the Lorentz group is the proper time interval.

Some symmetry groups can contain *subgroups*, which break up the group according to the type of transformations. For instance, the isometry group of the Euclidean plane, or the group of transformations that preserve the inner product and hence distance, consists of rotations of coordinate systems on the plane and translations of the origins of the coordinate systems. It has two subgroups, the *rotation* and *translation* groups. The rotation group preserves distance and a fixed point, which is the axis of rotations. The translation group preserves distance but not the point; in exchange, directions become fixed. When both translations and rotations are included in the isometry group, neither the point nor direction is fixed, and the only invariant feature is distance. Thus extra invariant features appear when a group is restricted to one of its subgroups.

A larger symmetry group singles out the more general invariant. Distance is invariant under all three groups of isometry, rotation, and translation. Yet it is the sole invariant of the largest isometry group. If we follow mathematicians in associating the profundity of a concept with its generality and fruitfulness, then the largest transformation group picks the most profound concept. *Symmetry transformations erase particularities*. A world with high symmetry characterized by a large symmetry group is rather featureless; it retains only the important features. Perhaps we can understand why Einstein wanted to enlarge the spatio-temporal symmetry group. He was striving to formulate the most general and truly universal concept that alone answers to the notion of space–time. Physicists are continuously seeking larger symmetry groups in their drive for unification; the larger group provides more encompassing concepts. For example, the electromagnetic interaction is characterized by the unitary group of order 1, $U(1)$, the weak interaction by the special unitary group $SU(2)$. The unification of the two in the electroweak interaction is achieved by the larger group $SU(2) \times U(1)$.

When the symmetry group of a physical system is cut down to one of its subgroups, we say the *symmetry is broken*. The result of symmetry breaking is the appearance of more invariants and more features. Thus two distinct interactions, electromagnetism and weak, result when the electroweak symmetry is spontaneously broken at low energies. The strong interaction is the strongest of the fundamental interactions, it also happens to have the highest symmetry. When physicists say the symmetry of the strong interaction is broken in the weak, they mean features absent in the strong interaction, such as parity distinction, appear in the weak interaction. The symmetries applicable to specific interactions are also called *dynamical symmetries*, as distinct from *spatio-temporal symmetries*, which are universal.

The spontaneous breaking of the electroweak symmetry is not the result of external forces. However, symmetries can be broken by artificial means. For example, the isotropy of a system can be broken by the application of a magnetic field as in the Zeeman effect, in which the two spin states of electrons, which are degenerate without the magnetic field, are split by the field. There is nothing mystical about broken symmetries, for specific symmetries, like any specific physical properties, are always contingent. A symmetry group may apply under certain situations and only its subgroup applies in others.

The *coordinate* transformations discussed previously are usually called *passive* transformations, where we keep the system fixed and transform the coordinates. Another way of expressing symmetries is by *active* or *point* transformations, where we keep the coordinates fixed and transform the system. Consider the permutation of two entities A and B, respectively labeled "A" and "B," which are their coordinates. In coordinate transformations we shuffle the labels and write "A" $\rightarrow B$, "B" $\rightarrow A$. In point transformations we shuffle the entities and write $A \rightarrow$ "B", $B \rightarrow$ "A." Physicists used to emphasize coordinate transformations, which is the way the principles of relativity are stated. Increasingly, physicists adopt the mathematician's preference for point transformations, which are mathematically more elegant. However, although point and coordinate transformations are mathematically equivalent, we should not mix up the physical meaning of the abstract concepts. Generally, entities are physical, whereas coordinates are conventional. Psychologically, that which is fixed is often deemed more real. Mental shuffling of entities maybe the psychological genesis of a container space; we may be led to the illusion that the fixed coordinates stand for some container in which we variously put things.

Transformations in physics are not arbitrary. *They are generated by dynamical variables*, through which symmetries are associated with *conservation laws*. For example, linear momentum is the generator of translations. The invariance of a system under transformations implies the conservation of the associated generator. If a system is invariant under temporal translations, its total energy is conserved; if it is invariant under spatial translations, its linear momentum is conserved; if it is invariant under rotations, its angular momentum is conserved. Conservation laws are already great generalizations sweeping across many branches of physics, but they become only one aspect of symmetries. The ascendancy of symmetries signifies intensive abstraction and generalization in physics.

Since symmetries reveal the overall structures of physical systems, they provide a powerful tool to extract information about a system without knowing the details of its dynamics. They come in handy because equations of motions are often difficult or impossible to solve. Symmetry considerations cannot replace solutions, but in the default of solutions they enable us to classify possible solutions, exclude certain classes as forbidden, and find selection rules for various transitions. Symmetry is even more useful when we want to find out some fundamental laws. We know that if the unknown law has certain symmetries, then its solution would exhibit certain patterns. The solutions of the unknown laws correspond to experimental data. We start by gathering data, then organize them into a large pattern whose symmetries provide a handle for us to guess at the fundamental law. The development of the quark model for the strong interaction is a good example of the synoptic power of symmetries pointing the way to the fundamental laws.[28]

Local Symmetries

"Symmetry" without qualification usually refers to *global* symmetry. The para-meters of global symmetry transformations are constants independent of spatio-temporal positions. The Poincaré transformations in special relativity are global; their ten parameters are position independent. The phase symmetry in nonrelativistic mechanics is global, it can be written as $\exp(i\theta)$, where θ is a constant. A system is transformed as a unit under global transformations.

One often hears physicists say "to gauge a symmetry group." It means to *localize* the group, or to make the transformation parameters vary spatio-temporally so that the system does not behave as a unit under the transforma-tions. Local transformations allow a large degree of individuality for each point of a system by endowing it with an autonomous internal structure. The Poincaré transformations become local in general relativity. The quantum phase transformation is localized and becomes $\exp[i\theta(x)]$ in quantum field theories. Local symmetries are ubiquitous in quantum field theories. The sym-metry groups $U(1)$, $SU(2) \times U(1)$, and $SU(3)$ for the electromagnetic, electro-weak, and strong interactions are all local transformations whose parameters are functions of spatio-temporal positions. These local symmetry groups are often called gauge groups and theories employing them gauge field theories. They determine the properties of the phase space at each spatio-temporal point individually.

A *local symmetry* is more complicated than a global symmetry because it demands the *global invariance* of the entire system under *local transformations*. Generally this requires the introduction of extra structures to reconcile the difference of various local transformations. The extra structures are usually interpreted as interaction potentials, as discussed in the next chapter.

Relativistic Invariance and Free Elementary Particles[29]

The universal symmetry in relativistic quantum theories is associated with continuous spatio-temporal transformations, which include spatial and tem-poral translations, spatial rotations, and the Lorentz transformations among inertial frames. Together they form a group, called the *proper Poincaré group*, elements of which are denoted by g. Let T_g be unitary operators on a Hilbert space such that if $g = g'g''$, then $T_g = T_{g'}T_{g''}$. The operators T_g realize the group elements g as definite transformations on a Hilbert space. Mathematically, the T_g form a *unitary representation of the Poincaré group*.

The representation describes the effect of relativistic invariance on the state vectors in the Hilbert space.

Let S and S' be coordinate systems and $|\phi_S\rangle$ and $|\phi_{S'}\rangle$ quantum states represented in the respective coordinate systems. If S and S' are connected by the Poincaré transformation g, $S' = gS$, then the quantum state transforms as $|\phi_{S'}\rangle = T_g|\phi_S\rangle$. The state transformation T_g is completely determined by the coordinate transformation g. Thus a correspondence is established between the transformations of coordinate systems and the transformations of quantum states. The state transformations preserve the inner product or the probability of the states.

A specially interesting case occurs when $|\phi_S\rangle$ are states of free particles. The idea of a free particle is an approximation or idealization, for particles form an interacting system. The idealization is useful; in scattering problems, the initial and final states of the interactants are assumed to be free. The unitary representations of the Poincaré group lead to the equations of motion for free particles, for the group transformations include temporal translations, realized as the temporal evolution of the particle's states.

Particles can be elementary or composite. An elementary particle cannot be decomposed into parts. We do not have an exact criterion for elementary particles. However, Wigner has proposed a necessary but not sufficient condition. A free electron should be a free electron in all relativistic frames. Any two states of a *free elementary particle* should be connectible by a transformation in the Poincaré group. Thus all its states are representable by superposition of states obtained by relativistic transformations of a *single* state. In other words, there must be no relativistically invariant subspaces for the state space of a free elementary particle, otherwise we would call the invariant subspaces elementary. The state space of a free elementary particle is the Hilbert space for an irreducible representation of the Poincaré group.

Wigner had worked out the irreducible representations of the Poincaré group. The group-theoretic analysis shows there are two characteristics that are invariant under relativistic transformations. These characteristics are identified as the *mass m* and *spin s*. The spin is the angular momentum in the rest frame, it determines the number of linearly independent states that have the same momentum four-vector. To each (m, s) there corresponds only one irreducible representation up to unitary equivalence. The masses of the particle can be zero or positive; the values of the finite masses are not determined by relativistic invariance. If a particle has finite mass, then s can be 0, $\frac{1}{2}$, 1, or $\frac{3}{2}$, ... If the mass is zero, the spin is either 1 or 2. Neglecting dynamical effects, the general conceptual framework for the kinematic characteristics of types of free particles can be obtained from spatio-temporal invariance. *Thus pure relativistic considerations single out mass and spin as indices for the classification of various free elementary particles and put certain constraints on their values.* The specific mass and spin values for a particular species of particles must be determined empirically.

Besides continuous spatio-temporal transformations, there are also the discrete transformations of *space inversion* represented by the parity operator P

and *time reversal* represented by the operator T. Quantum field theory ushers in a third discrete transformation called *charge conjugation C*, which replaces a particle by its antiparticle and vice versa. All relativistic quantum field theories are invariant under the combined CPT transformations; their physical contents are unchanged by the simultaneous applications of space inversion, time reversal, and charge conjugation. This is the CPT theorem, first proved by Pauli.[30] Physicists used to believe in the invariance under individual discrete transformations. In 1956, P invariance was observed to be violated in the weak interaction, which does not conserve parity. The weak interaction also has no respect for C and CP invariance. There are theoretical arguments that CP invariance is violated in the strong interaction. Because of the CPT theorem, CP violation means time-reversal symmetry also breaks down.

§ 7. The Principles of Relativity and Local Symmetry

Principles in physics suggest not only the content of but also the criteria for physical theories. There are more symmetries than symmetry principles. $SU(3)$ is the symmetry of strong interaction, but there is no "principle of $SU(3)$." The reason is that a symmetry principle stipulates not a specific transformation group but a type of transformation, or more correctly a general conceptual structure.

The three outstanding symmetry principles in physics are the *principle of special relativity, the principle of general relativity*, and *the principle of local symmetry* or the *gauge principle*. The first stipulates *the form of global symmetry*, the third *the form of local symmetry*. The meaning of the second is controversial and entangled in the meaning of invariance and covariance. Since general relativity is not the central concern of this work, I will be brief in offering a view that highlights the similarity it shares with other fundamental theories of interaction dynamics.

There are at present four known fundamental interactions of the physical world: gravity, electromagnetism, and the two nuclear interactions. The first is the topic of general relativity, and the latter three are the topics of quantum field theories. General relativity stands apart from the others. However, theories of all four interactions share a general form; they are all *field theories with local symmetries*. Each theory of a fundamental interaction has *two* symmetry groups, a *spatio-temporal group* and a *local group*, the latter is intimately connected with interaction dynamics. The possession of two groups distinguishes them from special relativity, which has only one symmetry group, the spatio-temporal group, and which is not concerned with interactions. I argue that general invariance concerns the spatio-temporal group and is conceptually similar to special relativity, while general covariance concerns the local group and is similar to gauge field theories.

Historically, general relativity is the first theory with local symmetry. In contrast to special relativity, where a single inertial frame applies across the world, in general relativity the orientations of the inertial frames are free to vary from point to point, and the differences in orientations carry information about the gravitation field. This is the idea that Weyl tried to generalize to the electromagnetic field and that, with the advent of quantum mechanics, blossomed into gauge field theories.

General Invariance

The idea of invariance is discussed in the last section. It is part of the symmetry structure of a theory; a group of symmetry transformations leaves certain features of a system invariant. An invariance principle recommending a specific symmetry group also recommends certain invariant features for a physical theory. The specification of the features imposes restrictions for physical laws. Einstein said of the principle of general relativity: "These points of intersection naturally are preserved during all transformations (and no new ones occur) if only certain uniqueness conditions are observed. It is therefore most natural to demand of the laws that they determine no more than the totality of space–time coincidences."[31] It is quite plain here that he took general relativity to be an invariance principle stating the employment of the system of identifiable points or the differentiable manifold as the fundamental spatio-temporal concept for physical theories. James Anderson and Andrzej Trautman also argued that general relativity is best construed as an invariance principle. It assigns the group of general coordinate transformations or the group of diffeomorphisms, $diff(M^4)$, as the spatio-temporal symmetry groups of physical theories.[32]

The principle of general invariance seems forceless compared to the principle of special invariance. Why? The main reason is that the symmetry group stipulated by special relativity is more restrictive and hence informative. Proper time, the invariant feature under the Poincaré group of special relativity, is peculiar and novel. In contrast, the symmetry group of general relativity is totally general and nonrestrictive, its invariant features carrying little information. There is a tradeoff between generality and informativeness; information means something specific. In the case of space–time, generality is more important; space–time is truly universal. However, the lack of information makes the general concept inconspicuous. The identifiable points of a manifold, the invariants under $diff(M^4)$, are common to all physical theories parameterized by spatio-temporal variables. The points embody the idea of position. Position has always been the natural parameter that appears explicitly in physical theories. Until now it has been expressed relative to some coordinate systems; it is unsatisfactory, but must be tolerated because there was no alternative. Now that we have a satisfactory definition of position, we need not be told to use it. As Einstein said: "the reasonableness of the thing is too obvious." The principles of special and general invariance are like two Olympian weight-lifting champions. The first champion breaks the world record by a wide margin and fills us with admiration. The second champion doubles the load and lifts it with so much grace that some spectators overlook the weight and boo.

General Covariance

The idea of covariance arises from the usual formulation of the relativity principles, which states that physical laws are the same in certain coordinate systems, or that the laws are covariant under the specified coordinate transformations. The exact meaning of "covariant" is unclear. The problem is especially acute with general covariance, which has been criticized to be vacuous. The most common interpretation is that a physical theory is generally covariant if it is written in the coordinate-free or tensor form. In this usage, covariant is purely descriptive, and the principle of covariance is no more than a pragmatic maxim recommending a certain style of writing physical theories. The maxim is valuable; man is a tool-using animal. However, it lacks the status of a principle. The inadequacy of the formulation is suggested by the existence of many alternatives. In one variant version, general covariance means that the metric tensor is the only quantity pertaining to space admissible in the laws of physics, but the meaning of "pertaining to space" is left vague.

Anderson distinguished between a theory's covariance and symmetry groups by making a distinction between its absolute and dynamical objects. Absolute objects are not affected by dynamical interactions; dynamical objects depend on absolute objects but not vice versa. The symmetry group is the largest subgroup of the covariance group that preserves the absolute objects. If a theory posits only dynamical objects, then the symmetry and covariance groups become the same.[33] Anderson's distinction does not clarify how general covariance is different from general invariance. For according to his formulation general relativity has no absolute objects except the spatio-temporal points, thus its covariance and symmetry groups are degenerate.

Anderson has suggested another way of looking at general covariance when he compared the covariance group of general relativity to the gauge groups for other physical theories. Also, Steven Weinberg said: "The Principle of Covariance is *not* an invariance principle, like the Principle of Galilean or Special Relativity, but is instead a statement about the effects of gravitation, and about nothing else."[34] Weinberg did not say it explicitly, but his meaning is clear from his subsequent elaboration. The principle of general covariance states something similar to what field theorists have come to call the gauge principle or the principle of local symmetries.

I suggest that general covariance is best understood as a *local symmetry principle*. It makes the Poincaré or Lorentz group into the local symmetry group of general relativity and stipulates the global invariance of a dynamical system under transformations of the local group. As in gauge field theories discussed in § 10, the institution of a local symmetry is a two-step process, which figuratively speaking breaks the world up into individual entities and then piece them back together.

In general relativity, the laws of special relativity are localized to each point in the manifold with the help of the equivalence principle; the other basic postulate of the theory of general relativity. This step essentially expels the effect of gravity from the laws in the infinitesimal displacement around each point. The second step of enforcing global invariance leads to the reintroduction of the gravitational potential, which reconciles the local laws on various points. Mathematically, the two steps can be seen in the distinction between tangent spaces and the affine connection (D3 and D4 in § 5). The tangent spaces, to which the special relativistic laws are localized, are disjoint. The connection, which represents the gravitation potential, relates the features on various tangent spaces to present a global order. The logic of the local symmetries in general relativity and its similarity to gauge field theories are most apparent in the fiber bundle formulation (Appendix B).

The *equivalence principle* has many formulations. A weak form of the principle is the equivalence of inertial and gravitational masses. Einstein's original version focuses on how the effects of a homogeneous gravitational field can be canceled by a nonrotating uniformly accelerating reference frame. Physicists soon realized that the spatio-temporal variation of gravitational fields is basic to the equivalence principle, which must be restricted to infinitesimal regions. Einstein's 1916 paper and Pauli's 1921 article both say that for every infinitely small world region there always exists a coordinate system in which gravitation has no influence either on the motion of particles or any other physical processes.[35]

The term "coordinate system" is confusing in theories of local symmetry because each of a theory's two symmetry groups has its coordinate systems. To avoid confusion I will use the mathematical term and call the coordinate systems for the local groups *tetrads*. Infinitesimal displacements at a point are collectively represented by its tangent space. On the tangent space over each point we can define orthonormal frames, technically called tetrads. The tetrads are free to rotate. They should not be confused with the spatio-temporal coordinate systems in which the points are identified, (D1 in § 5). The coordinate systems are defined on the manifold, the tetrads on the tangent spaces. (For the tetrad formulation of general relativity, see Appendix B.)

The equivalence principle says that in a gravitation field, on the tangent space T_x over each point x in the four-dimensional manifold M^4, a tetrad can always be defined such that the gravitational field has no effect. In the specific tetrad, the laws of special relativity hold. A tangent space is a linear vector space, on which the Poincaré group is naturally applicable. Since each point is considered independently, no extra constructions such as the connection are involved in the local statement of special relativistic laws. *The equivalence principle localizes the laws of special relativity to the tangent space over each point and makes the Poincaré group the local symmetry group of general relativity.*

As a result of the equivalence principle, there exists on each point in the spatio-temporal manifold a tiny Minkowski space. Unlike the case of special relativity, in which all the tiny spaces are fixed and fused into a global Minkowski space, the tiny Minkowski spaces in general relativity are disjoint and free to rotate independently of each other. We can express the orientations of the tetrads and hence the tiny Minkowski spaces in terms of the components of the coordinate patch. When we perform a coordinate transformation, each tetrad transforms differently from the other, and we have a position-dependent or local transformation. *The principle of general covariance demands that the entire system of physical laws should be invariant under such position-dependent transformations.* To satisfy the principle an extra structure is required to compensate for the rotations at various position. This structure is the *connection*.

The importance of the principle of general covariance does not lie in the formalism of the connection but in its decree that the new structure is given full *physical significance*. The connection represents the *gravitational potential*. The physical significance distinguishes general relativity from the formal exercises wherein Newtonian mechanics is written in fancy mathematical terms involving the connection, which is only to be forced to vanish by some artificially imposed constraints. The physical interpretation of the connection makes general covariance into a physical principle, and brings it into close association with the gravitation field, as Weinberg argued.

General covariance is a local-symmetry principle. It is thus distinct from general invariance, which is a global-symmetry principle specifying $diff(M^4)$ as the spatio-temporal symmetry group. The distinction between local and global symmetries is not important in special relativity because there the Poincaré transformations are global; therefore, the domains of transformations and invariance are the same. This is not the case with general relativity; the local character of its transformations makes an issue of what is invariant.

The interpretation of general covariance as a local-symmetry principle puts gravity on the same footing as the other three fundamental interactions. At first sight, general relativity and gauge field theory appear quite different. The reason is they place different emphasis on their two symmetry groups. The Poincaré group, which is the symmetry group of special relativity, turns out to be the space–time group of gauge field theories and the local group of general relativity. It is well known and yields attention to the novel groups. Thus gauge field theories highlight the local symmetry groups $U(1)$, $SU(2) \times U(1)$, and $SU(3)$, while general relativity highlights the spatio-temporal group $diff(M^4)$. The difference in emphasis is not due to novelty alone. Gauge field theories describe fully interactive systems, and their focus is on interaction dynamics, which is local in character. General relativity is not fully interactive, and its focus is on the global structure of the world. However, despite the slant, the logics of the theories are similar, for the logic must encompass both groups and underwrite the entire theoretical structure.

I do not dwell on the principle of local symmetry here. It will be discussed in Chapters 5 and 7 in the context of gauge field theory. More specifically, its meaning in general relativity and gauge field theory will be compared in § 26.

4

Quantum Field Theory

§ 8. Quantum Fields and Elementary Particles

James Clerk Maxwell synthesized the works of Michael Faraday and others and originated the theory of electromagnetism in 1865. It is our first field theory; it introduced a new form of matter, the electromagnetic field. At first, people accustomed to mechanical models thought that the field needed a medium of propagation, the ether. Soon they found the mechanical scaffold was theoretically superfluous and experimentally dubious. Any lingering predilection for the ether was dashed by the failure of Albert Michelson's heroic effort to detect it. The compatibility of the behavior of the electromagnetic field with the motions of its sources, which are ponderous bodies, became the outstanding problem in physics. It drew the efforts of many physicists, including Hendrik Lorentz and Henri Poincaré, and was solved by Einstein's theory of special relativity.

Dirac united nonrelativistic quantum mechanics and special relativity and founded quantum field theory. However, the results of the theory yielded infinities, which are plainly not physical. The infinities cast doubt on field theories, and many alternatives were tried. The problem was solved in the late 1940s by Freeman Dyson, Richard Feynman, Julian Schwinger, and Sin-Itiro Tomonaga. The solution is known as *renormalization*. The resultant quantum electrodynamics, or QED, is the first and most successful quantum field theory. Its predictions of fundamental constants agree with experimental results to better than one part in a million. The agreement testifies to the marvel of precision experiments as well as to the accuracy of theoretical predictions.[36]

The attempt to extend the success of quantum electrodynamics to the nuclear interactions met with many difficulties. Field theory again plunged into a dark age, while physicists sought alternative solutions in all directions. Then it rebounded. During the development of quantum field theories for the nuclear interactions, the concept of local symmetry came to the fore. Theories centering on local symmetries are often called gauge field theories. The root idea of local symmetries went back to general relativity, in which the orientations of local inertial frames vary. Weyl tried to generalize the idea. He suggested that the "scale" of local frames should also be allowed to vary, so the frames would be enlarged or reduced as we go about in the manifold. The variation of the frame's scale would be reconciled by the electromagnetic field, just as the variation of their orientation is reconciled by the gravitational field.

Weyl called this *Eichinvarianz*, which was translated into English in the 1920s as "gauge invariance." Weyl's idea did not work. Einstein pointed out that the proposed scale change makes the rate of a clock dependent on its history, which is unacceptable. The scale of the vectors is not variable. It is determined by the masses of elementary particles, which are universal constants.

With the arrival of quantum mechanics, it became apparent that what varies from point to point is not the scale but the phase of the electron wavefunction. Fritz London worked out the local phase invariance of electromagnetism, and derived the electromagnetic coupling from local symmetry considerations. However, the old name sticks, and "gauge invariance," "gauge fields," and "gauge theories" become prevalent. The names are half right because the potential is an equal partner in the local symmetry, and the gauge transformation of the potential is the standard term in electromagnetism. But as far as matter fields are concerned, the symmetry is with respect to their phase, not some length scale. Chen Ning Yang said several times: "If we were to rename it today, it is obvious that we should call it phase invariance, and the gauge fields should be called phase fields."[37]

The electromagnetic interaction has been known since the Maxwell equations. The first gauge theory made little impact; local symmetry needs harder testing stones to show its mettle. Opportunities abound in the nuclear interactions, which differ from electromagnetism in a striking way. The electromagnetic field is not self-interacting; its quanta, the photons, do not carry the electric charge, so that photons do not stick to each other to form a light ball. Mathematically, this feature is manifested in the fact that the local symmetry group of electromagnetism is commutative or Abelian. In contrast, the symmetry groups of the nuclear interactions are noncommutative or non-Abelian. The quanta of the strong and weak interaction fields carry the coupling charges and are self-interacting. For instance, the gluons of the strong interaction carry the color charges and can form a "glue ball." The self-interaction makes the non-Abelian fields much more complicated than electromagnetism. The non-Abelian physical theories are also more restrictive and powerful; in them the values of the interaction charges cannot be scaled arbitrarily as is the case for Abelian theories. In 1954, Yang and Robert Mills extended the idea of local symmetries to noncommutative systems in their study of the strong interaction, and the power of local symmetries was revealed. Physicists often refer to gauge field theories as Yang–Mills theories.

The application of quantum field theory to the nuclear interactions has to overcome many other difficulties. Whereas the field quanta of the electromagnetic field are massless, the field quanta of the weak interaction are massive. The finite masses account for the short range of the weak interaction; zero-mass fields such as the electromagnetic field have infinite range. The problem of masses was solved with the concept of spontaneous symmetry breaking. In the 1960s, Sheldon Glashow, Abdus Salam, and Steven Weinberg unified the electromagnetic and weak interactions in the gauge field theory of the electroweak interaction. They were shown to be renormalizable by Gerard t'Hooft in 1971, and the preeminence of gauge field theories was established. Whereas the strength of other interactions attenuates with increasing distance between the interactants, the strong interaction gets stronger when the interactants move apart. In the early 1970s, several authors introduced the idea of asymptotic freedom, which leads to the idea of quark confinement. The quarks are like pebbles in a bottle, relatively free in small confines of radius smaller that 10^{-14} cm but cannot break loose as independent particles. They are described by quantum chromodynamics or QCD.[38]

Quantum field theory is successful with three out of four fundamental interactions of the physical world, but it is unable to incorporate gravity. In the past decade, much attention has been paid to superstring theory, which offers the hope of unifying all physical interactions and superseding field theory. However, so far superstrings have not received any experimental support. Quantum field theory is still our best experimentally tested theory.

The Standard Model of Particle Physics

According to the current standard model of elementary particle physics based on quantum field theory, the fundamental ontology of the world is a set of interacting fields. Two types of fields are distinguished: *matter fields* and *interaction fields*. Their general properties, including spin statistics, differ widely. The quanta of matter fields, called *fermions*, have half-integral spins. The electron is a fermion; its spin quantum number is $\frac{1}{2}$. The fermions are exclusive; only a single fermion occupies a particular state. This is the Pauli exclusion principle, which is the basis of structured matter; the entire periodic table of chemical elements is built upon it. The quanta of interaction fields, called *bosons*, have integral spins. The photon is a boson; its spin quantum number is 1. The bosons are gregarious; many bosons can occupy one state. The coherent radiation of a laser beam comprises billions of photons oscillating in a single state.

There are 12 matter fields, and each has its antifield. The 12 matter fields are divided into three generations; the higher generations are replicas of the first except their quanta have larger masses (Table 4.1). Recent Z_0 linewidth measurements done in the LEP collider in Geneva suggest there are only three light neutrinos. Thus there may not be a fourth generation of matter fields. All stable matter in the universe is made up of only three matter fields in the first generation: electron, up quark, and down quark fields. The quarks are constituents of nucleons such as the proton and neutron. The neutrinos interact weakly with everything and are not part of stable matter. The members of the higher generations have very short lifetimes and are observed only in high energy cosmic rays or in accelerators.

There are four fundamental interactions. Gravity holds our feet on earth and the earth in orbit; it is responsible for the large-scale properties of the universe. Neither a quantum theory nor a fully interactive theory is available for gravity. Electromagnetism binds electrons and nuclei into atoms and atoms into molecules; it is responsible for all physical and chemical properties of solids, liquids, and gases. The strong interaction binds quarks into nucleons and nucleons into

Table 4.1 The fundamental fields of the physical world

Matter fields					
	First generation		Second generation		Third generation
Quark (q)	up	(u)	charm	(c)	top	(t)
	down	(d)	strange	(s)	bottom	(b)
Lepton	electron	(e)	muon	(μ)	tau	(τ)
	neutrino	(ν)	neutrino	(ν_μ)	neutrino	(ν_τ)

Interaction fields				
Interaction	Quanta	Strength	Range (cm)	Acts between
Strong	gluons	1	$\simeq 10^{-13}$	quarks
Electromagnetic	photon	1/137	infinite	charged particles
Weak	$W^\pm$, Z^0	10^{-6}	$\simeq 10^{-15}$	quarks, leptons
Gravitation	graviton?	10^{-39}	infinite	all particles

Fundamental interactions
(The gravitation interaction is not as well understood.)

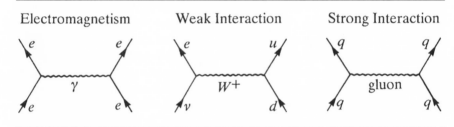

Electromagnetism   Weak Interaction   Strong Interaction

atomic nuclei. The weak interaction is responsible for the decay of certain nuclei. The ranges of the nuclear interactions are very short; thus they are not significant in dimensions of ordinary experience.

In fully interactive field theories, the interaction fields are permanently coupled to the matter fields, whose charges are their sources. The electric charge is the source of the electromagnetic field, and the color charges of the quarks are the source of the strong interaction. There is no sourceless electromagnetic field, and the electron always comes fully dressed in the electromagnetic field. The interacting electron and electromagnetic fields form one integral dynamical system, the system of quantum electrodynamics. The nuclear interactions are similar.

Fundamental interactions occur only between matter and interaction fields, and they occur at a point. The form of the point coupling is the same for electromagnetic and the nuclear interactions. In Feynman diagrams, a matter field is often represented by a straight line and an interaction field by a wavy line. For example, in Fig. 4.1, the lines represent electron fields, and the wavy line an electromagnetic field. The coupling of the two is represented by the vertex in the diagram. The strength of the coupling is determined by the

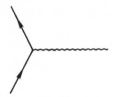

Figure 4.1 Coupling between electrons (solid lines) and a photon (wriggled line).

electronic charge e, which is characteristic of the interacting electron field. The three-way coupling depicted in Fig. 4.1 is the basic form of interaction; it is shared by the strong and weak interactions. The diagrams in Table 4.1, which represent the interactions of two matter fields mediated by an interacting field, involve two basic couplings.

§ 9. Fields, Quantum Fields, and Field Quanta

"Field" has at least two senses in the physics literature. A field is a continuous dynamical system or a system with infinite degrees of freedom. A field is also a dynamical variable characterizing such a system or an aspect of the system. Fields are continuous but not amorphous; a field comprises discrete and concrete point entities, each indivisible but each having intrinsic characteristics. The description of field properties is local, concentrating on a point entity and its infinitesimal displacement. Physical effects propagate continuously from one point to another and with finite velocity. The world of fields is full, in contrast to the mechanistic world, in which particles are separated by empty space across which forces act instantaneously at a distance.

To get some ideas of fields, consider an illustration found in many textbooks. Imagine the oscillation of N identical beads attached at various positions to an inelastic thread whose ends are fixed. For simplicity the bead motion is assumed to be restricted to one dimension, and the thread is weightless. We index the beads by integers $n = 1, 2, \ldots, N$ as in Fig. 4.2a. The dynamical variable $\psi_n(t)$ characterizes the temporal variation of the displacement of the nth bead from its equilibrium position. The dynamical system is described by N coupled equations of motion for the $\psi_n(t)$s, taking into account the tension of the thread and the masses of the beads.

Now imagine that the number of beads N increases but their individual mass m decreases so that the product mN remains constant. As N increases without bound, the beads are squashed together, and in the limit we obtain an inelastic string with a continuous and uniform mass distribution as in Fig. 4.2b. Instead of using integers n that parameterize the beads, we use as indices real numbers z, $0 \leq z \leq L$ for a string of length L. A particular value of z designates a specific point on the string, just as a value of n designates a specific bead. Instead of $\psi_n(t)$, the dynamics of the string is characterized by $\psi(t, z)$, called the *displacement field* of the string. Sometimes $\psi(t, z)$ is said to be the displace-

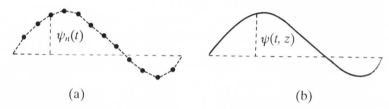

(a)   (b)

Figure 4.2 (a) Beads on a vibrating thread are designated by discrete indices *n*. (b) Points of a vibrating string are designated by continuous indices *z*. The displacement of the string at any time *t*, $\psi(z, t)$ is called the displacement field of the string.

ment at each space point *x*; however, *z* denotes not position in space but position *on the string*; *z is simply a continuous index having the same status as the discrete n*. We can call *z* a spatial index but not an index of space points. The continuity of the parameter *z* should not mislead us into thinking that its values are not distinct. The real line continuum is rigorously a point set satisfying the condition of completeness (see note 167). The *N* beads on the thread form a system with *N* degrees of freedom. Similarly, the string is a dynamical system with infinite degrees of freedom, characterized by the dynamical variable $\psi(t, z)$.

In the case of the string, the parameter *z* is a number; so we have a one-dimensional field. The same meaning of dynamical variables with continuous indices applies in cases with more dimensions, where the spatial parameter becomes an ordered set of numbers. Consider a vibrating membrane such as the surface of a drum. Physicists describe the displacement of the membrane by a field $\psi(t, \mathbf{x})$ with $\mathbf{x} = (x_1, x_2)$. Again x_1 and x_2 are parameters indexing points on the drum, not points in space.

In the examples of the string and the drum, the dynamical variables ψ represent spatial displacements of matter, and are explicitly called displacement fields. Fields generally are *not* spatial displacements. To arrive at the general concept of fields, we abstract from the spatial characteristic of ψ. With the abstraction we depart from locomotion, which dominates classical physics, and consider other kinds of variations. Nonspatial changes are common; imagine a spot changing in color.

A field $\psi(t, \mathbf{x})$ is a dynamical variable for a continuous system whose points are indexed by the parameters t and **x**. Unless explicitly specified as in displacement fields, the spatial meaning of a field is exhausted by the spatial parameter **x**, which has the same status as the temporal parameter *t*. With **x** fixed, ψ varies but generally does not vary spatially. The field variable ψ need not be a scalar as it is in the previous examples; it can be a vector, tensor, or spinor.

The string and the membrane have names; most continuous systems do not and are generically called fields. A fundamental field of physics is a freestanding and undecomposable unity by itself; it cannot be taken apart materially, unlike the string that can be decomposed into beads. The parameter **x** allows us to distinguish points, but they are points-in-the-field and are meaningless when detached from the field.

The classic example of fields is the electromagnetic field $F(x)$, where x is the four-dimensional spatio-temporal parameter whose coordinates are $\{x^\mu\}$ = $(x^0, x^1, x^2, x^3) \equiv (ct, \mathbf{x})$. The parameter $\mathbf{x}$ exhausts the spatial connotation of the field. The field F is not "waving" in some kind of ether, and it needs no support of a propagating medium. F is a tensor.

Lagrangian Field Theory[39]

Being dynamical variables of systems with infinite degrees of freedom, fields can be treated within mechanics. Besides the Newtonian formulation, in which the force is specified, there are the Hamiltonian and Lagrangian formulations of mechanics, which are more powerful for systems of particles. The *Hamiltonian* of a system is its total energy; the *Lagrangian* is the difference between its kinetic and potential energies. They carry the substantive content of a dynamical system and play the same role as the force in the Newtonian formulation. Nonrelativistic quantum mechanics mostly uses the Hamiltonian version, where the basic variables are generalized coordinates and momenta. Its disadvantage is that it is not manifestly relativistically invariant because time is singled out. Relativistic field theories mostly use the manifestly relativistic Lagrangian formulation, where the field equations are derived from a Lagrangian density via a variational principle. Quantities of the dynamical system, such as energy–momentum and charge, follow as the invariants of the system under certain symmetry transformations.

The Lagrangian for a continuous system is parameterized in the same way as the field variable. We are only interested in *local fields* whose Lagrangians $\mathscr{L}(x)$ depend only on the properties in the infinitesimal neighborhood of the point x. In such cases the Lagrangian is a function only of the field variable $\psi(x)$ and its first partial derivative and can be written in the form $\mathscr{L}(x)$ = $\mathscr{L}(\psi(x), \partial_\mu \psi(x))$, where $\partial_\mu \psi = \partial \psi / \partial x^\mu$.[40] The field $\psi(x)$ can be regarded as the generalization of the configuration variable in point-particle mechanics to continuous systems. The generalization of the momentum variable, called the *canonical conjugate* $\pi(x)$ of the field $\psi(x)$, can be readily derived by taking the proper derivatives of the Lagrangian. Given the Lagrangian, the equation of motion for the field $\psi(x)$ can be derived from variational principles.

The Lagrangian can have certain symmetry properties; that is, it is invariant under certain groups of transformations of x^μ and ψ. Using the variational principle, it can be shown that if the Lagrangian is invariant under a group of continuous transformations expressible by an infinitesimal parameter, then there is a conserved quantity. This is known as the *Noether's theorem*. The conservation of energy, momentum, angular momentum, and charge can be derived this way.

Field Quantization and Quantum Fields

The Schrödinger equation is not relativistic. There are several relativistic quantum equations of motion; the most common ones are the Klein–Gordon equation for spin 0 systems and the Dirac equation for spin-$\frac{1}{2}$ systems. In one interpretation, they are relativistic wave equations for single particles, so that their variables $\psi(x)$ are single-particle wavefunctions in strict analogy with nonrelativistic quantum mechanics. This interpretation runs into two difficulties. First, unlike the Schrödinger equation, the Klein–Gordon equation does not give a positive-definite probability. Second, its solutions include states with negative energies. The negative-energy states make the system unstable in interaction, for the particle would continue to cascade toward the state with infinite negative energy, emitting infinite radiative energy in the process. These difficulties force us to abandon the single-particle interpretation of the Klein–Gordon equation. The Dirac equation gives positive probability, but it too contains negative-energy states. To avoid the catastrophe of infinite decay, Dirac postulated that the negative-energy states are completely filled by other particles, which led to the prediction of a "hole" in the filled sea. The prediction of the "hole," or the *antiparticle* as it came to be known, was a great triumph. However, it demands a many-particle picture in contradiction to the original single-particle interpretation.

The difficulties disappear if the variables $\psi(x)$ of the relativistic equations are interpreted not as single particle wavefunctions but as dynamical variables for continuous systems, or simply, fields. The field interpretation became the consensus among physicists. The discussions of two possible interpretations in textbooks do not imply the underdetermination of theories; the single-particle interpretation does not work.

We go from classical to quantum mechanics by replacing the classical variables of position and momentum by quantum operators Q and P that obey the commutation relation $[Q, P] = QP - PQ = i\hbar$. Similar canonical quantization procedures can be used to go from classical to quantum field theories. Quantum fields are obtained by replacing the classical field variable $\psi(x)$ and its conjugate $\pi(x)$ by operators obeying certain commutation relations. However, the procedure is mainly mathematical. Physically, many quantum fields have no classical counterparts. We straightforwardly posit *quantum fields*, or continuous systems with quantum characteristics, independently of classical analogues.

Continuous quantum systems constitute the basic ontology of the world, according to contemporary physics. A field system is a complicated whole that can be characterized by several field variables and their interactions, for instance, an electron field, an electromagnetic field, and their coupling. Often the coupling is weak and can be neglected in the first approximation, so that the various fields can be treated ideally as *free fields*, for instance, the free-electron field. A free field is a dynamical variable describing an aspect of an interacting field system.

Consider an idealized free field $\psi(x)$ with Lagrangian $\mathscr{L}(\psi, \partial_\mu \psi)$ and conjugate momentum $\pi(x) = \partial\mathscr{L}/\partial\dot\psi$. The field is *local*, which ensures the point-to-point transmission of physical effects. It is *relativistic*; its parameter x with four-component coordinates $\{x^\mu\}$ satisfies the symmetry of special relativity, and is invariant under the Poincaré group of coordinate transformations. Here is the first big difference between quantum field theory and quantum mechanics. In quantum mechanics, time is a parameter, while position is an observable. The preferential treatment of space and time violates relativity. In quantum field theories, position is no longer an observable; it is a parameter having the same status as time. The Lagrangians of closed systems do not depend explicitly on x.

Being *quantum* fields, $\psi(x)$ and $\pi(x)$ are quantum-dynamical variables represented not by numerical functions but by *operators* in a Hilbert space, which is the state space of the entire continuous system. As discussed in § 3, operators are transformations of the states of the field, for instance, excite it in a particular mode at a particular point x. Operators generally do not commute. Since a transformation may have an effect on its successor, the order of transformations cannot be arbitrarily changed. However, spacelike separated transformations do not affect each other, for causality demands that physical effects propagate from point to point with finite velocity. Thus all spacelike separated field operators commute. This condition is known as *microcausality* and is a basic postulate of quantum field theories. It agrees with the local characteristic of fields.

As discussed in § 6, relativistic invariance considerations classify fields according to their spins. There are two major types of quantum fields: those with integral spins and those with half-integral spins. They have radically different commutation relations. The commutation rules specify the consequences of changing the order of two operations and represent the quantum conditions of the fields. It can be proved that if microcausality is to be preserved, then integral-spin fields and half-integral-spin fields must obey their respective commutation rules.

Normal Modes, Field Quanta, and Particles[41]

So far we have not introduced the notion of particles. Particles in field theories are nothing like tiny pebbles; they are normal modes or quanta of excitation for the field. To get some idea of them, consider the motion of a violin string with length L. Since the ends of the string are fixed, only those wavelengths of vibration that are integral divisions of $2L$ are allowed. These wavelengths correspond to the harmonic frequencies ω_n, each of which is a *normal mode* of the string. For instance, the curve marked A_1 in Fig. 4.3a is the first normal mode with wavelength $2L$, and the curve A_2 is the second mode with wavelength L.

Normal modes have peculiarities that make them extremely useful in the analysis of vibrations, waves, and fields. Each mode oscillates in a definite frequency. The modes do not couple to each other in the first approximation, where the amplitude of oscillation is not too large. Furthermore, the normal modes form a complete set of solutions for the wave equation governing the string. The general motion of the string can be expressed as a superposition of harmonics with various amplitudes, each oscillating in its own frequency ω_n. Not all modes of a system need be excited in all times; for instance, the even

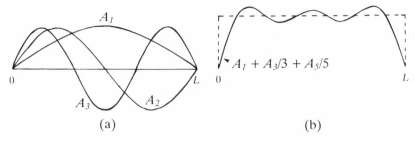

A_1

$A_1 + A_3/3 + A_5/5$

0 L 0 L

A_3 A_2

(a)   (b)

Figure 4.3 (a) The first three normal modes of a string with length L are $A_1 = \sin \pi z/L$, $A_2 = \sin 2\pi z/L$, and $A_3 = \sin 3\pi z/L$. (b) A square wave (dotted line) is approximately represented by the superposition of the first, third, and fifth normal modes. The even modes do not contribute in this case. The approximation will improve when more normal modes are included in the superposition.

modes are not excited in the square wave depicted in Fig. 4.3b. The energy of the string is the sum over the energies of excited modes. Thus an arbitrary vibration pattern of a violin string can be analyzed as a system of an infinite number of harmonic oscillations, whose motions are independent of each other if the string is not vibrating too violently. The independence of the normal modes in the first approximation greatly eases analysis and is the source of the name "particle."

In a similar way, a free quantum field can be analyzed in terms of its normal modes. Instead of the parameter n that designates the modes of the string, the modes of a quantum field are usually indexed by a continuous parameter $\mathbf{k}$, which accounts for the wave vector, polarization, and other factors. The mode $\mathbf{k}$ has frequency $\omega_{\mathbf{k}}$, and the energy $\hbar\omega_{\mathbf{k}}$ is called the *energy quantum* of the mode. As a quantum system, the field energy cannot change continuously but only in steps of energy quanta. Quanta of interaction fields usually have names; the quantum of the electromagnetic field is called a *photon*, that of the strong interaction field a *gluon*. Quanta of matter fields are usually by called the name of the field; thus the quantum of the electron field is an electron. Field quanta are usually called *particles*.

A state of a free quantum field can be written as $|n(\mathbf{k}_1)$, $n(\mathbf{k}_2), \ldots n(\mathbf{k}_i), \ldots \rangle$, which is the eigenstate of an observable $N(\mathbf{k}_i)$. The occupation number $N(\mathbf{k}_i)$ introduces the *occupation number representation* of quantum fields. A measurement of $N(\mathbf{k}_i)$ finds $n(\mathbf{k}_i)$ quanta in the mode $\mathbf{k}_i$. The higher the number $n(\mathbf{k}_i)$, the more excited the field is. A measurement of the field energy yields $\Sigma_{\mathbf{k}} n(\mathbf{k})\hbar\omega_{\mathbf{k}}$.[42] The *ground state* of the field, or the state with the lowest energy, is one in which no mode is excited. Thus the ground state is $|0, 0, \ldots \rangle \equiv |0\rangle$.

The field operator $\psi(x)$ is a combination of transformations called creation and annihilation operators, which excites and deexcites various modes. The *creation operator* $a^{\dagger}(\mathbf{k}_i)$ excites a quantum in the mode $\mathbf{k}_i$; that is, it increases the number $n(\mathbf{k}_i)$ in the state $| \ldots n(\mathbf{k}_i) \ldots \rangle$ by 1. Similarly, the *annihilation operator* $a(\mathbf{k}_i)$ deexcites a quantum in the mode $\mathbf{k}_i$ and decreases the number

$n(\mathbf{k}_i)$ by one. The creation and annihilation operators are related to the occupation number, $N(\mathbf{k}_i) = a^{\dagger}(\mathbf{k}_i)a(\mathbf{k}_i)$. However, unlike $N(\mathbf{k}_i)$, which is an observable, the creation and annihilation operators are dynamical variables but not observables. Field operators $\psi(x)$ generally are not observables. This does not mean that field operators have no physical significance. One manifestation of their significance is that the classical electromagnetic field is the limit of the expectation value of the creation and annihilation operators and not the observable $N(\mathbf{k})$.[43]

An arbitrary state $|n(\mathbf{k}_1), n(\mathbf{k}_2), \ldots \rangle$ in the occupation number representation can be built up from the ground state $|0\rangle$ by successive applications of various creation operators. Since the ordering of operations is important in successive creations, the commutation rules of field operators play a crucial role in deciding what states are attainable. For integral-spin fields, the commutation rules place no restriction on the values of $n(\mathbf{k}_i)$; it can be 0, 1, 2, . . . Any number of quanta can be excited in one mode. Thus we are led to the *Bose–Einstein statistics*, where the field quanta are called *bosons*. This is not the case with half-integral-spin fields, whose commutation rules dictate that repeated operations of the same creation operator yield zero. Hence it is impossible to create more than one quantum in any mode, and $n(\mathbf{k}_i)$ can take on only two values, 0 or 1. This is called the *Pauli exclusion principle*. We are led to the *Fermi–Dirac statistics*, where the field quanta are called *fermions*. The correlation between spin and statistics based on microcausality and commutation rules is a theorem in axiomatic quantum field theory.

Normal modes, field quanta, and particles are good concepts for describing continuous systems only when the coupling between them is negligible. The condition is not always satisfied. For instance, the modes of a violin string cannot be regarded as independent of each other when the vibration is violent enough to become anharmonic. Similarly, when quantum fields interact, quanta can be excited and deexcited easily so that the static picture of free fields depicted above no longer applies. That is why field theorists say particles are epiphenomena and the concept of particles is not central to the description of fields.

The Vacuum or the Ground State[44]

In fundamental physics, the *vacuum* is the *ground state* $|0\rangle$ or the *state with the lowest energy* of the physical world. For an individual free field, the vacuum is the state with zero expectation of finding a field quantum in any mode. The vacuum average value of a field operator vanishes, $\langle\psi\rangle_0 \equiv \langle 0|\psi(x)|0\rangle = 0$. However, the square of the field operator has nonzero average, $\langle 0|\psi^2(x)|0\rangle \neq 0$. The nonvanishing square averages, or *vacuum fluctuations*, lead to observable physical effects. For example, they contribute to the Lamb shift, which splits the degeneracy of the $2S_{\frac{1}{2}}$ and $2P_{\frac{1}{2}}$ states of the hydrogen atom.

In an interacting field system, it is not necessary that all fields have zero average energy in the system ground state. Some nonvanishing scalar fields in

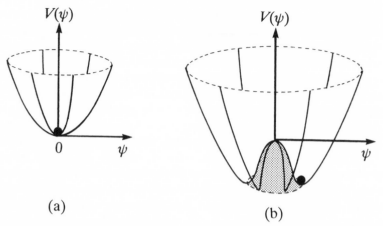

(a)   (b)

Figure 4.4 (a) The effective potential $V(\psi)$ has rotational symmetry and a unique minimum or *vacuum*, which is realized in the stable ground state of a system depicted by the black dot. (b) The effective potential also has rotational symmetry, but its vacuum is degenerate; there are many possible minimum-energy states. When one of the states is realized by the stable state of a system, the rotational symmetry of the potential is *hidden* or *spontaneously broken*.

the system ground state are crucial to explain the finite masses of some interaction fields. Consider a system with an effective potential energy $V(\psi)$ as illustrated in Fig. 4.4a. The minimum in the potential occurs at $\langle \psi \rangle_0 = 0$. The average energy of the field vanishes at ground state, which is nondegenerate. This is the case for most fields.

There are also cases where the effective potential is like that illustrated in Fig. 4.4b. Here the minimum occurs not at $\langle \psi \rangle_0 = 0$ but at some finite average field value, $\langle \psi \rangle_0 = \psi_0$. The ground state is now degenerate; that is, there are many possible states satisfying the minimum energy condition. If we examine the Lagrangian for the system, we find that it is rotationally symmetric with respect to ψ. However, the rotational symmetry is not manifested by the field as it realizes one of the possible ψ_0s in a stable configuration. This phenomenon, in which the symmetry of the Lagrangian is not displayed in the actual stable state of the field system, is called *spontaneous symmetry breaking*. The scalar field responsible for symmetry breaking is usually called the Higgs field. In an interacting field system, the Higgs field restructures the system ground state and changes the masses of the fields with which it interacts.

§ 10. Interacting Fields and Gauge Field Theories

Fields form an interacting dynamical system, of which free fields are approximations and idealizations. Systems of interacting fields constitute the most fundamental structure of the world as we know it today. In the past couple of decades, physicists have made considerable effort to find a unified field theory for all four known fundamental interactions. This aspiration has yet to be fulfilled. However, we now know that all interacting field theories share a common structure; they are all *field theories with local symmetries*. The idea of local symmetries is present but not conspicuous in classical electromagnetic theory.[45] It becomes prominent in general relativity for the gravitational interaction, which resists quantization. Its significance is fully exploited in the quantum theories for the nuclear interactions. These theories are often called *gauge field theories*.

A symmetry exhibits certain features that are invariant under certain transformations. The features are not necessarily spatio-temporal. An interacting field theory has two symmetry groups, the space–time group and the local symmetry group. Here we consider only the latter. In quantum theories, non-spatio-temporal features are generically called *phases*, which are defined in a *phase space* or *state space* appropriate to the dynamical system under study. The local or non-spatio-temporal symmetry transformations act on the state space.

Consider isospin, with which the serious development of gauge field theory began. The nuclear interactions are blind to the electric charge. To them the charged proton and the chargeless neutron are merely different states of the same particle, the nucleon. What distinguishes the neutron state from the proton state is the orientation of the nucleon's isospin. The isospin orientation determines the relative "neutron-ness" and "proton-ness" of the superposed state. This is a standard quantum superposition. All possible orientations of the isospin form a three-dimensional state space called the isospin space. The three elements of symmetries, coordinates, transformations, and invariance, are present. "Proton" and "neutron" name the coordinate axis of the isospin space. The symmetry transformations are isospin rotations. The invariant is the nameless feature that characterizes the nuclear interactions.

Although many phases are symmetric and equivalent, only one of them is realized in a definite state. In calculations we have to pick a convention and set up a definite representation of the state space to describe a particular situation. In global symmetries, the same convention applies over all spatio-temporal positions. If we have chosen to designate a certain state as proton at one spatio-temporal point, we are not free to designate another state as proton elsewhere. The global convention requires that all field operators share a common state space. It violates the spirit of local field theories, in which descriptions concentrate on a point and its infinitesimal vicinity. The relaxation of the global requirement is the starting point of gauge field theories. The localization of the symmetry effectively installs a separate state space on each point in the field. The independent state spaces underlie the ideas of discrete events and point interaction.

Quantum Electrodynamics (QED)[46]

To see the meaning of non-spatio-temporal symmetry transformations and their localization, consider the experiment in which an electron field passing through two slits forms an interference pattern on a second screen (Fig. 4.5a).

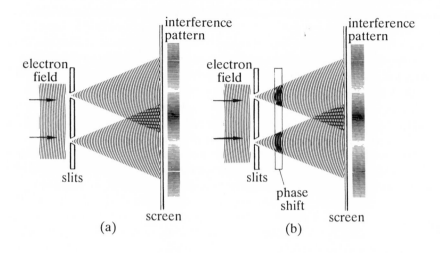

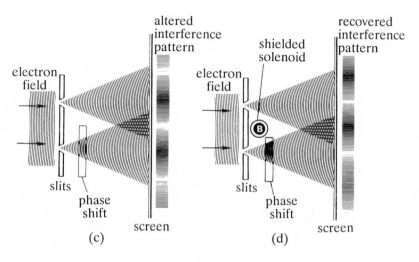

Figure 4.5 (a) An electron field passing through two slits is intercepted by a second screen, forming an interference pattern. (b) Shifting the phase of the entire field by the same amount leaves the interference pattern unchanged. The electron field is invariant under global phase transformations. (c) Shifting the phase of part of the field changes the interference pattern. The free-electron field is not invariant under local phase transformation. (d) The effect of the local phase shift is compensated by the potential A_μ of an electromagnetic field. This is a variation of the Aharanov–Bohm effect. A whisker of a shielded solenoid is placed where the electron field intensity is almost zero. A current in the solenoid generates a magnetic field B inside the whisker and a vector potential A_μ outside. As the current varies, the interference pattern changes. At some value of A_μ, the original interference pattern is recovered. This demonstrates that the interacting electron–electromagnetic field system can be made invariant under local phase transformations.

The electron field $\psi(x)$ is determined up to a phase factor θ; its Lagrangian is invariant under the phase transformation

$$\psi(x) \rightarrow \psi'(x) = e^{i\theta}\psi(x) \tag{4.1}$$

The phase transformations form a group, the unitary group $U(1)$, for electromagnetism. The invariance under phase transformations is readily demonstrated by inserting a phase-shifting plate in the field as it emerges from the slits (Fig. 4.5b). The phase transformation, as represented in Eq. (4.1), is *global*, because the phase change θ is a constant independent of the position x in the field. A global symmetry means any phase transformation must apply across the entire electron field. The whole field is described by a single state space and must be treated as a unit.

A field can be extensive. It is unreasonable to expect that phase changes are always global. Global transformations fail to describe situations such as that in Fig. 4.5c, where a phase shift occurs only in part of the field. For symmetry transformations that cover general situations, we need a phase parameter $\theta(x)$ that can vary as a function of the position x

$$\psi(x) \rightarrow \psi'(x) = e^{i\theta(x)}\psi(x) \tag{4.2}$$

Going from Eq. (4.1) to (4.2), *the symmetry U(1) is localized* to each point x in the field. The position-dependent phase $\theta(x)$ effectively sets up a state space for each point, in which the symmetry group acts independently of its action in the state spaces for other points. Thus each point in the field becomes an entity by itself; each acquires its own internal structure so that it can choose its own phase convention independently of its neighbors.

In the two-slit experiment, the interference pattern changes with a local phase change. This means that the free electron Lagrangian is *not* invariant under local phase transformations. The failure of invariance is unsatisfactory. In the experiment, the local phase shift is artificially induced, but there is no artificial change in the general characteristics of fields. Generally, phase changes are merely changes in the convention of description; active phase transformations are mathematically equivalent to passive transformations of the coordinate systems of the state space. We do not expect the physical content to change with our choice of phase coordinates, just as we do not expect it to change with spatio-temporal coordinates. As discussed in § 6, the whole point of symmetries is to suppress the undesirable effects of coordinates. Apparently local phase transformations alone are unable to accomplish the task.

The additional factor required to preserve the invariance of the system under local phase transformations is seen in the Aharanov–Bohm effect, first observed in 1960 (Fig. 4.5d). The experiment shows that the effect of an artificially induced local phase shift can be canceled by the application of an electromagnetic potential $A_\mu(x)$. Generally, if a term

$$-\bar{\psi}(x)e\gamma^\mu A_\mu(x)\psi(x) \tag{4.3}$$

is added to the free-electron Lagrangian, then the resultant Lagrangian is invariant under the joint local transformation of $\theta(x)$ *and* $A_\mu(x)$, Eq. (4.2) and

$$A_\mu(x) \rightarrow A'_\mu(x) = A_\mu(x) - \partial_\mu\theta(x)/e \tag{4.4}$$

The partial derivative term $\partial_\mu\theta(x)$ compensates for the variation due to the position dependence of the phase factor. $A_\mu(x)$ transforms in the same way as the electromagnetic potential, and is identified with it. The Aharanov–Bohm experiment shows that the potential $A_\mu(x)$ is physically significant and its introduction is not merely formalism.

In the preceding, we start with a free-electron field, apply local symmetry transformations, and "generate" a coupling term, Eq. (4.3). The invariance under Eqs. (4.2) and (4.4) can be satisfied if we stipulate that $A_\mu(x)$ has the form

$$A_\mu(x) = \partial_\mu\Lambda(x) \tag{4.5}$$

which implies that the electromagnetic field vanishes. If we acceded to the restriction on A_μ, then the exercise in local transformations would be much ado about nothing. The crux of the *principle of local symmetry* is the assertion that the potential need *not* satisfy Eq. (4.5) or that the electromagnetic field need *not* vanish. It does not stipulate an interaction field but rules against its *a priori* exclusion imposed by prescriptions such as Eq. (4.5). Once the physical significance of the potential is acknowledged, the electromagnetic field can be derived by taking appropriate derivatives of it. We add the kinetic term of the electromagnetic field to the invariant Lagrangian. The result is the full quantum electrodynamic Lagrangian for an interacting electron–electromagnetic field system, where the term Eq. (4.3) represents the interaction between the electron and the electromagnetic fields.

We started with a free-electron field, and end with an interacting electron–electromagnetic field system. The electron field is no longer free; it is coupled to the electromagnetic potential $A_\mu(x)$, which rotates its phase $\theta(x)$ at each point. A change in the local phase $\theta(x)$ simultaneously alters the electron field $\psi(x)$ and the electromagnetic potential $A_\mu(x)$. The form of the coupling term Eq. (4.3), which is illustrated in Fig. 4.1, is dictated by local symmetries. It is strictly a *point interaction*; the coupling occurs at a point befitting local field theory.

The Logic of Gauge Field Theories

The derivation given of the quantum-electrodynamics Lagrangian via the argument of local symmetries is common to gauge field theories. For the nuclear interactions, the interaction potentials and their coupling to matter fields are unknown. The line of reasoning provides a powerful approach to find the unknown characteristics. Roughly speaking, we start with a free matter field and derive the interacting field system in these steps:

  1. *The localization of symmetry transformations.* The symmetry transformations of the phase factor of a matter field are characterized by certain

parameters. Making these parameters functions of spatio-temporal position localizes the transformations to each point x in the matter field. The points neighboring to x undergo different phase transformations. Effectively, an individual state space is set up for the local field $\psi(x)$ for each x.

2. *The enforcement of global invariance and the derivation of the term for reconciliation.* To ensure that the autonomy granted to the localities does not lead to anarchy, it is required that the Lagrangian of the field as a whole is invariant under the local transformations. A term is found to preserve the global invariance. It reconciles the phase factors at various points in the field.

3. *The introduction of an interaction field.* The principle of local symmetry endows the extra term with full physical significance by interpreting it as the coupling between the phase of the matter field and the potential of an interaction field. The potential compensates for the phase change at each point, and their coupling is identified as the point interaction between two fields. The form of coupling so derived is called *minimal coupling.* It is universal and is the same for any matter field with the same symmetry properties. The theory then proceeds to find the characteristics of the interaction field by exploring the structures of the coupling terms. The result is the description of an interacting matter–interaction field system.

In quantum electrodynamics, the phase transformations involve only one parameter θ, and the phase exponent contains only numbers, which commute. In the nuclear interactions, the transformations involve several parameters, and the phase exponent contains matrices, which do not commute. The non-commutative cases are much more complicated. However, the major ideas are similar.

The Fiber Bundle Formulation of Dynamical Theories

The conceptual similarity of the theories for all four fundamental interactions is rigorously shown in their fiber bundle formulations. The fiber bundle is an important concept in differential geometry. It includes manifold or differentiable structures, group or symmetry structures, and affine or connection structures. It provides a comprehensive framework within which we can compare various theories.

Fiber bundles can represent many dynamical systems, both interacting fields and classical systems with ponderous bodies. Different dynamical systems involve different types of fiber bundles as listed in Table 4.2, but all systems share the general notion of the fibrillation of a whole. The fiber bundles representing classical dynamical systems differ from those representing field theories; they lack local symmetry groups. The possession of a local symmetry group, which characterizes field theories with local symmetries, sets general relativity and the gauge field theories apart from classical dynamics.

Table 4.2   The fiber bundle formulations of dynamics

Dynamical system	Physical theory	Fiber bundle
Gravitation interaction	General relativity[47]	Orthonormal frame bundle
Strong, weak and electro-magnetic interactions	Gauge field theories[48]	Principal fiber bundle
Classical mechanics	Lagrangian formulation[49]	Tangent bundle
	Hamiltonian formulation[49]	Cotangent bundle

Crudely speaking, each point in a matter field is mathematically represented by a distinct fiber in a vector bundle. The fiber has an intrinsic structure determined by the local symmetry group, and is indexed by a unique point in the spatio-temporal parameter space. The potential of the strong, weak, or electromagnetic field is represented by a connection on a principal fiber bundle, the potential of gravity by a connection on an orthonormal frame bundle. The connection is related to the phase, which is an element in the associated vector bundle representing the matter field. The entire dual bundle is invariant under local symmetry transformations. The specifics of the bundles differ, but a common general conceptual structure is unmistakable. (See Appendix B.)

To find a unified theory for the basic structure and dynamics of the world is a persistent aspiration of physicists. Einstein left an unsuccessful unified field equation on the blackboard of his office. The touch passes on. Physicists have not yet found a satisfactory unified theory, but they have made major strides. t'Hooft summarized his review of the fundamental interactions of the physical world: "All four forces are now described by means of theories that have the same general form. Thus if physicists have yet to find a single key that fits all the known locks, at least all the needed keys can be cut from the same blank."[50]

5

Object of Experiences: Quantum State–Observables–Statistics

Her [nature's] fundamental laws do not govern the world as it appears in our mental picture in a very direct way, but instead they control a substratum of which we cannot form a mental picture without introducing irrelevancies. The formulation of these laws requires the use of the mathematics of transformations. The important things in the world appear as invariants (or more generally the nearly invariants, or quantities with simple transformation properties) of these transformations. The things we are immediately aware of are the relations of these nearly invariants to a certain frame of reference, usually one chosen so as to introduce special simplifying features which are unimportant from the point of view of the general theory. The growth of the use of transformation theory, as applied first to relativity and later to quantum theory, is the essence of the new method in theoretical physics.

P. A. M. DIRAC
Principles of Quantum Mechanics, Preface

§ 11. Properties and Quantum Properties: Amplitudes

Interpretation usually applies to texts. Unlike discourses, which are rooted in concrete situations and surroundings to which demonstratives and other indexical terms refer, texts are to some degree decontextualized. A text addresses unknown audiences and frees itself from the conditions of its production, thus opening the possibility for various readings. Reading is receptive but not passive, and genuine reading is interpretive. The reader contributes by adopting, fixing the references, finding meaning for textual descriptions in his own comprehension, entering the world jointly opened by the text and his anticipation, thereby gaining an understanding of the text. Understanding, interpretation, and application are inseparable elements of a hermeneutical act.

No application of a theory is possible without some interpretation. Quantum mechanics is a physical theory, not pure mathematics draped in meaning assigned by correspondence rules. Therefore there is no pure quantum formalism, no "neutral" presentation of the structure of quantum mechanics. The quantum theories taught in classrooms and used in research are already interpreted, so that at least some mathematical symbols have some physical or empirical significance. The working interpretation has been incorporated in the structure of quantum mechanics presented in § 3. It harbors a notion of the objective world by invoking general concepts such as physical properties and measured results. These concepts enable physicists to apply the theories in specific concrete situations.[51] Like our everyday understanding of the world, which is embedded in practice, the working interpretation of quantum theories is inarticulate and dull. However, it is more fundamental than fancy academic interpretations, for living and working are the primal modes of our being in the world.

The philosophical task of this work is to articulate a sense of the objective world that accounts for the working understanding and the success of quantum theories. I do not speculate on what really occurs in quantum measurements, because no satisfactory theory exists and as Feynman said, the physics we are studying is not twisted up with the measurement process.[52] The measurement problem is important, but at issue is some physical process that can only be illuminated by future physics. Physical theories are typically explicit about processes and mechanisms. Thus the silence of a theory on them should be respected; silence is an authentic mode of communication. Philosophical interpretations need not fill the silence with noise, and they should not mistake noise for foundations. My concern is how we have managed to understand the quantum world to the extent that we do without a satisfactory quantum measurement theory.

I follow a method of mathematical scientists that Kant described and adopted in his analysis of experiences. "The true method, so he [the first man who demonstrated the properties of the isosceles triangle] found, was not to inspect what he discerned either in the figure, or in the bare concept of it, and from this, as it were, to read off its properties; but to bring out what was necessarily implied in the concepts that he had himself formed *a priori*, and had put into the figure in the construction by which he presented it to himself."[53] Kant brought out what are implied in the general concept of empirical objects, which we have formed and put into our experiences so that we present what we experience to ourselves as objective reality. I try to do the same with physical theories. We examine quantum theories and try to bring out the general conceptual structure by which we acknowledge what they describe as the objective world.

To articulate the structure of quantum theories, I compare it to the logical structures of common sense. By common sense I mean the theoretical attitude of our everyday thinking, in which we step back from the tool context of our activities, abstract from the values we attach to objects, and regard the objects as mere presence. This attitude is most apparent when some tool breaks and we try to figure out what went wrong. As Martin Heidegger argued, it is an essential but not the primary mode of our being in the world. My shoes function most perfectly when I am oblivious of them, not when they pinch and cause me to look at them as objects.[54] I concentrate on the theoretical attitude because it is the part of everday thinking that the sciences develop and refine.

Endangered Quantum Properties

To pick out some entity and characterize it, as "the meteor moves at 1,000 miles per hour," is perhaps the most common and fundamental form of human thought. A predicate such as "1,000 mph" stands for a physical property that is independent of our observations, although it has ramifications that are empirically detectable. One thinks that physical properties are the major topics of physics. However, both the Copenhagen school and some of its opponents opt for abandoning the concept of objective properties in quantum physics. Niels Bohr said: "Atomic systems should not even be thought of as possessing definite properties in the absence of a specific experimental setup designed to measure these properties." Many philosophers, even those who disagreed with Bohr's instrumentalism, subscribed to the first part of Bohr's statement. In a lengthy recent review, R. I. G. Hughes concluded that "quantum systems do not have properties."[55] The dismissal of the concept of properties does not mean that a certain quantity has zero value, for colorless and odorless are definite properties; nor that the property varies, for time and change are not under consideration; nor that the property cannot be measured accurately, for the problem arises precisely when the physical system is not measured. It means that the question "what is it like?" and the statement "it is thus and so" are illegitimate. Abner Shimony called this "objective indefiniteness."[56] The adjective is limp; the concept of objects itself becomes unclear without the concept of definite properties.

Positivists are happy to let the reality of the microscopic world evaporate and to talk exclusively about "it appears so when observed in this specific way." Some interpreters, who reject the phenomenalist doctrine, try to find replacements for the concept of properties. Several alternatives have been suggested: "potentiality," "propensity," "eventuality," and "latency." Also introduced are the notion of "event" or "process," in which the potentiality is "actualized" so that a definite property obtains.[57] Potentiality and its like are categorically different from ordinary potentials or dispositional attributes. Potentials in physics, such as electric potential, potential energy, or latent heat, are physical properties. Dispositional attributes such as soluble or flammable are usually shorthand descriptions of definite properties, for example, the property of interacting with a certain liquid in a certain way or igniting under certain conditions. The interactions and conditions can be substantiated by definite physical mechanisms, at least in principle. This is not so with potentiality, which is seldom elaborated, except that it has something to do with probability, which is again left vague. Howard Stein gave as good an exposition as I can find, saying that "eventuality" is introduced "to suggest the 'potential' character of our attributes—which are phantoms in limbo until they are called forth by suitable test procedures."[58]

The abandonment of definite properties is mainly based on the alleged violation of two assumptions, which Bas van Fraassen stated as: "*Value definiteness*: Each physical parameter always has some value, one of the values which may be found by measurement. *Veracity in Measurement*: Measurement of a parameter faithfully reveals the value it really has."[59] It is argued that the postulates are violated by quantum systems, therefore quantum systems have no definite properties. The meanings of "parameter" and "value" are not clear. If the values are physical properties, one wonders why their definition should include the stipulation "may be found by measurement," which makes them dependent on observation.

In quantum interpretations, the "values" are usually taken to be eigenvalues or spectral values, which can be directly revealed in experiments, although the revelation may involve some distortion so that the veracity postulate does not hold. It is beyond a reasonable doubt that quantum systems generally have no definite eigenvalues. However, this does not imply that they have no definite properties. The conclusion that they have none arises from the fallacious restriction of properties to classical properties, of which eigenvalues are instances. Sure, quantum systems have no classical properties. But why can't they have quantum properties? Is it more reasonable to think that quantum mechanics is necessary because the world has properties that are not classical?

The no-property fallacy also stems from overlooking the fact that the conceptual structure of quantum mechanics is much richer than that of classical mechanics. In classical mechanics, the properties of a system are represented by the numerical values of functions, which assign real numbers to various states of the system. In quantum mechanics, functions are replaced by operators, which are structurally richer. A function is like a fish with only one swaying tail, its numerical value; an operator is like an octopus with many legs. Quantum mechanics employs the octopus with good reason, and we miss something important if we look only at the one leg that reminds us of the fishy tail. Quantum systems generally do not have definite eigenvalues, but they have other definite values. The stipulation that the values must be directly revealable in measurements confuses the empirical and physical meanings of properties.

I argue that we cannot give up the notion of objective properties. If we did, the quantum world would become a phantom and the application of quantum mechanics to practical situations sorcery. Are there predicates such that we can definitely say of a quantum system, it *is* such and so? Yes, the wavefunction is one. The wavefunction of a system is a definite predicate for it in the position representation. It is not the unique predicate; a predicate in the momentum representation does equally well. Quantum properties are none other than what the wavefunctions and predicates in other representations describe.

I hope physicists will not yawn and throw this book down. This seemingly trivial answer is dismissed by many interpreters, not without reason. It is not enough to assert the physicality of quantum properties; the assertion is futile if we cannot claim knowledge of the properties. How is the claim of knowledge justified when the properties described by the wavefunctions are never found in experiments? At once the relation between objects and experiences and the battery of philosophical questions it entails spring into focus. Here is the crux of the difficulties in quantum interpretation. The treatment of these questions will take up the whole of this chapter.

States, Property Types, and Values

To rescue the intangible quantum properties, we must clarify the meaning of properties in general. In ordinary speech, a thing's state refers to its overall mode or condition, such as a bad state or a state of excitement. This sense is captured in the technical meaning of a *state* as the abstract summary of a system's characteristics. The concept of states serves two important logical functions. First, it separates the qualities and the numerical identity of a system, so that we can think of the system's change through time. Second, it distinguishes the objective state of affairs from its definite descriptions, which, as we will see, always involve certain conventions.

We often think of an object in a state different from the one it is in; we think of its possibilities when we envision its past and future. The thought of *possibilities* frees us from what is momentarily given through the senses and opens us to the world. Kant entitled the category of qualities "anticipation of perceptions." We anticipate because we see the possibilities. In physics and many other sciences, the intrinsic relation between the concepts of objective states and possibilities is made rigorous. The set of all *possible* states of a system, called its *state space*, is the most basic postulate of a theory and is the fundamental framework underlying all descriptions of the system. It is important to note that the concept of possibilities is not a superstructure but a primitive element in our thoughts about objects.

The "space" in the state space, like many mathematical spaces, refers to an algebraic system or abstract set endowed with certain structure; it has nothing to do with the space that is often mentioned with time. A state space usually has certain intrinsic structures, so that the possibilities of the system to which it belongs are limited in certain ways. As will be discussed in § 19, the state space of an entity can be interpreted as the *kind* of things it belongs and the notion of kinds is integral to the notion of entities. Various kinds of state spaces specify various kinds of entities. The Hilbert space and the differentiable manifold distinguish quantum and classical kinds, and various types of Hilbert spaces distinguish various kinds of quantum systems.

Properties are more specific than states; with the concept of properties we analyze the state of a system into various aspects and ascribe specific features to the system. Generally, the object's state is not committed to any property types. If a property is chosen, then the meaning of states becomes more specific and "state" is qualified, as in "energy state" or "momentum state."

In ordinary language, properties are expressed by common terms, which include common nouns, adjectives, verbs, and adverbs. Common terms have three usages. They can be predicates, as in "the apple is red," subjects, as in "red is a color," or assume a classificatory role, as in "the flower and your dress are similar in color." The classificatory role is crucial, for often we ascribe the same property to different things based on their similarity in certain respects. The flower and the dress are similar only in color, not in texture or other respects. Thus the concept of properties must incorporate both a covering element to account for respects such as color, and a specific element to account for specific features such as red.[60] Under the covering element, the specifics become "individuals," which is why essentially predicative terms such as "red" can stand as the subject in certain sentences. The duality in the concept of properties becomes clear in questions: "What texture does it have?" "How heavy is it?" The conceptual complexity of properties agrees with the argument of Kant, who said a quality must admit variable degrees.[61] Thus its concept has two components: One specifies what quality, the other what degree.

An attributive concept has a covering aspect and a specific aspect. I call the former a *property type* and the latter a *value*. Property types such as color and weight are rules that systematically assign values to things and are often called parameters, variables, or dynamical variables in scientific theories. In ordinary speech where values are less finely differentiated, the rulelike nature of property types can be seen from comparatives and superlatives. Since the value attributed to a thing is an instance of a rule, the attribution goes beyond the assigned value and implies the alternatives under the type. Thus the concept of properties calls upon the concept of possibilities and agrees with the concept of state spaces.

The values are definite properties; linguistically, they are the predicates in attributive statements. Rough is a value of texture and 60 mph of speed. A vanishing value is a definite value, for example, something at rest is said to have zero velocity. Values are usually numerical in physical theories; numbers provide powerful means to assign systematic, accurate, and finely discriminatory predicates. However, generally values need *not* be numerical. The broad sense of "value," which includes number as a special case, is common in mathematics, where values are generally the end variables of transformations. For instance, the value of a projection-valued measure is a projection operator. The broad sense is also common in ordinary speech; love has intensities or values not measurable by numbers or dollar amounts. The type to which a value belongs is manifested in the unit if the predicate is numerical and is often understood if it is not.

A property type may be inapplicable to a thing; for instance, the types color and taste are inapplicable to atoms. If a property type is applicable, then a definite value follows. It makes no sense to say that something falls under a property type but has no definite value, as something is colored but has no definite color. We may not know the value, but conceptually it is definite.

A value is a partial description of a system; the system can have many aspects covered by many property types. Often many properties are attributable to an entity, but not all are required to characterize it completely. We say a thing can be described in many ways, although the things we handle every day are usually so complicated that the descriptions are rarely complete. The completeness and redundancy of descriptions are more apparent in basic physics, where the systems are simple. For instance, a classical particle can be completely characterized by its three components of position and momentum or by its position, energy, and two momentum components or by its position, energy, and angular momentum. Thus there is a choice of property types in describing objects. The specific choice is conventional, but there are transformation rules relating the descriptions in various types, so that the objectivity of the description is preserved.

States and Properties in Classical Mechanics

States, possibilities, types, and values constitute a conceptual framework that is implicitly understood in the attributive statements we make every day. It has been incorporated into physical theories.

Consider a classical particle. Its state space is a six-dimensional differentiable manifold, each point x of which represents a possible state. The actual states of the particle at various times trace a trajectory in the state space. The description is highly abstract. To describe the particle more definitely we introduce property types such as position, momentum, angular momentum, or energy. The property types are represented by *functions*, which are maps from the state space to the real line or its products. A function assigns to a state of the particle a number or an ordered set of numbers as the value of a property type, and we usually call the value the particle's property at that moment of time.

For example, suppose the function f represents position; f assigns to each possible state x of the particle a triplet of numbers $\{x^\mu\} \equiv \{x^1, x^2, x^3\}$

$$f(x) = \sum_{\mu=1}^{3} x^\mu \hat{\mathbf{x}}^\mu \tag{5.1}$$

The entire set of numbers $\{x^\mu\}$ is the *value* of the particle's position and we often simply call it the position. $f(x) \rightarrow \{x^\mu\}$ reads "the particle at state x has position $\{x^\mu\}$," analogous to "the position of Boston is 42°N 71°W."

The position function generates a *coordinate system*, and the real number x^μ is called the position coordinate for the basis $\hat{\mathbf{x}}^\mu$. The basic idea of coordinatization is the institution of bases of "scales" that make possible the systematic assignment of values. The idea can be generalized to other property types that are not spatial, for instance, the momentum. Such generalized coordinate systems are called *representations*.

Other properties of the particle are defined in similar ways. The many property types capable of describing the system are related. Properties such as energy can be defined either as a function on the state space, or as a function on the values of other properties such as the momentum.

Quantum States and Properties

The state space of a quantum system is a complex Hilbert space. The state of the system is represented by a vector $|\phi\rangle$ up to a phase factor $e^{i\theta}$, or more exactly by the equivalence class of vectors differing only by a phase factor. We will consider only a quantum system at a particular time, for the evolution of the system according to the Schrödinger equation is not controversial. Thus I often abbreviate "the physical system in the state $|\phi\rangle$" by "the state $|\phi\rangle$."

Penrose said: "It seems to me to be perfectly clear that there is (if we accept standard quantum mechanics) a completely objective meaning to $|\psi\rangle$ [the state vector]."[62] I think the assessment is common in the working understanding of quantum mechanics. When physicists say $|\phi\rangle$ is the state of a quantum system, they are not speaking metaphorically.

The property types of quantum systems are generically called *observables*. For instance, the momentum and the spin observables are two aspects of electrons. Some people prefer to call them physical quantities. Whatever they are called, they are property types and not things. Thus unless one is committed to the Platonic view of universals, it is not appropriate to say "an observable A has a certain value." Suppose A represents momentum. We say a particle or a car has a certain value of momentum, not the momentum has a certain value.

Observables are represented by *operators*. Like functions, operators are transformations or mappings. Unlike functions, the range of operators is not the real line but the state space itself. What are the values of quantum properties? The most obvious answers are the eigenvalues associated with the operators; they are real numbers and realizable in experiments. This answer does not work. Its failure is responsible for the abandonment of properties by many interpreters. The eigenvalues will be discussed in § 13. Here we look for alternatives.

When Dirac introduced the concept of observables in his classic textbook, he cited the criterion that an observable must have a complete set of eigenstates. A complete set of eigenstates is a basis of the Hilbert space, so that every state vector can be expressed in terms of it. A basis is a kind of coordinate system. The idea of a coordinate system brings quantum mechanics into analogy with classical mechanics and suggests the candidate for the value.

In classical mechanics, a particle's position and momentum assume definite values in certain coordinate systems. The same holds for quantum systems. But there are differences. The state space of quantum mechanics has a richer intrinsic structure.[63] The Hilbert space has a builtin metric structure embodied in the inner product, which enables the quantum state space to internalize coordinatization as a kind of relation among states. Thus quantum properties can assume definite values in bases or coordinate systems defined within the state space itself.

The most interesting bases are associated with observables. Let A be an observable with nondegenerate eigenvalues $\{a_i\}$ and eigenstates $\{|\alpha_i\rangle\}$, which is a basis for a state space. A state $|\phi\rangle$ can be written in terms of the basis

$$|\phi\rangle = \sum_i c_i|\alpha\rangle_i, \qquad c_i = \langle\alpha_i|\phi\rangle \tag{5.2}$$

where the set of complex numbers $\{c_i\}$ is the *amplitude*. The basis of A constitutes a *representation* of $|\phi\rangle$. The state $|\phi\rangle$ and the amplitude are both determined only up to a phase factor. Except for an arbitrary phase $e^{i\theta}$, $\{c_i\}$ is the definite property of the quantum state as $\{x^\mu\}$ is the property of a classical particle, or we can say that the value is the equivalence class of amplitudes differing only by an overall phase factor. There are types of amplitudes associated with different observables, such as position amplitude, momentum amplitude, or spin amplitude. The position amplitude is called the *wavefunction*, which is the value for the property type position. It is important to note that "position" here is a quantum property, which is totally different from "position" in classical mechanics. The classical position is only a partial description of a classical system; the position amplitude or wavefunction is a complete description of a quantum system without spin.

There is an alternative definition of values that makes the logical analogy with classical mechanics closer and brings in the eigenvalues. In the basis of the observable A, A acts as a multiplication operator (Fig. 2.1b). It maps $|\phi\rangle$ into another state $A|\phi\rangle$; each of its component c_i is multiplied by a unique real value a_i,

$$A|\phi\rangle = \sum_i a_i c_i |\alpha_i\rangle \tag{5.3}$$

There is a one-to-one correspondence between $A|\phi\rangle$ and the sequence of complex numbers $\{a_i c_i\}$, which we can call the *A-amplitude*.[64] Equation (5.3) bears a formal analogy to Eq. (5.1). The base state $|\alpha_i\rangle$ plays the role of the classical coordinate axis $\hat{\mathbf{x}}^\mu$. The observable A and the function f are both property types. f is a rule that systematically assigns to each classical state x a unique set of coordinates $\{x^\mu\}$. Similarly, A specifies the rule $|\phi\rangle \to \{a_i c_i\}$ that assigns a unique A-amplitude to each state $|\phi\rangle$. The set of coordinates $\{x^\mu\}$ is the value and x_i the component value of the state x for the classical property type f. The A-amplitude $\{a_i c_i\}$ is the *value* and $a_i c_i$ the *constituent value* of the state $|\phi\rangle$ for the quantum property type A.[65] As a set of three real numbers is required to specify the classical position, so the entire set of complex numbers $\{a_i c_i\}$ is invoked to specify the quantum state's property. If A is the position operator, we say "the quantum system has positional value $\{a_i c_i\}$" just as we say "the classical particle has position coordinates $\{x^\mu\}$." The quantum positional value $\{a_i c_i\}$ involves all classical positions a_i. It precisely depicts the nonlocal and entangled quantum characteristics shown in correlation experiments. It is strange, but that is the way it is.

As the classical positional value is defined relative to a coordinate system, so the quantum amplitude is defined relative to a basis associated with an observable. Another basis associated with another observable gives another representation of the state. In the classical case, different coordinate systems are connected by symmetry transformations. In the quantum case, different observable representations are connected by unitary transformations. In either case, the choice of a particular system or representation is conventional or pragmatic.

The amplitude $\{c_i\}$ or the A-amplitude $\{a_i c_i\}$ is intimately related to the characteristic of the state vector $|\phi\rangle$ and fully compatible with it. It is predictable; it evolves deterministically accordingly to the Schrödinger equation. Thus I interpret *the amplitude or the A-amplitude as the definite property or the value of a quantum system in a certain state for the property type A.* The interpretation is based on the *general* way in which we think about properties and a *formal* analogy with classical mechanics. It does not rely on any substantive similarity on classical and quantum properties.

This interpretation raises many questions. What the amplitudes and *A*-amplitudes describe are nothing like classical features; they are literally unimaginable. What empirical evidence do we have of them? By what right do we ascribe them to quantum objects? What is the meaning of the eigenvalues measured in experiments? How do we correlate the amplitudes and the eigenvalues? How do we account for the many observables, especially incompatible ones? These questions are examined in the following sections. The next section consolidates the physical significance of the amplitudes by showing that their specifics can be determined from experimental data. Eigenvalues, probability, and their empirical connotation are discussed in § 13. Sections 14 and 15 investigate the meanings of observables and representations and examine a basic conceptual structure shared by quantum mechanics and other physical theories. In the remaining two sections, various versions of the observer and the phenomenalist claims are refuted.

§ 12. The Form of Observation and the Reality of Quantum States

Amplitudes, *A*-amplitudes, and transition amplitudes are the quantities with which physicists work. However, there is strong resistance among philosophers to accept them as definite quantum properties. The amplitudes and quantum states are complex quantities of which we cannot even form a mental picture. In some formulations of the definite value principle, for example that by Richard Healey, the values are specifically restricted to "real values" so that the amplitudes are excluded. Michael Redhead noted that interpreters often regard "picturability" to be a criterion of understanding.[66] Some interpreters argue that an adequate interpretation must invoke a certain kind of hidden variables, which show how the world could be the way quantum mechanics describes it. Hidden variables are picturable.

The insistance on visualizability is not without justification; it is related to the fundamental limitation of our ability to observe. Quantum states and amplitudes are suspect because they are never directly found in experiments. I argue that the suspicion is based on a confusion. In talking about value definiteness, interpreters often tacitly assume that values must be susceptible to direct measurements. Thus the concept of physical properties is made to depend upon the notion of observations. The dependence is unwarranted; physical and empirical are distinct notions, as we say there are things too small to be seen. We must have empirical evidence before we can claim knowledge of physical properties, but the evidence need not be a direct revelation in experiments. The relation between the physical and the empirical does not entail a conflation of the concepts.

In this section I make a clear distinction between the meaning of being physical and being empirical. I clarify the visualizability criterion and argue that it is violated by quantum states and amplitudes. Finally I show that the specific features of quantum states and amplitudes can be determined, although indirectly, from experimental data.

The Form of Observations: Real Number Representability

Kant and empiricist philosophers of science both argued that observations have conditions that cannot be taken for granted. However, they differed in major ways about the nature of the conditions. In the positivist philosophy of science, the criterion that separates observational and theoretical terms is specific and substantive. It is the sensitivity of our perceptual organs, perhaps aided by some primitive tools.[67] The physiological criterion is usually extrinsic to physical theories; a categorical distinction between the visible range and the rest of the radiation spectrum is totally foreign to the theory of electromagnetism. Consequently the observability criterion is imposed externally and performs no function in the theories. The meaning of "observation" it entails is incompatible with the common usage by experimental physicists. Anyone who has visited laboratories or read experimental papers would notice that what physicists say they observe are not what meet the eyes, for example, oscilloscope traces or computer printouts. Consider the experimental reports: "We report the observation of a heavy particle J," and "we have observed a very sharp peak in the cross section for $e^+e^- \rightarrow$ hadrons." One of the experiments involves the scattering of protons, the other the collision of electron–positron beams.[68] According to the positivist observability criterion, both the object, the J/ψ particle, and the tools of measurement are unobservable; the experiments are like looking for black cats in dark rooms.

Kant argued that the general form of possible experiences is not decided by the grossness of our senses. Instead of a substantive criterion for observability, he introduced an expanded general structure that accommodates a categorical distinction between the sensual and conceptual aspects of objective knowledge. Unlike Leibniz's confused and distinct ideas or David Hume's faded and vivid impressions, the Kantian distinction is not one in degree but one in kind. Ultimately, knowledge rests on our sensual contact with the world, and our sensibility has its limits. The sensual limits Kant called the *forms of intuition;* intuition includes perception and imagination. He argued that the forms are space and time, which are not substantive. Although we can think generally, what we observe are always particular items. Space and time are the forms we have used in differentiating particulars as outside and alongside each other. Since space and time are inalienable elements in experiences, the forms of intuition play crucial roles in empirical knowledge.[69]

How do the philosophical theories answer to physical theories? There is no mention of any condition of observation in classical mechanics. I guess there would be none in quantum mechanics if we could observe amplitudes and phase correlations with the help of our experimental equipment. Unfortunately we cannot. The predicament forces quantum mechanics to make explicit the criterion of observability. The criterion does not invoke human physiology. Like Kant's thesis, it is abstract, intrinsic, does active work, and forms part of a vastly expanded conceptual structure. Unlike Kant's thesis, it is not particularly stuck to space and time.

In quantum mechanics, property types or dynamical variables are represented by operators. However, not all operators representing dynamical variables are observables. The most obvious counterexamples are found in quantum field theory. The creation and annihilation operators, and field operators generally, are dynamical variables with clear physical meaning, but they are *not* observables. For instance, the classical electromagnetic field is the limit of a dynamical variable that is not an observable (see note 43). There are criteria for observables which are associated with quantities that in principle can be measured. Quantum theories dictate that observables must have eigenvalues that are *real numbers*. Dirac explains: "When we make an observation we measure some dynamical variable. It is obvious physically that the result of such a measurement must always be a real number, so we should expect that any dynamical variable that we can measure must be a real dynamical variable."[70] (From now on I use "real" and "complex" in the technical sense as pertaining to real and complex numbers, "reality" and "complexity" in the ordinary sense.)

It is a fact that physical experiments return numbers, strictly speaking, rational numbers. There is a limit to accuracy; even if we can measure to the hundredth significant figure, the result is still a rational number. However, it can very well be idealized as a real number. Perhaps we can interpret the idealization in the Leibnizian sense, where phenomena are results of some kind of "blurring." In quantum mechanics, measurements return eigenvalues of observables that are real. *In stipulating that quantities admissible as measured results must be real numbers, quantum theories make explicit a general limit to human empirical capabilities: The general form of our sensual capability is representability by real numbers.*

I call a characteristic *visualizable* only if in principle some function can be found that maps the object structure into the real number system or its direct products such that the usual addition, multiplication, and ordering features of the numbers are preserved. In formal measurement theory, similar conditions are expressed by representation theorems.[71] We do not see the numbers; we can see those qualities that are representable by them. The paradigm entity represented by the real numbers is a continuous line. Visualizable qualities are in principle analogous to the subsets of the real line, its direct products, and subsets or patchwork thereof. These qualities are spatial in a very general sense. The visualizability criterion means that we are essentially spatial beings. Only visualizable characteristics are realizable in experiments within our capability.

The visualizability condition extends far beyond Cartesian spaces and includes manifolds. For example, an undulating two-dimensional surface is a subset of a three-dimensional Cartesian space or a patchwork of tiny flat surfaces. There is no restriction to the number of dimensions; one can visualize a four-dimensional cube. All spatial things, temporal events, and geometric configurations, including general relativistic ones, admit real representations and are visualizable; time is visualized spatially as a curve. The real-number representation admits many abstract quantities and qualities that are not spatio-temporal. We spatialize them in visualization. Visualization includes perception and imagination, which are "faithful" mental pictures, and should not be confused with symbolic representations. For example, Feynman diagrams are not visualizations but informative symbols of microscopic processes.

As the form of the ultimate contact between us and the world, visualizability is important. The physically significant terms in classical mechanics, electrodynamics, and thermodynamics are exclusively real quantities. When complex quantities are used for calculational convenience, it is always understood that only the real and imaginary parts of the quantities have physical meaning. This is possible because the classical equations of motion do not couple the real and imaginary parts; they do not explicitly contain the imaginary number $i = \sqrt{-1}$. A complex quantity is visualizable if it can be decomposed into real and imaginary parts, which are separately represented by real numbers.

Classical characteristics are representable by real numbers; their visualizability, not narrow instrumentalism, explains the indispensability of classical concepts in objective knowledge. Classical concepts are present in quantum mechanics; the eigenvalues of the position and momentum observables—not the observables themselves—are essentially the classical notions. Bohr rightly emphasized this point. The question is whether visualizable features have exhausted physical characteristics.

The Irreducible Complex Nature of Quantum States

Quantum characteristics are irreducibly complex, they cannot be decomposed into real and imaginary parts. i appears explicitly in the Schrödinger equation and other fundamental relations. The canonical commutation relation between the position operator Q and the momentum operator P is $QP - PQ = i\hbar$. Suppose we write a state vector as $|\phi\rangle = |\phi_1\rangle + i|\phi_2\rangle$. We find that $|\phi_1\rangle$ and $|\phi_2\rangle$ are coupled by the Schrödinger equation. Thus the entire complex vector $|\phi\rangle$ is significant in quantum mechanics, and its meaning is destroyed if we try to separate the real and imaginary parts.

I do not mean that the symbol i must be present in all formulations of quantum mechanics. There are many mathematical ways of expressing the same thing. For example, i can be represented by a real matrix whose square is -1. The general definition of complex numbers is the ordered pairs of real numbers obeying complex multiplication rules.[72] Alternatively, we can consider a complex wavefunction as a pair of coupled real functions. However, all these formulations involve some extra structure, and it is the extra structure that interests us. The extra structure is most conveniently represented by the complex number system, which includes the real number system as a substructure.

The complex nature of the quantum state space is neither an accident nor a mathematical convenience. George Mackey and E. C. G. Stueckelberg had independently casted quantum mechanics in a real Hilbert space. The methods they used are different, but the results are essentially the same. To describe the physics as supported by experiments, the real formulation must contain an observable J that commutes with all observables and $J^2 = -1$. Without this operator there is no uncertainty relation. The operator J is just another way of representing i. Thus the real formulation reverts to the complex. The quantum states must be complex if we are to obtain a satisfactory probability concept with nontrivial dynamics.[73]

A pure quantum state is represented by a unit vector up to an arbitrary phase factor. This may give the impression that phases are unimportant. The impression is wrong. The state vector is an abstract summary. To describe the state more definitely we have to use representations, which reveal the state's relations with other states. The relation is contained in the complex relative phases. Due to the builtin metrical structure of the Hilbert space, a state $|\phi\rangle$ can be expanded in terms of a set of basis vectors $\{|\alpha_i\rangle\}$

$$|\phi\rangle = e^{i\theta_1}(|c_1\||\alpha_1\rangle + |c_2|e^{i(\theta_2-\theta_1)}|\alpha_2\rangle + \ldots) \tag{5.4}$$

where the *amplitude* $c_i = \langle\alpha_i|\phi\rangle$ is written as $|c_i|e^{i\theta_i}$; $|c_i|$ is called the *modulus* and θ_i the *phase*. The absolute phase in front, $e^{i\theta_1}$, has no physical significance. However, the relative phases $e^{i(\theta_2-\theta_1)}$ are all important; they account for the peculiarly quantum-mechanical interference phenomena. That is why we need complex numbers c_i instead of their moduli $|c_i|$ in the expansion of $|\phi\rangle$. A similar condition holds for the A-amplitudes.

The set of moduli $\{|c_i|\}$ is readily measured, but the relative phases are somehow destroyed in experiments, exactly how we do not know. Measurements return real numbers, into which complex numbers cannot be homomorphically embedded. It is like trying to squeeze a three-dimensional thing into a two-dimensional plane: Some damage is unavoidable. This is not due to our clumsiness. The source is more basic; it is due to the fundamental limitation on our form of observation. The complex quantum states and amplitudes cannot be visualized. Dirac said: "In the case of atomic phenomena, no picture can be expected to exist in the usual sense of the word 'picture,' by which is meant a model functioning essentially in classical lines."[74]

Empirically Determinable Properties

Nonvisualizable properties pose great epistemological difficulties. What distinguishes quantum states and amplitudes from angels, when the characteristics they describe cannot be realized in any experiment we can perform? We are only interested in knowable physical properties, and to claim knowledge we must be able to supply empirical evidence. I call a characteristic *observable* or *perceivable* if it is visualizable and can be linked to our sensual perceptions by whatever instruments we can devise with the aid of whatever established physical laws. I think this criterion of observability is descriptive of the practice of natural scientists.

Liberal as it is, observability is a criterion, and criteria can be overstepped; otherwise they would be idle. For unobservable characteristics, we are forced to seek indirect evidence. A characteristic has *empirical ramification* if it is either observable or "kickable." Kickability is Alfred Landé's term.[75] Something is *kickable* if it can be kicked and kicks back, or it can be somehow physically manipulated and the manipulation produces observable effects. Presumably the property is remote and obscure if we must resort to the indirect kickability criterion alone. Thus kickability can only work in a well-developed theory in which the property is clearly defined, for we must be able to say specifically what we are kicking and how it is supposed to kick back. Suppose I wave a wand saying "hocus pocus" and something happens. No empirical ramification obtains because I have no idea what I have manipulated; I cannot even tell if the happening has anything to do with my action. In contrast, empirical ramification of radiation frequency obtains when I turn a knob on my radio and the music it produces changes. For we have a good theory informing us specifically that I have tuned the frequency. The existence of a theory means that a kickable characteristic can be manipulated in many ways and has multiple manifestations. Different instruments involving different processes concerning the kickable characteristic would agree on the observable effects of the characteristic. Conversely, from a diversity of observed phenomena, we can precisely extract the working of the kickable characteristic by considering the different processes involved.

A characteristic is *physical* if it has empirical ramifications and its specifics can in principle be determined, no matter how indirectly, from experimental data. A physical characteristic is the *value* or *property* of an object if its description can be abstracted from the particular empirical ramifications, and its ascription does not conflict with other properties of the object. The abstraction separates us from the phenomenalist position in which the physical are logical constructions upon measured data and are dependent on them. The kickability criterion is important in addition to empirical determination because in remote cases, specific manipulability alone ensures that the characteristic belongs to the physical system and is not some spurious effect arising in complicated experimentation. The experiments may have disturbed or even destroyed the system, the data may be statistical, and much analysis may be required to piece them together. However, as long as we can somehow determine the original kickable characteristic, we admit that the property is physical.

The crux of the preceding explication is to lift the unreflectingly assumed degeneracy between being physical and being observable without forsaking the requirement of empirical evidence for knowledge of physical characteristics. The degeneracy in meaning has been a stumbling block in understanding the quantum world view. Quantum properties are not visualizable, but this will no longer prevent them from being physical.

The Empirical Determination of Quantum Amplitudes

The kickability of quantum phases can be seen in the familiar two-slit experiment (Fig. 4.5). A version of the two-slit experiment is the Aharanov–Bohm effect, in which the relative phases of the fields from the two slits are changed by an electromagnetic potential, resulting in predictable changes in the interference pattern. Quantum theories describe exactly how the potential produces the observable effects by coupling to the phases of the electron field. The deeper physical significance of the coupling between phases and potentials will be discussed in § 27.

Phase manipulation is routine in all kinds of experiments, especially interferometries, which are conceptually similar to the two-slit experiment. Take, for example, neutron interferometry. A beam of monoenergetic neutrons is coherently split into two subbeams, represented by states $|\phi_1\rangle$ and $|\phi_2\rangle$, respectively. The subbeams are analogous to the waves that pass through the two slits. Along the path of the subbeam $|\phi_1\rangle$ is placed some mechanism that induces a phase shift in the passing neutrons, analogous to the phase shifter in Fig. 4.5c. For example, a magnetic field induces neutron spin precession, as a result of which the neutron state vector $|\phi_1\rangle$ suffers a phase change $\delta\theta(\mathbf{B})$, which is proportional to the magnetic field strength $\mathbf{B}$. Since the state $|\phi_2\rangle$ of the other subbeam is not affected by the magnetic field, $\delta\theta(\mathbf{B})$ represents a relative phase difference between the two beams. The two subbeams are then combined and their interference patterns recorded. Quantum mechanics predicts that the intensity of the recombined beam in the interference region should vary with the relative phase shift, and hence with the magnetic field. When the magnetic field is experimentally varied, the intensity changes as theory predicts. With only one kind of experiment, we can argue that all they show is the correlation between the changes in the magnetic field and the interference pattern. However, the same changes in interference can be produced by other means, including nuclear and gravitational interactions. If we ask what in the system is changed by all these means, quantum mechanics tells us it is the relative phase. Thus the relative phase of quantum states is manipulated with measurable effects.[76]

The preparation and measurement of quantum states are two separate empirical procedures. There is no difficulty in preparing systems in a definite quantum state. Practically, we have various kinds of filtration and selection arrangements, the most familiar one being the Stern–Gerlach setup in which particles in pure spin state are prepared. The measurement problem is much harder; given an arbitrary microscopic system, we want to determine its state including all phase relations.

To determine a quantum state $|\phi\rangle$ completely, we need to find out its amplitudes in one set of bases; only one is enough. Once it is known, we can transform to other bases using the transformation functions (2.4). The empirically relevant bases are eigenstates of observables. Consider a complete set of compatible observables summarily represented by A with spectrum $\{a_i\}$. $|\phi\rangle$ is expressed in terms of the eigenstates $\{|\alpha_i\rangle\}$ of A as in Eq. (5.2). We want to determine $\{c_i\}$ and $\{a_i c_i\}$ including all phase information.

A single observation on a single system tells us nothing; we need to observe repeated transitions of the system or use ensembles. That is true of all quantum experiments. So we take an ensemble of systems all in the pure state $|\phi\rangle$. Even with ensembles, the empirical determination of the complete state is not easy, for it requires the measurement of at least two *incompatible* observables. A measurement of the observable A on the ensemble yields $\{|c_i|^2\}$ as the fraction of data points with eigenvalue a_i. However, it misses all phase information; the θ_i are not determined. Measurements of observables compatible with A yield no further information, for they are mere functions of A. To determine the phases, we need to measure observables incompatible with A.

The mention of measuring incompatible observables may raise a mental block. Much attention has been paid to the fact that the eigenvalues of incompatible observables cannot be determined simultaneously to arbitrary accuracy. Much effort has been spent to find joint distributions for incompatible observables. However, these considerations do not concern us because we are *not* interested in simultaneous eigenvalues or joint distributions. We seek neither the precise values of the system's position or momentum, nor the joint position and momentum distribution for some ensembles. Our aim is to determine exactly the quantum amplitude $\{c_i\}$, not the eigenvalues of the observables. We want to use the information obtained from the position and momentum data to determine the amplitude in one of the representations. Such information can be obtained by performing separate measurements on identical ensembles. Various methods of simultaneously measuring incompatible observables, such as heterodyne and parametric amplification, are devised and applied, not only in physics laboratories but in engineering systems.[77]

The general problem of experimentally determining the state is complicated. Let us consider some specific cases. The simplest case is for spin-$\frac{1}{2}$ systems, which can be found in standard textbooks. To completely determine an arbitrary spin-$\frac{1}{2}$ state, we need to measure all three spin components.[78] A harder case is whether we can uniquely determine the states of spinless particles from the experimental data of their position and momentum distributions. Pauli pondered on this question early on. J. V. Corbett and C. A. Hurst showed that bound states can be uniquely determined, but scattering states cannot. Busch and Lahti showed that scattering states can be determined if the notion of observables is suitably generalized.[79]

The empirical determination of quantum amplitudes and phases is not an idle problem. It is not a major issue in fundamental physics whose attention is on universal laws, not individual systems. However, the behavior of individual systems is important in many applied areas, for instance, in communication engineering, which aims to squeeze as much information into the electromagnetic field as possible. Quantum phases carry information; therefore, phase manipulation and determination are important practical problems. The utility of quantum phases in real-life engineering systems is the ultimate proof of their physicality.[80]

The preceding presentation is brief, for a more detail discussion is too technical for this book. The interested reader can find a large amount of supportive technical material in the bibliography of the review articles cited. From the material we can conclude that *quantum amplitudes and complex phases are kickable and empirically determinable with sufficient labor. They are physical.*

To sum up, the general concept of properties is not changed by quantum mechanics. If a property type applies to a physical system, then the system at a particular time has a definite value as its definite property. The value is unambiguously predictable, precisely describable in mathematical terms, and empirically determinable. This holds for quantum as well as classical systems. Unlike classical properties, quantum properties are not visualizable, but visualizability is not a requirement for physical properties. The peculiarity of quantum systems lies in the specific features of their properties, not in the violation of the concept of properties.

§ 13. The Gap in Understanding: Eigenvalue and Probability

Consider a complete set of observables of a quantum system summarily represented by the self-adjoint operator A and regarded as a single observable. The observable A is associated with a set of eigenstates $\{|\alpha_i\rangle\}$ and a set of eigenvalues $\{a_i\}$. Any state $|\phi\rangle$ of a quantum system can be described in terms of its amplitude $\{c_i\}$ or A-amplitude $\{a_i c_i\}$, where $c_i = \langle \alpha_i | \phi \rangle$. When a measurement of A is made on the system, the eigenvalue a_i is observed with a probability $|c_i|^2$. In § 11, I interpret A as a property type and $\{c_i\}$ or $\{a_i c_i\}$ as the value or definite property of the system in state $|\phi\rangle$. The significance of the eigenvalues is still to be found and the meaning of probability clarified.

Eigenvalues are not Properties of Quantum Objects

A specific eigenvalue a_i has the unique significance of being the one that can be found in an experiment on a single system. It is also classical in nature and perceivable by us. However, there are great objections to ascribe it as the property of the quantum system. The nature of eigenvalues conflicts with that of state vectors, which claim to be the summaries of quantum properties; the one is real and the other complex. A state vector has its governing equation of motion, a specific eigenvalue found in an experiment does not. The eigenvalue found for a single system is not kickable within quantum mechanics. We cannot manipulate a quantum system to obtain a specific eigenvalue in an experiment. Hidden variable theorists have tried to amend quantum mechanics so that specific eigenvalues are attributed to the states. Many such theories have been refuted by experiments. Many studies have shown the impossibility of systematically and consistently assigning definite eigenvalues to quantum systems under certain reasonable constraints.

One important work of great generality that blocks the eigenvalue's claim to being the property of a quantum system is done by S. Kochen and E. P.

Specker. I have used and will use the symbol A to denote both an observable and its representative operator, except in this paragraph, where the observable represented by the operator A is denoted by O_A. Let us assume there is a one-to-one correspondence between self-adjoint operators A in a Hilbert space and observables O_A. Now suppose there is a classical-style function, *val* that assigns an eigenvalue a to each observable O_A for each state $|\phi\rangle$, $val_\phi(O_A) = a$. The eigenvalue a would be interpreted as the property or value of $|\phi\rangle$. In the classical case, the structure of values is preserved by functional relations. We will demand the same for *val*. If $f(O_A)$ is another observable, then its value is $f(a)$; $val_\phi[f(O_A)] = f[val_\phi(O_A)] = f(a)$. Kochen and Specker have proved that these conditions lead to contradictions, if the dimension of the Hilbert space is greater than 2. It is generally impossible to assign eigenvalues to quantum observables after the classical style.[81]

The contradiction in the Kochen–Specker proof can be avoided if we allow an operator to represent many observables. However, if we reject the arbitrary multiplication of concepts, we have to admit that quantum systems do not generally possess definite eigenvalues. This is no surprise to physicists. As David Mermin remarked in his paper on no-hidden-variable theorems, quantum physicists must be shocked that such a commonplace needs such elaborate proofs.[82]

Abstractly, the eigenvalues of an observable can be regarded as labels of the eigenstates, which are the axes of a coordinate system in the state space that enables us to determine definite amplitudes. Physically, the labels are realized in classical objects we can measure. How they are realized no one knows. Eigenvalues are analogous to symptoms of a disease, which are disturbances of the body that show up and indicate something that does not show up. Just as a cold persists though its symptoms are suppressed, a quantum system has a definite amplitude though it has no eigenvalue. Unlike the amplitudes, the occurrence of eigenvalues needs the extra condition of classical realization. Theoretically, the condition is represented by selecting one term from the series of eigenstate expansion. This step occurs at the end of a calculation and marks the end of a quantum process, when the Born postulate is invoked. Practically, an indicator is somehow triggered in measurements and experiments. Unperformed experiments have no results, but this does not imply that the quantum system on which the experiment might be performed has no properties.

Suppose $|\phi\rangle$ represents the state of a hypothetical quantum cat. Let the observable A be a property type whose two possible symptoms are a_1 = alive and a_2 = dead. Let B be a property type whose possible symptoms are b_1 = agile and b_2 = immobile. The quantum cat's definite property is its A amplitude $\{a_1c_1, a_2c_2\}$, or equivalently its B-amplitude $\{b_1d_1, b_2d_2\}$. The complex numbers c_i and d_i are related by the transformation function Eq. (2.4). Each amplitude is a complete description of the cat's property. The choice between them is like saying a pen is 6 inches or 15 centimeters long. We have no words for the amplitudes, but that does not imply they are illusory. It is legitimate to say the quantum cat *is* $\{a_1c_1, a_2c_2\}$, or it *is found* alive or dead. It is illegitimate to say the quantum cat *is* alive or dead or a superposition of both. Even if the quantum cat happens to be in the eigenstate $|\alpha_1\rangle$ labeled by the eigenvalue alive, it is still only legitimate to say that it *is* $\{a_1, 0\}$. However, here we can *abbreviate* and say it has property a_1 or it is alive. In the abbreviation, a_1 refers not to an eigenvalue but to the sole nonvanishing element in an amplitude. This is similar to the case of classical positions, where we briefly say of something with position $\{0, 1, 0\}$ that it is due north. The abbreviation "the cat in the eigenstate $|\alpha_i\rangle$ has value a_i" does not involve measurement outcomes or other empirical concepts. It is a plain attributive statement.

The quantum predicates of amplitudes describe characteristics more complicated than the classical predicates of eigenvalues can handle. However, in special cases, a quantum system may be "aligned" in such in a way that a classical predicate becomes a good abbreviation. As an analogy, imagine two-dimensional beings whose language can describe only two-dimensional phenomena. Suppose their two-dimensional world is embedded in a three-dimensional world; they would be puzzled by events that are inexplicable in their language. However, certain special phenomena can be described in a two-dimensional language. These special cases are analogous to the eigenstates.

How about those cats with possible properties of being alive or dead? They are classical cats, not quantum cats. The quantum formalism, when applied directly, is not valid for classical entities. Of course, quantum mechanics can be used to construct theories of complicated systems that exhibit classical properties, but such properties are not represented by state vectors.

The so-called wave-particle paradox is similar to the paradox that arises from attributing life or death to the quantum cat. A system is often said to be a particle if a position eigenvalue is observed and a wave if a momentum eigenvalue is observed, hence it is said to present a paradox. The paradox is the result of fallacious attribution. Both eigenvalues are classical quantities, and neither can be attributed to the system as its property. The property of the quantum system is its wavefunction in the position representation and momentum amplitude in the momentum representation. Either one of them offers a complete description, and the descriptions can be shown to be equivalent by rigorous transformations. Paradox arises only when interpreters insist on considering exclusively classical quantities that we can observe. The insistency closes the mind to quantum properties. The rentention of classical dynamical concepts in quantum mechanics worried Schrödinger, who complained that one is always constrained to speak only about the probability for the measurement of some classical quantity. He said: "But in so doing one gets the feeling that only with difficulty can precisely the most important statements of the new theory be forced into this Spanish boot."[83]

A Remark on Realism

According to Michael Dummett, the principle of bivalence marks the difference between realism and anti-realism. Realism asserts that every statement is determinately true or false, and the singular terms in the statements denote objects in the relevant domain of discourse.[84]

Many philosophers argue that realism is refuted by quantum mechanics because "a quantum system has a certain eigenvalue associated with the observable A" is neither true nor false, since generally it is impossible to assign eigenvalues to quantum systems. The above analysis shows that "the quantum system has the eigenvalue a_i" is like "Caesar is prime (as in prime number)." Strictly speaking, they are not false, because their negations, "the quantum system does not have the eigenvalue a_i" and "Caesar is not prime," are not true. Perhaps we can say they are "deep falsities" in parallel to Bohr's "deep truths" whose negation also contains deep truths.[85] Usually we call them nonsense. Caesar is simply not a number to which the predicates prime and not prime apply. Erroneous application of property types can always lead to senseless statements; the quantum case is not peculiar in this respect. However, nonsense does not imply the breakdown of bivalence or the end of realism. Now the philosophical problem becomes how to rule out senseless propositions. The answer demands a proper concept of the individuals that stand as subjects. It is provided with the help of the notion of state spaces; a state space represents the kind of things the individual belongs and delimits its possible characteristics, as discussed in § 19. Here we note that the eigenvalues are ruled out because their general nature is incompatible with the possibilities allowed by the state spaces of quantum systems.

Because of the difficulties in the theories of truth, especially the correspondence theory of truth, I am wary of defining realism in terms of truth.[86] However, if we adopt Dummett's differentiation, then the realistic interpretation applies to quantum mechanics if it applies to our common sense notions of ordinary things. The statements "the quantum system in the state $|\phi\rangle$ has A-amplitude $\{a_i c_i\}$" and "it has wavefunction $\{c_i\}$" are like the statement "the car travels at velocity $\{v_i\}$", they are either true or false. We can find out whether they are true by experiments in an indirect way, but their being true or not is independent of our measurements.

The Unresolved Quantum Measurement Problem

In the working understanding of quantum mechanics, the eigenvalues have a strong empirical connotation; they are what are observed or found in experiments. The meaning of observers and experiences will be discussed in more detail in the following sections. Here we note that the legitimate empirical significance is exhausted by the abstract statement of our form of observation. Eigenvalues are numerical and fall within the bounds of our form of perception, whereas quantum amplitudes do not.

It appears that the world requires two general classes of predicates, quantum and classical, for description. Why it is so no mortal can answer. Since both classes describe physical properties, we expect that we can find some physical theories that properly relate them. Many people also expect that the theory can be found within the framework of quantum mechanics. The efforts to find such a theory are customarily lumped under the heading of the quantum measurement problem. The frustration is that the solution is so far elusive.

The eigenvalues describe classical characteristics, which do not exhibit the phase entanglement peculiar to quantum amplitudes. A position eigenvalue of the position observable is the familiar classical position, and it can be legitimately regarded as the property of a classical system. Quantum mechanics includes both classical and quantum characteristics but treats them differently. Quantum properties described by amplitudes are given full dynamical description; classical characteristics described by eigenvalues do not have an equation of motion. Furthermore, quantum mechanics gives no substantive or causal relation between the two classes of characteristics. The two-line explanations of the "collapse of wavefunctions" or the "jump" into an eigenstate can only be metaphorical by the standard of physics. The projection postulates, which attempt to substantiate the metaphor, remain inconclusive and controversial, (§ 4). Physicists have tried many other ways to establish substantive criteria for classicality within the quantum framework to give a better treatment of classical characteristics. They include superselection rules, decoherence effects, systems with infinite degrees of freedom, and generalized observables. None seems satisfactory.[87]

There are many philosophical analyses of the measurement problem, which I do not go into.[88] I leave the unsolved problem as a gap in our understanding and note that physicists understand quantum mechanics to a significant extent despite the gap. The editors of the first anthology on quantum measurements, which appeared as late as 1983, remarked that no textbook on the subject existed. The projection postulate is seldom covered in standard quantum textbooks. These omissions show that a substantial account of the relation between amplitudes and eigenvalues is not required for the working understanding of quantum theories. My philosophical task is to find the categorical framework that is responsible for both the working understanding and the recognition of deficiency in the understanding.

The Levels of Description

Quantum mechanics does provide some means to account for classical characteristics, but not in the way as formulated in the quantum measurement problem. Classical objects are macroscopic and composed of many microscopic constituents whose characteristics are governed by quantum laws. When we try to derive their characteristics from that of their quantum constituents, we introduce additional postulates and concepts to account for the effect of composition. By using quantum mechanics and some statistical methods, we can show that when we average over the behavior of the constituents of a large system, often the quantum phases are averaged out so that the system as a whole is essentially classical. There are also decoherence theories showing that the phase coherence of a quantum system is washed out in its interaction with the environment (see note 87). Consequently, the resultant classical systems are characterized not by amplitudes but by a new set of predicates that fall in the same class as the eigenvalues. Unfortunately, such calculation of classical characteristics does not help the quantum measurement problem, for it has already used quantum mechanics with its amplitude–eigenvalue duality.

The quantum measurement problem highlights a general problem in the philosophy of science. *The world exhibits many levels of scale and complexity that require radically different descriptions.* The classical and the quantum are two such levels, and there are many more beyond physics. Ontologically, we can agree that all systems, no matter how large and complicated, are made up of subatomic constituents. However, this does not imply that the theory for the subatomic constituents is applicable to all systems as integral units. Those who assume that it does have neglected the effect of composition and confused the properties of the parts with that of the whole. Generally, predicates of the parts of a system do not apply to the system as a whole. Consider "a crowd of hungry people." It is erroneous to say that the crowd itself is hungry, although each of its constituents is hungry. Similarly, the universal applicability of quantum mechanics means only that everything physical can be decomposed into atomic constituents, which can all be represented by appropriate quantum states. This does not imply that a macroscopic thing as a whole can be so represented. Quantum mechanics simply does not apply in the classical world. As Anthony Leggett said in his review article: "It is well known that the extrapolation of the conventional formalism of quantum mechanics to the macroscopic level leads to highly paradoxical consequences."[89]

The emergent properties of complicated systems can be investigated in their own right. We have some account of how the levels are related, but often there are scant substantive connections among concepts in our theories about different levels. Our knowledge is patchy and harbors substantive gaps. Some philosophers claim there is a privileged level to which the theories of all other levels can be reduced. The dogma is challenged by the quantum measurement problem, where it is shown that despite much effort, we are unable to reduce classical characteristics. There is no universally unified theory that accounts for all levels of complexity.

The Concept of Probability in Quantum Mechanics

The Born postulate prescribes the probability that a certain eigenvalue shows up in a measurement of the observable A on a system with a certain amplitude. The lack of a substantive relation between amplitudes and eigenvalues as exemplified by the quantum measurement problem agrees well with the notion of probability. The genuine sense of probability also involves a conceptual gap, which separates the notions of probability and statistics (§ A3). The corresponding gaps make quantum mechanics a probabilistic instead of a statistical theory.

The ideas of probability, randomness, and indeterminism are often intermingled. However, although quantum mechanics predicts a random phenomenology, the concept of randomness has no function in its theoretical structure. The absence is significant, because random processes are physical and physical theories are usually explicit in their accounts. For instance, in the kinetic theory of gases, the randomness of molecular collisions is expressly postulated by the factorization of the distribution functions and the postulate contributes to the Boltzmann transport equation. In contrast, quantum theories include no random mechanism. The Schrödinger equation is deterministic, and no stochastic process underlying quantum measurements has been found despite much effort. Probability just pops up in the Born postulate, and it is immediately converted to statistics. Perhaps "God throws a die" is the best description of the situation; at least it is honest to the fact that we cannot even tell what we have missed. Thus to understand quantum theories, we need a concept of probability that is not closely tied to randomness and indeterminism. We also need a notion that is distinct from statistics, so that we can separate the meaning of the Born postulate in terms of single systems and ensembles.

A detailed discussion of probability is found in Appendix A. The probability calculus involves two major strands of ideas, measure and independence, and the former dominates. *Independence* is related to the common notion of randomness, and it is crucial to the proof of many peculiarly probabilistic limit theorems such as the laws of large numbers. However, independence is not a primitive concept in the probability axioms. It is only a definition stipulating the special factorization characteristics of certain magnitudes. The axioms of the probability calculus are *measure-theoretic*.

The *measure* of a set is the natural generalization of such concepts as the length of a line segment or the volume of a figure. It extends the notion of magnitudes to abstract spaces and enables us to rigorously ask "how much" about abstract quantities such as possibility or evidence. Measure theory has become the foundation of analysis or calculus. Measures are used extensively in the analysis of Hilbert spaces, where they have no probabilistic connotation.[90]

Mathematically, *probability* is by definition a normalized measure, or a relative magnitude of a part in a whole whose total magnitude is 1. Thus the mathematical meaning of "probability" is precisely a ratio or a proportion. In many probability textbooks, probability is explained by analogy to the relative mass of bodies.[91] The analogy is abstractly strict. It warns against hasty interpretation of the probabilistic formalism employed in physical theories, for something akin to mass may be more appropriate.

In common usage, probability means more than a ratio; it invokes the notion of an *instance*, which is not included in the notion of relative magnitudes. Unlike relative magnitudes, which apply to sets that are abstractly interpreted as properties, probabilities apply to single instances with certain properties. To avoid confusion, I use "probability" only when "relative magnitude" or its synonym is inapplicable, for example, in the Born postulate (Q4 in § 3) but not in its statistical reformulation (Q4′).

The *concept of probability* is a formal structure that highlights the tension between an individual and a system with certain classification schemes. It comprises five elements: an individual ω in an aggregative system, a classification scheme of the system, the range of the classification parameter x, a specific parameter, and a verdict conveying the judgment that the individual falls within the class designated by the specified parameter. In most probability problems, the classification is made according to individual properties. However, this is not a necessary requirement. The range of parameters can be crude; it need contain only the possibility that the specified parameter does not obtain. For example, "the probability that it rains tomorrow is r" contains the weather classification with the alternatives rain or shine, an incident specified by the date, and the judgmental value r for the parameter rain. In contrast, the element of the individual is absent in *statistics*, where the value r has the meaning of a proportion.

The five elements in the concept of probability are found in quantum mechanics. The individual is the state of the single physical system that is the topic of quantum mechanics. The classification scheme is provided by the eigenvalues associated with a relevant observable. Here the scheme is based not on the physical properties of the quantum system but on their empirically accessible manifestations. To an A-amplitude $\{a_i c_i\}$ there correspond many possible eigenvalues a_i, and no substantive rule exists to pick the specific one that is realized. This dichotomy makes quantum mechanics probabilistic.

Probabilistic statements are judgments and not empirically verifiable. To make the Born postulate empirically meaningful, it is converted to a statistical statement about an ensemble of systems. We say when a measurement of A is performed on a large ensemble of systems all in the state $|\phi\rangle$, it is overwhelmingly likely that a fraction $|c_i|^2$ of the data points has the eigenvalue a_i. In the conversion, the notion of the single physical system is lost, and we are dealing with relative magnitudes of data points. The important point is that the statistics applies only to data points and cannot be ascribed to the quantum systems in the ensemble as their properties. Quantum mechanics is essentially the physics of single systems, thus it is essentially probabilistic.

It is often said that quantum mechanics implies "objective probability." The term has the poetic sense of a dice-playing god or a chancy nature. Karl Popper construed objective probability as propensity, which is some kind of property that endows the object with a certain tendency to produce certain results.[92] If the propensity is physical and law-governed, then the notion of probability vanishes once the underlying law is made explicit. If there is in principle no law for propensity besides the probability calculus, then it is only a new name. I admit I can only make metaphorical sense of propensity. Conceptually, the only sense I can make of objective probability is that our most fundamental physical theory irreducibly contains both a single-system description and a statistical description of ensembles and no substantive connection between the two.

§ 14. Coordinatization of the Quantum World: Observables

We have examined the meaning of various elements of the observable in the previous sections. *An observable coordinatizes the quantum world in a particular way with its eigenstates, and formally correlates the quantum coordinate axes to classical indicators, the eigenvalues.* An observable introduces a *representation* of the quantum state space by coordinatizing it.

Within the representation, a quantum state acquires a definite description in terms of amplitude. The state vector $|\phi\rangle$ and amplitude $\{c_i\}$ in quantum mechanics play the same formal roles as the coordinate-free state x and its coordinates $\{x^\mu\}_\alpha$ in relativistic mechanics. They realize, each in its own specific way, the general conceptual distinction of physical states and their definite representations in terms of specific property types.[93]

An amplitude is ascribed to a quantum system only with the choice of a representation, just as coordinates are assigned to a classical particle only in a coordinate system. More generally, a definite value is specified only within a particular property type; we cannot say the flower is red without the idea of colors. This condition means only that definite predications are necessarily conceptualized. It is completely general and purely conceptual, and has nothing to do with observations, measurements, or instruments. The empirical verification of the attributes is a separate question. Therefore, the interpretation of observables as quantum coordinatizations is different from Bohr's instrumentalism that says that a quantum system has no definite property without specific measurement arrangements. The instrumentalist's requirement is physical, tied to measurements, and peculiar to quantum systems.

Whenever we use coordinate systems, we can either transform the state holding the axes fixed, or transform the axes holding the state fixed. The feature is preserved in quantum mechanics. In the Schrödinger picture, the state evolves while the observable or equivalently the quantum coordinate system is fixed. In the Heisenberg picture, the observable evolves while the state is fixed.

The Conventionality of Observables and Property Types

The theories of relativity have taught us that coordinate systems are conventions. So are quantum representations. As discussed in § 11, an object can be described by the values of many property types. The choice of a specific description is arbitrary. Similarly, a quantum state can be characterized in terms of the amplitudes of many observables. Physicists choose a representation for expediency. They can work in the position representation (expanding in the eigenstates of the position observable) or the momentum representation (expanding in the eigenstates of the momentum observable), depending on which is more convenient for the solution of a particular problem.

All observables are represented by self-adjoint operators on the state space, but not all self-adjoint operators represent observables. Prominent exceptions are made by superselection rules that forbid superposition of and transition among certain groups of states. A superselection rule denies physical significance to certain self-adjoint operators. Even without superselection rules, some self-adjoint operators, notably projection and density operators, are interpreted as states. Among the infinitely many self-adjoint operators on a Hilbert space, physicists use only a handful, if we consider all observables related by a symmetry group to be the same. Usually the observables are chosen so that their eigenvalues can be measured in experiments. Wigner showed there are limitations to the measurability of arbitrary operators, and argued that only those operators that are actually measurable represent observables.[94] The empirical restriction is not preposterous if observables are quantum coordinate systems; we classify in the ways that suit us.

The arbitrariness of choosing observables is also seen in the Bell experiments. Suppose an atom decays and emits two electrons whose spins are antiparallel to each other. The two electrons travel apart, and their spins are separately measured in two experimental stations. Each station can freely set a parameter on its measuring apparatus, here the orientation of the spin measurement. John Bell proved that under certain locality conditions, the correlation of the measurement outcomes satisfies certain inequality that disagrees with the prediction of quantum mechanics. Many such measurements have been carried out, and the results uphold the quantum prediction.

J. P. Jarrett showed that Bell's locality condition is equivalent to the conjunction of two conditions. The first condition, which Shimony called "parameter independence," holds when the distribution of outcomes obtained in one measuring station is independent of the parameter setting on the apparatus in the other station. The second condition, called "outcome independence," holds when the distribution of outcomes obtained in one station is independent of the distribution of outcomes in the other station. Experiments show that one of the conditions is violated, and interpreters generally agree that it is outcome independence. The surviving condition means that the measured outcome is independent of the choice of the parameter setting in the distant station.[95] Quantum mechanically, the choice of parameter settings is a choice of observables, for instance, a choice between the spin observables S_x and S_y. The Bell experiments show that the choice is unimportant.

To say that observables are conventional does not imply that they are phantasmal. A concept is different from its realizations. Rigid rods, light rays, Greenwich, the fixed stars, and other things that implement coordinate systems in classical mechanics are concrete things by their own right. Conventionality lies only in the choice that make them into reference bodies. Similarly, observables are conventional only in their conceptual role as providers of quantum coordinate systems. Once an observable is chosen, its eigenstates that realize the coordinate bases are as physical as any other state, and the classical quantities that realize its eigenvalues are as concrete as a rigid rod.

The conventionality of representations does not lead to relativism. Quantum mechanics prescribes rigid *transformation rules* among various representations that leave the quantum state *invariant*. The significance of the transformations and invariance is discussed in the next section.

Incompatibility, Indeterminacy, and Statistical Dispersion

A salient feature of quantum theories is their admission of incompatible observables that have no simultaneous eigenstates and cannot be simultaneously measured to arbitrary accuracy. Incompatible observables are represented by operators that do not commute. How does the interpretation of the eigenstates of observations as quantum coordinate systems account for incompatible observables?

The most familiar account of the indeterminacy relation between incompatible observables is in terms of the variances of measured statistics.[96] Let A and B be observables obeying the commutation relation $(AB - BA)|\phi\rangle = i\hbar|\phi\rangle/2$. Let $\langle\delta A^2\rangle_\phi$ and $\langle\delta B^2\rangle_\phi$ be the statistical variances of the measured data of A and B on ensembles in the state $|\phi\rangle$. A straightforward calculation within the quantum formalism shows that

$$\langle\delta A^2\rangle_\phi\langle\delta B^2\rangle_\phi \geq \hbar/4 \tag{5.5}$$

This relation is a factual statement about data statistics. It has been beautifully verified in quantum optics, where minimum uncertainty states satisfying the equality in (5.5) are produced. The equality is shown to hold as the variances $\langle\delta A^2\rangle_\phi$ and $\langle\delta B^2\rangle_\phi$ change.[97] As a feature of data statistics, Eq. (5.5) is phenomenological. It is independent of the interpretation of observables as coordinatizations, which demands indeterminacy to be an intrinsic property of the quantum system.

The meaning of the indeterminacy relation has attracted much attention. Some interpreters argued that the statistical reading is the only viable meaning. However, Busch and Lahti have clearly distinguished two meanings of the indeterminacy relation arising from different mathematical sources, which they called *support* and *dispersion* properties.[98] Mathematically, the dispersion property is the source of (5.5), and the support property underlies the condition that incompatible observables have no common eigenstates. The support property is related to the coordinatization interpretation of observables.

Suppose the observable A has eigenvalues $\{a_i\}$ and eigenstates $\{|\alpha_i\rangle\}$, in terms of which the state $|\phi\rangle$ possesses A amplitude $\{a_i c_i\}$. Observable B has eigenvalues $\{b_i\}$ and eigenstates $\{|\beta_i\rangle\}$, in terms of which $|\phi\rangle$ has B amplitude $\{b_i d_i\}$. Suppose $|\phi\rangle$ coincides with $|\alpha_1\rangle$ so that $\{a_i c_i\} = \{a_1, 0, \dots\}$. If A and B are *incompatible* and *share no eigenstates*, then $|\phi\rangle$ cannot coincide with any of the $|\beta_i\rangle$, so that $\{b_i d_i\}$ must contain at least two nonzero elements. This condition can be manifested in the statistics of measured data; $|c_1|^2 = 1$ and $|d_i|^2 < 1$ for all i, because $\Sigma_i |d_i|^2 = 1$ and the sum contains at least two terms. The probability is 1 that a measurement of A yields a_1, and probability is less than 1 that a measurement of b yields any specific b_i. Thus incompatibility is a condition on the way in which two quantum coordinate systems pick subsets of the state space as bases. This meaning of incompatibility highlights the "out-of-jointness" of two quantum coordinate bases.[99]

Under the support property, the indeterminacy relation is applicable to *single systems*, in contrast to (5.5), which applies only to ensembles. It agrees with Heisenberg's original explication of the indeterminacy principle, which is for single systems. However, it deviates from Heisenberg's argument that relies on procedures of measurements.[100] In the previous interpretation, the indeterminacy relation expresses the conceptual limitation on the ways various property types or observables can be defined and does not depend on substantial means of measurements.

Recently, E. B. Davies, J. T. Lewis, and others introduced a generalized observable with builtin indeterminateness so that the indeterminacy principle naturally applies to single systems.[101] The generalization is amenable to the interpretation of observables as quantum coordination systems correlating real numbers to sets of states, which has not assumed any specific characteristic for the measured values. Generalized observables characterize a class of property types or quantum coordinate systems that is coarser grained and more tolerant than ordinary observables.

§ 15. The General Concept of Objects in Physical Theories

Bohr argued for the "*impossibility of any sharp separation between the behavior of the atomic objects and the interaction with the measuring instruments.*" A microscopic object cannot be specified without an account of the entire measuring apparatus because of the unavoidable and inexplicable interaction between them. Hence "an independent reality in the ordinary physical sense can neither be ascribed to the phenomena nor to the agencies of observation." Heisenberg said: "We can no longer talk of the behavior of particles apart from the process of observation."[102] Other interpreters generalized their doctrine that quantum objects are inseparable from observations to all objects. It is not difficult to find papers on quantum interpretation with concluding remarks such as "the moon is *not* there when nobody looks" or "perhaps an unheard tree falling in the forest makes no sound after all."[103] Someone asked the participants of a conference on quantum interpretation whether glass is transparent in the dark. A quarter of those responded could not decide.[104] Such sayings are so prevalent that Feynman deemed a criticism appropriate in his lectures on physics. Feynman argued that the interpreters have confused *sound* and the *sensation* of sound. Sound is a disturbance of air, which is always produced by falling trees in natural surroundings, whether a hearer is around to sense it or not.[105]

What Feynman said is common sense, which is deemed vulgar by some philosophers. The doctrines of quantum interpreters are sophisticated and have the backing of the long tradition of empiricist philosophy. Einstein remarked that it comes to the same thing as Berkeley's principle, "to be is to be perceived."[106] Perhaps a stronger influence came from the phenomenalism of logical positivism, which was powerful at the time when the quantum debate was at its height and provides an underpinning for the Copenhagen interpretation.[107] To counter the idealist and phenomenalist doctrines, we have to meet them on general philosophical grounds.

We can look at the philosophical problem this way. Suppose we amass many experiences and theories, abstract from their substantive contents, and discard various concepts until we reach a bare skeleton beyond which the impoverished conceptual structure fails to support the kind of intelligible experiences we have. Let us call the skeleton the *minimum conceptual complexity of experiences*. It is the presupposition of empirical knowledge because without it experiences collapse even if our sense organs are somehow stimulated. It is formal and devoid of information because we have abstracted it from all empirical contents. When we try to take the minimal structure apart, finally we come upon conceptual nexuses that must be taken as units. The nexuses we call *primitives*, which can be analyzed but not decomposed; the components of a primitive derive their meaning from being part of the whole and cannot be taken out and given independent meaning. If a primitive is simple and unanalyzable, we can regard what it represents as *given* and absolute; the given is what stops analysis.

The ultimate evidence of empirical knowledge is what we directly perceive at an instant. Therefore, the logical form of momentary perceptual experiences, abstracted from all the substantive details, should be the most basic primitive. It is less than the minimal structure of experiences because we may need additional concepts to account for temporal awareness. Its conceptual complexity lies at the center of the philosophical debate.

In our everyday thinking, we say we immediately access objects in perceptions; we see tables and chairs. Empiricists from John Locke onward argue that we are wrong. They note that the perceptual image varies incessantly whereas the object is supposedly immutable. The round table appears as an ellipse with ellipticity of 0.5 in some perceptions and ellipticity of 0.7 in others and so on. Furthermore, we can dream and hallucinate, seeing things that are not there. Based on these arguments from illusions, and more importantly on their premise that the immediately perceived must be given and unanalyzable, empiricists arrive at the conclusion that what we perceive are not objects but ideas, sense impressions, or sense data. We see not the round table but various elliptical images, which are somehow impressed on the passively receptive mind. Berkelian idealists say things are made up of ideas; positivist phenomenalists say they are logical constructions out of sense data. Both argue that they are fictitious and not necessary for experiences, although they do provide a certain economy of thought.[108] The empiricist theory is called the "representative theory of perception," in which we are separated from the objective world by a veil of perception. Many attempts to breach the veil and reach the world fail. We are trapped with sense impressions, which are all we have and which vanish when not being perceived. Sound dies, sensation remains and usurps the name of sound.

The argument from illusions has a strong point. The arbitrariness in perceptual images is genuine, hence we cannot simply identify what we see with things that are supposed to be given out there. To answer the empiricists, it is not enough to insist that we do see things directly; we have to explain how. We must find an appropriately broad categorical framework that accommodates both the changing appearances and the invariant object.

Kant argued that the trouble starts from the assumption of unanalyzable primitives. "Representation" in its full significance contains an intrinsic relation—it is the representation *of* something—therefore, it should be conceptually analyzable. Empiricists fail to account for the relation because they take representations as given. Kant argued that the given is incomprehensible and does not constitute experiences. Even the most immediate experiences are not without preconditions and structures. We do perceive tables and chairs directly, and that is possible because the primitive of perceptions involves a web of concepts. Our experiences are intelligible, and they are structured by a conceptual scheme with a certain minimal complexity, which he tried to analyze. He called the scheme the *concepts of objects in general*, for only through them can we recognize anything as objects and realize that we are having direct experiences of the world. The scheme is the presupposition of all experience and knowledge. It introduces the notion of objective regularities and admits the arbitrariness of subjective states. It dictates that our judgments are *objectively valid* only if "they are combined *in the object*, no matter what the state of the subject may be."[109]

Logically, representation is a general concept that includes but is not restricted to representations in sensual experiences. Arbitrariness similar to that in perceptual images is found in the representations in physical theories. As discussed in the previous section, the choice of observables in quantum mechanics is arbitrary and conventional. If physical theories include conventional representations, and if they are to convey objective knowledge, then we expect them to include means for controlling the fallout of the arbitrariness and maintain objectivity, in effect to address the arguments from illusions. Can we discern a primitive conceptual structure in physical theories that takes account of the distinction between sound and the experiences of sound? I argue yes, it is the *representation–transformation–invariance structure*, of which the quantum observable is an instance. The structure makes explicit the concept of being objective. The structure is a primitive but not the minimal conceptual structure of experiences; much more is required to support objective knowledge, as discussed in the following chapters.

A Common Primitive Structure in Physical Theories

Dirac said: "The growth of the use of transformation theory, as applied first to relativity and later to quantum theory, is the essence of the new method in theoretical physics. Further progress lies in the direction of making our equations invariant under wider and still wider transformations. This state of affairs is very satisfactory from a philosophical point of view, as implying an increasing recognition of the part played by the observer in himself introducing the regularities that appear in his observation, and a lack of arbitrariness in the ways of nature."[110]

By transformations in quantum mechanics and quantum field theory, Dirac meant the unitary transformations among the representations of various observables. He noted the logical similarity between them and the symmetry transformations of relativity. Relativity and quantum mechanics each has a primitive conceptual structure that incorporates various representations, transformations among the representations, and states that are invariant under the transformations. We are interested in the shared structures of the primitives.

Since its debut in the special theory of relativity, symmetry quickly outgrew its spatio-temporal characteristic and proliferated into every corner of physics and beyond. Its leading idea, *invariance under transformations of representations*, is found in many rigorously formulated scientific theories. For instance, we find it in the utility theory of economics, where the numerical representation of value is subject to monotone transformations. Also, in formal measurement theory, the representation theorem specifies the representational rules that map various relational structures into the real line, and the uniqueness theorem specifies the transformations among the representations. The two theorems together lay the foundation of measurements.[111]

Symmetry is so prevalent now that we tend to forget that group theory, which underlies it, was developed only in the second half of the nineteenth century.[112] The elements of symmetry are separately intuitive and were around for some time. The origin of the idea of something unchanged or invariant lay way back in history. Coordinate systems were introduced by René Descartes in the first part of the seventeenth century. Nineteenth-century physicists routinely made coordinate transformations. The novelty of symmetry is to unite the elements in an integral framework, and more important, to bring out its physical significance. Lorentz had worked out the transformation rules that bear his name, but Pauli remarked that "the relativity principle was not at all apparent to Lorentz. Characteristically, and in contrast to Einstein, he tried to understand the contraction in a causal way." The major novelty of symmetry is not a substantive feature but a formal structure. As Pauli said, it changed "the general way of thinking of the physicists of today."[113] Since formal structures are empty and can accommodate any content, we can retrofit it into older theories and talk about the Galilean group, thus brining Newtonian mechanics into the symmetry ambit. The retrofit shows that what has changed is the logical form and not the substantive content of the theories.

I argue that the representation–transformation–invariance structure, which is shared by symmetry and the transformation theory of quantum mechanics, embodies the first-cut distinction in the general concept of objects that is the basic presupposition of empirical knowledge. The structure separates the objective state of affairs from its various representations, which may include perceptual representations. The notion of objects is intuitive and has always been in use in our everyday thinking. It is not always included in scientific theories, which are supplemented by common sense. Modern physical theories have made it explicit.

When a concept as basic as that of objects is incorporated in scientific theories, its power and ramification is fully revealed and at once recognized.

Weyl said: "All *a priori* statements in physics have their origin in symmetry." Weinberg said: "Symmetries took on a character in physicists' mind as *a priori* principles of universal validity, expressions of the simplicity of nature at its deepest level."[114] The specific symmetry groups that hold in specific situations, for instance, the Galilean, the Lorentz, or the unitary groups, must be determined empirically. What is *a priori* is only the formal structure common to all theories employing symmetry groups. And it can be *a priori* only because it is the explicit statement of a tacit presupposition of objective knowledge.

The Concept of Objects in Everyday Thinking

Before we consider the details of the representation–transformation–invariance structure, let us reflect upon the common-sense way of thinking about objects. If the structure is the statement of some basic presupposition and not an exotic invention of scientific theories, it must agree with our common understanding.

When we look around, we see objects, books and pens. The presence of objects is immediate, we do not infer them from sense data; I have never seen a sense impression. Nevertheless, for all its imminence, we are aware that we may be in error and things may be different from how they appear. We make mistakes about objects and doubt our assertions; we hallucinate and acknowledge the hallucination. Macbeth said: "Is this a dagger which I see before me,/ The handle toward my hand?"[115] The image is vividly given to him, yet he doubted. The doubt can arise because Macbeth, a man of good sense as well as imagination, distinguished between the object and the given. Then he said: "Come, let me clutch thee!" He was trying to transform his perspective to get a better "look" at the dagger. In the attempt he had presupposed that objects are "sensible to feeling as to sight," or they follow certain rules. The transformation and the rule application are not given in the image; they express the spontaneity of intelligible experiences. When Macbeth finally judged that the image is "a false creation/ Proceeding from the heat-oppressèd brain," he acknowledged a defective mode of experience in which the object is absent.

In most if not all situations, we operate in a mode of partial knowledge, in which we have some understanding and yet realize that some substantive content is missing or faulty. *The partial mode that recognizes its own deficiency is an essential feature of common sense.* It not only accounts for our blunders and fumbles; it is vital to our ability to doubt, see possibilities, and learn. The notion of errors is crucial to that of objectivity, for objective judgments are falsifiable. Thus a satisfactory categorical structure of objective knowledge must have enough complexity so that some nexus can be defective or left blank without the total collapse of intelligibility. Only in such a structure can illusion and hallucination make sense. The structure is not simple, it implies the *reflexivity* of knowledge and the ability to think beyond what meets the eye.

Imagine two persons seeing something. One says it is a sea of electrons in an ionic lattice. The other says: "What? It's a plain old metal desk," and mutters, "crazy physicist, but I guess he knows what he's saying." Such disagreements occur frequently in daily life. In all cases, we always immediately understand that we are talking about the *same object* we separately see and judge, otherwise there is nothing to quarrel about. We may not say so explicitly, but we understand there are objects, which admit various representations and yet impose certain agreement. If one learns enough physics, he would not merely acquire a set of fancy terminology and be admitted into the exclusive club of physicists. He would come to agreement about the object, so that all talk objectively about the world. Usually people accede even without knowing the physics, for it is our basic understanding of objects.

Within our everyday understanding of perceptual experiences, we have tacitly distinguished a *general concept* and its *specific embodiments,* or a *categorical framework* and *a substantive conceptual scheme,* both of which are simultaneously involved in definite observational statements. The general concept, that of empirical objects, is embodied in all empirical statements and guarantees a minimal mutual understanding. The embodiments, for instance, as a desk or a lattice of atoms, can differ in various substantive schemes. The substantive schemes, which provide various specific representations, depend on the psychological, sociological, and cultural background in which they are employed. They are conventional.

In the current debate centering on conventionism, "conceptual structures" usually mean substantive schemes. The substantive schemes have a strong influence on our thoughts. We often stubbornly adhere to our own tradition and are reluctant to look into alternative schemes; to get acquainted with an unfamiliar scheme usually requires considerable effort. However, their hold is not absolute. We do not think of people holding alternative schemes as inhabitants of a different world, even if we refuse to find out the details of their views. The sense of tolerance is made possible by the categorical framework in which the physicist and the writer agree that they are both talking about the same object. The agreement, made in the ignorance of the details of the other's scheme, is another manifestation of common-sense functioning in a partial mode.

The categorical framework, in which we acknowledge empirical objects, includes at least three elements: first, the possibility of various substantive schemes, which are acknowledged to be essentially biased and conventional; second, the possibility of translating among the schemes; and third, the idea of the object as something invariant under the translations. The framework acknowledges that as human beings, we are radically situated, physically and intellectually, and our particular representation is conventional. It also admits the thought of alternative representations and the object's independence from them. I somehow understand other people because I have spontaneously attributed to what I see the general properties of being capable of being variously represented and yet independent of the representations.

Mutual understanding on substantive terms requires much more than a simple substitution of terms such as "a sea of electrons" for "a desk." It requires the transformation of the entire representation or a change in the frame of mind, which is often difficult. Often it is not clear if we can actually find substantive schemes that can be rigorously transformed into each other. For instance, the physicist's sea of electrons may not coincide with the ordinary desk because it excludes the formica top. The object framework provides the condition of the possibility of transformations but does not guarantee their success. Yet the very thought of the possibility is crucial to the common sense of tolerance. Let us see how such thought is expressed in physical theories.

The Representation–Transformation–Invariance Structure

"Represent" and its cognates have a broad and a narrow meaning. Broadly speaking, all ideas, images, impressions, words, and symbols of things are representations; things are neither in the mind nor on paper. Thus we say a vector in a Hilbert space represents a quantum state. The broad-sense representations can be very abstract and general. The representations we are concerned with have a narrower meaning. They are schemes with sufficient structures for us to attribute definite predicates to particular things. The representations in quantum mechanics, the coordinate systems in relativity and other symmetry theories, and the languages of daily discourse are narrow-sense representations. Our everyday thinking may be vague, but it is not abstract; we think in terms of one representation or another. I will use "representation" only in the narrow sense but "represent" in both senses.

Modern formulations of physical theories incorporate both a state space that represents the object states and various representations of the state space. The representations usually have some empirical significance. There are many equivalent representations; "equivalent" means that although the definite predicates or values assigned by different representations are different, they are equally good in describing the same state of the system. The choice of a particular representation is conventional. Various representations are connected by transformation rules, and the objective state is invariant under the transformations. This categorical structure is schematically depicted in Fig. 5.1.

In Fig. 5.1, M is the state space of a physical system, and x one of its states. $f_\alpha(M)$ and $f_\beta(M)$ are the representations associated with the property types f_α and f_β. $\{x^\mu\}_\alpha$ and $\{x^\mu\}_\beta$ are the values or definite predicates for the state x in the respective representations. $f_\beta \cdot f_\alpha^{-1}$ is the rule for transforming from the representation $f_\alpha(M)$ to $f_\beta(M)$. In quantum mechanics, M is a Hilbert space, x a state vector, f_α an observable, $f_\alpha(M)$ a representation, $\{x^\mu\}_\alpha$ the amplitude of x in the basis of the observable, and $f_\beta \cdot f_\alpha^{-1}$ a unitary transformation. In relativistic mechanics, the modern formulation of classical mechanics, and theories employing symmetry groups, f_α is a coordinate function, $f_\alpha(M)$ a coordinate system, $\{x^\mu\}_\alpha$ the coordinate of the invariant x in the system, and $f_\beta \cdot f_\alpha^{-1}$ a symmetry transformation. Like symmetries, coordinate systems include but are not limited to spatio-temporal ones. They include the coordinatization of various state spaces; for instance, "protonlike" and "neutronlike" can be the coordinate axes of the state space of nuclear interactions.

Some philosophers erroneously interpret M and $f_\alpha(M)$ separately as two ways of writing physical theories. The former is coordinate-free and modern;

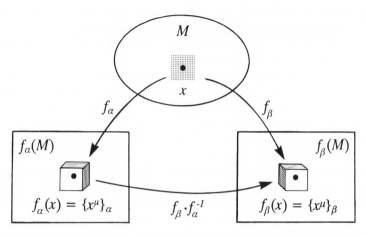

Figure 5.1 The general concept of objects in physical theories. M is the state space, which encompasses all possible states of an object. The representative rules f_α and f_β assign definite predicates $\{x^\mu\}_\alpha$ and $\{x^\mu\}_\alpha$ in the representations $f_\alpha(M)$ and $f_\beta(M)$ for the state x. The transformation $f_\beta \cdot f_\alpha^{-1}$ does not merely relate the predicates in different representations; it is a composite map that points to the objective state x. The objectivity of x is ensured by its invariance under transformations among all representations, and thus abstracts from them all.

the latter depends on coordinates and is old-fashioned. They argue that we can map a manifold into itself without using any coordinates. They have forgotten that manifold mappings are based on the manifold, whose definition depends on coordination (compare Fig. 5.1 to the definition of the differentiable manifold in Fig. 3.1). In more abstract spaces, we have implicitly assumed some identifying means that allows us to differentiate the initial and final states of a transformation. Even if specific coordinate representations need not be invoked, still the general idea of representations is indispensable in transformations.

I contend that specific representations are necessary for *empirical* knowledge of the objective world. The indispensability of observables and the representations they introduce is apparent in quantum mechanics. The representation-free form M is too abstract and by itself is insufficient for physical theories, for the theories must predict the behavior of particular objects so that they can be supported by experiments. M, being a total abstraction from all particularities and observational conditions, is never observed; observations are always of particulars or representations of M. That is why Wigner said we do not observe the symmetries of physical laws in the world surrounding us. What we observe in experiments, characterized by initial and boundary conditions expressed in coordinates, are irregular and not symmetric at all. Robert Wald said: "All experiments in physics measure numbers, so all quantities of physical interest must eventually be reducible to numbers. However, many quantities of interest—such as the magnetic field or the stress tensor . . . require the additional specification of a basis of vectors in order to produce numbers."[116] The additional basis or representation is extraneous to physical states, just as our observations are, but they are not extraneous to physical theories, for they secure our empirical access to the physical states.

The abstraction that brings us the representation-free form M also severely curtails what we can say of a specific object's definite features. We can generally talk about length as the invariant of isometric transformations. However, we cannot talk about the definite length of a particular line segment without the use of coordinates. We state the length in a particular representation and then supplement the statement with transformation conditions to erase the peculiarity of the representation. "This pen is 6.2 inches long, which is equivalent to 15.7 centimeters and . . . . " In doing so, we have invoked the whole structure in Fig. 5.1. In most daily discourse, the transformation supplement is understood and suppressed.

The representation–transformation–invariant structure is an integral unit. Even as the spotlight of explication falls on a component, the rest of the structure contributes in the shadow or semishadow. The recognition of it as a primitive structure is what changed the general way of thinking of physicists. In the following discussion of various categorical frameworks that interpret scientific theories, the issue is the primitive. Thing, phenomenon, and object are primitives in various frameworks. We are concerned only with them and not with the superstructures that can be constructed upon them.

The Given

Let us start with the substructures of Fig. 5.1. Cover up $M, f_\alpha, f_\beta, f_\beta(M)$, and $f_\beta \cdot f_\alpha^{-1}$ in Fig. 5.1, leaving only the box $f_\alpha(M)$. Since $f_\alpha(M)$ is now the only item, its subscript α and functional connotation are superfluous. The structure consisting only of $f_\alpha(M)$ has no room for the thought of any theoretical contribution to the attribution and any alternative description. Thus $\{x^\mu\}_\alpha$ is regarded as absolute, self-evident, and theory-free. It is given, either in observations or in reality. This simple conceptual structure does not allow any partial mode of knowledge. If $\{x^\mu\}_\alpha$ is given, it is certain; if it is not given, all is blank.

The structure $f_\alpha(M)$ can be interpreted as the primitive of idealism and phenomenalism. Positivists divide the terms in a theory into observational and theoretical. The observational terms are privileged because they are free from all theory. They represent phenomena that are certain, self-evident, and given in observations. All else are logical constructions upon them. They can be represented by $\{x^\mu\}_\alpha$. Positivists tried to reduce theories into ideal forms that contain only observation terms; the effort failed. Their doctrine of theory-free observational terms has been much criticized.[117]

The structure $f_\alpha(M)$ can also be interpreted as the primitive of metaphysical realism. Hilary Putnam said: "On this perspective [metaphysical realism], the world consists of some fixed totality of mind-independent objects. There is exactly one true and complete description of 'the way the world is.' Truth involves some sort of correspondence relation between words or thought-signs and external things and the sets of things. I shall call this perspective the *externalist* perspective, because its favorite point of view is a God's Eye point of view."[118] Here $\{x^\mu\}_\alpha$ represents the one true description corresponding to some given thing.

Whether a philosophy is committed to the notion of the given is seen not in its external embellishment but its intrinsic structure. Consider a tax code embodied in, say, the Form 1040. The tax form includes items called exemptions and deductions, which show that the code represented by the form has the complexity to delimit taxation explicitly by stipulating what is *not* taxable. If there is no place in the tax form for taxpayers to enter exemption or its like, the politician's sweet talk of restricting taxation amounts to nothing. Analogously, we can ask: Does the categorical structure of a philosophy provide places for us to enter the theoretical aspect of observations or real facts? Does it provide separate places for us to enter what is not theoretical in them? Many philosophies allow elaborate theoretical superstructures to be constructed upon the given primitives but no conceptual complexity for their primitive elements, be they sense data or real things. There is no place in their primitives to express the notion of not-givenness; hence they cannot free themselves from the notion of the given. It is symptomatic that the criterion of observability is usually prescribed extratheoretically in empiricist philosophies and the meaningfulness of names described extrasemantically in realist philosophies. The problem is an old one. Kant complained that rationalists intellectualized appearances and empiricists sensualized concepts because they were stuck to the idea of things-in-themselves.

Conventions

Uncover $f_\alpha(M)$ and $f_\beta(M)$ but not $M, f_\alpha, f_\beta,$ and $f_\beta \cdot f_\alpha^{-1}$ in Fig. 5.1. This adds the idea of a *plurality* of boxes $f_\alpha(M)$ and $f_\beta(M)$. The meaning of $\{x^\mu\}_\alpha$ and $\{x^\mu\}_\beta$ remains unanalyzable, but now there are a host of them. The plurality brings to relief the theoretical contribution to descriptions; various theories give radically different representations, usually called conventions. The role of conventions in scientific theories is well known since the historical study of Thomas Kuhn.[119]

The framework of conventions spawns relativism, which argues that the conventions are incommensurate and the choice among them is arbitrary. The conventions are given, not in reality or observation, but by the culture in which one happens to be born. Relativists deny that scientific investigation is superior to other ways of thinking, arguing that it is only one of many incommensurate social customs. Relativism is often criticized for being incoherent and self-defeating. For applied consistently, it implies that relativism itself is one of the conventions that anyone not accustomed to it can reject.

"Convention" loses weight if people holding different theories are really like inhabitants of different worlds, as Kuhn asserted. The word is meaningful because we have tacitly assumed something common in the theories, that they are all about the objective world. This assumption is implicit in Kuhn who, after declaring that the theories of Ptolemy and Copernicus are incommensurate, went on to describe and compare them in detail. However, the notion of the objective world is officially excluded in the framework of conventions, which lacks the notion of what is *not*-conventional, that is, the notion of objects detached from conventional descriptions. Consequently conventionists deny the objectivity of theories, scientific or not.

The Empirical Object

"Object" is used in two senses in the following. The narrower sense is the *physical object* whose state is represented by x in Fig. 5.1. The broader sense is the *empirical object*, the topic of knowledge. "Empirical" here includes only the conceptual aspect of experiences, which are recognized as a kind of representation; it does not include the sensual aspect, which was considered in § 12. An empirical object is an object-variously-representable-but-independent-of-representations. The concept of empirical objects is represented by the full structure in Fig. 5.1; it includes the physical object as a conceptual element.

The introduction of the physical object whose state is x not only adds an element in our conceptual structure; it enriches the elements discussed earlier. For the first time, the idea of representations *of* something is made explicit. The physical object reinforces the common-sense notion of things that are independent of our representations. On the other hand, since representations are associated with observations of things, the idea of phenomena becomes more weighty. The multiplicity of representations of the same object forces us to acknowledge the idiosyncrasies of particular representations; hence it clarifies the meaning of conventionality.

Since the object x is categorically different from any of its representations, the meaning of $\{x^\mu\}_\alpha$ is no longer unanalyzable. It is now $\{x^\mu\}_\alpha = f_\alpha(x)$, reading "the value $\{x^\mu\}_\alpha$ of x for the property type f_α," "the predicate $\{x^\mu\}_\alpha$ of x in the representation $f_\alpha(M)$," or "the appearance $\{x^\mu\}_\alpha$ of x from the perspective $f_\alpha(M)$." Various representations can be drastically different, but they represent the same object. The same electromagnetic configuration that is a mess in the Cartesian coordinates can become simplicity itself when represented in the spherical coordinates. However, the two representations are equivalent.

Since f_α and f_β are imbedded in the meaning of $\{x^\mu\}_\alpha$ and $\{x^\mu\}_\beta$, the transformation $f_\beta \cdot f_\alpha^{-1}$ connects the two representations in a necessary way dictated by the object x. $f_\beta \cdot f_\alpha^{-1}$ is a composite map. It not only pairs the two predicates $\{x^\mu\}_\alpha$ and $\{x^\mu\}_\beta$, it identifies them as representations of the *same* object x, to which it refers via the individual maps f_α^{-1} and f_β. Since $f_\beta \cdot f_\alpha^{-1}$ always points to an object x, the representations they connect not only enjoy intersubjective agreement; they are also objectively valid. To use Kant's words, the representations are no longer connected merely by habit; they are united *in the object*.

The objective state x is called coordinate-free or representation-free. This is not an arbitrary designation but an active negative concept that signifies a *lack* of representation. The invariant x explicitly articulates the common-sense notion that physical objects are independent of our conventions and free from the arbitrariness of our perceptual conditions. A negation is a distinction between what is and what is not; for instance, what is given and what is not, what is conventional and what is not. A theory must have certain minimum conceptual complexity to internalize a distinction. The negativity, being free from or independent of, drives a wedge between the physical state and its representations, which become truly significant in the larger conceptual framework. Since modern physical theories have internalized the distinction signifying detachment, they themselves can assert objectivity for their objective statements, a task of which older theories are incapable.

The representation–transformation–invariance structure can also represent momentary perceptual experiences. The content of an experience is represented by $\{x^\mu\}_\alpha$, for observations are always specific. The conceptual complexity of $\{x^\mu\}_\alpha = f_\alpha(x)$ implies that we directly access the object x in our experiences and do not indirectly infer it from some given sense impressions. The object is not a transcendent reality but is immanent in experiences. Looking at the other side of the coin, the phenomenon $\{x^\mu\}_\alpha$ is not a semblance or mere appearance that stands for something else; it is what the object shows itself in itself. The idiosyncrasy in $\{x^\mu\}_\alpha$ is ascribed to the conditions of experience. The conceptual complexity implies that our experiential content goes beyond mere sensory stimulation. When we observe a particular representation $\{x^\mu\}_\alpha$, we simultaneously observe the invariance-under-transformations-of-representations. Suppose x represents a round table and $\{x^\mu\}_\alpha$ and $\{x^\mu\}_\beta$ various elliptical profiles. When we see the table from an angle, we see *in* the particular profile its invariance when seen from alternative angles. That is how we distinguish a round table from an ellipse.

The categorical framework of objects is a unitary whole. The physical object x is neither posited in advance nor constructed out of its representations afterwards. It is defined simultaneously and encoded in all its representations in the integral structure. Neither the representation-free x nor the representation $\{x^\mu\}_\alpha$ alone is sufficient to characterize the primitive unit of empirical knowledge. Both and their interrelation are required; x realizes the *general conditions for the possibility of objects* and $\{x^\mu\}_\alpha$ the *general conditions for the possibility of experiences of objects*. The two arise together in objective knowledge, as Kant argued. Representation–transformation–invariance is an integral structure that realizes the general concept of empirical objects in physical theories.

Since the concept of empirical objects has enough complexity to endow the content of experiences with meaning beyond what meets the eye, it can account for doubts, errors, illusions, and partial knowledge. There are enough elements in the categorical structure so that some can be left blank without a total collapse of comprehension. We may know $\{x^\mu\}_\alpha$ but not $\{x^\mu\}_\beta$, or we may know both but not the transformation relating them. In each case the ignorance exists along with some understanding. Furthermore, the general idea of the possibilities of alternative representations and transformations enables us to recognize our own deficiency and to learn if we please. Examples of partial knowledge are given in § 21, where the structure of empirical objects is applied in the linguistic context of reference.

Philosophically, the importance of the representation–transformation–invariance structure lies in the conceptual complexity of the general structure and not in the details of the various elements. It is the adoption of something like it instead of the simplistic structures of the given and the conventional that differentiates common sense from phenomenalism, metaphysical realism, and conventionism. The conceptual structure points out the possibilities of various representations and transformations but neither prescribes the procedure rules for formulating them nor guarantees they can be successfully formulated. In mathematical physics, the representations are rigorous and the transformations explicitly performed. In our everyday thinking, the representations are sloppy and often defy exact transformations. However, this does not warrant a lapse to conventionism, which denies the general idea of transformations because specific transformations fail. On the contrary, the imperfection of specific representations makes the general conceptual complexity more important, for it alone allows the thoughts of approximations, idealizations, and improvements in objective knowledge.

The representations may be partial in the sense that they characterize only one aspect of the objective state. For instance, the momentum representation does not include the spin of an electron. Representations of different aspects cannot be connected by transformations. Einstein was dissatisfied with special relativity, saying "what has nature to do with our reference frame?" He expanded the theoretical framework so that more representations are included and connected in general relativity. Dirac said the general direction of research is in making our equations invariant under wider transformations. Physicists acknowledge that even frontier physical theories offer only partial representations of the world. The world of our daily activity is much more complicated than the world of basic physics. Often various "world constructions" highlight various aspects and are therefore not mutually translatable. They should not distract from the objectivity of knowledge in general, for the important idea is the recognition that they are representations and representations can be partial.

The representation–transformation–invariance structure discussed pertains to a single system, which we have assumed to be somehow individuated. The entity may be the universe, in which case the representations would include schemes of individuation if we also consider entities within the universe. The individuating scheme is discussed in the next chapter. In Chapter 7 we see that even more concepts are required to reconcile the requirements of individuation and representation.

Conditions of Objects and Conditions of Experiences

Kant's thesis that empirical objects are not given and even immediate perceptual experiences have conceptual complexities has great philosophical consequences. It means that our notion of nature is structured by the general concepts that are preconditions of experiences. Kant was emphatic that these concepts are completely formal and devoid of substantive contents; the contents of the formal structure must be empirically determined. Thus his theory should not be confused with some talks of "constructing reality," which asserts that the substantive contents are inventions. For Kant, spontaneous intellectual contribution is limited to something like the general structure depicted in Fig. 5.1, which makes possible the empirical determination of specific features $\{x^\mu\}_\alpha$ or $\{x^\mu\}_\beta$. It is similar to the case of symmetry; we must first adopt a group-theoretic way of thinking before we can decide whether the Lorentz or the Galilean group agrees with experimental data. Only in this sense can we say symmetry is *a priori*. The group structure is our construction, and it expresses a general way of thinking about nature. This is completely different from saying that the Lorentz and Galilean groups are various constructions of reality.

The analysis of the general concepts of objects constitutes the bulk of the constructive part of the *Critique of Pure Reason*.[120] Here we are only interested in the first step, found in a long section usually called the "Transcendental Deduction of the Categories." Kant reasoned in the first person singular case, for empirical knowledge ultimately rests on individual experiences. Very crudely, he argued that the most basic structure of experiences is the unity of consciousness, which distinguishes the object from my experience of it. It enables me to break free from the immediate presentation and to think of the object's other possible representations, including those in the past and future. Thus it paves the way for rulelike combinations of various representations. These rules will be the categories: quantity, quality, modality, and relations, which are analyzed subsequently. Hence the object–experience distinction is the presupposition of the categories. (Kant used "deduction" in the old legal sense of justification. A transcendental investigation is concerned not with specific knowledge of objects but with the general features of the pure scheme that makes possible experiences and knowledge of objects.)

In our notation, perhaps I can say that starting from $\{x^\mu\}_\alpha$, which represents the content of my experience, we have to argue that it must have a conceptual complexity that goes beyond itself and implies a broader framework that includes the notion of the physical object, and further to argue that the object satisfies certain general criteria so that various $\{x^\mu\}_\alpha$ form a coherent picture characteristic of intelligible experiences of the object. The first step is to find the framework that can instill the required conceptual complexity into $\{x^\mu\}_\alpha$. I think that the resultant framework is something like that depicted in Fig. 5.1. I will only mention some of Kant's results instead of going into his involved argument, which makes heavy use of the notion of time.

Consider the simple experience "I see that this is a tree." Its content, "this is a tree," involves a concept, tree. The "this" points out the *it*, the particular item that is brought under the concept of trees in the observation. In our activities of seeing, the "this" is usually implicit, but it is there, it is hidden in the directedness of my attention. That "tree" appears in experience means I am attending to the tree. The attention is captured by the "this." "This" is what Bertrand

Russell called an egocentric particular.[121] It reflects on the attention that points, and it is only meaningful when properly anchored; "this is such and such" scribbled on a piece of paper carried by the wind means little. Thus "this is a tree" always extends beyond itself and is always accompanied by an I-think, which expresses the intellectual component of "I see that." Since the I-think is outside "this is a tree," it makes "this is a tree" stand out as a unit, an experience. This unit, the notion of experience, is nowhere to be found inside the that-clause, but it makes what falls inside the that-clause a content. It is true that *it* occupies most of my attention, but it is also undeniable that *my* self-awareness always stays in the periphery as a kind of horizon. The awareness distinguishes my observation from a camera's detection.

Kant argued that an account of intelligible experiences must be complicated enough to include the self-awareness expressed by the I-think that accompanies all my experiences. With this he at once pinpointed the fundamental human conditions of activity, reflexivity, and finitude. "I think" informs the thought that it is a thought, prevents my attention from being totally absorbed by the object, lets me know that I am having an experience of an object, and drives a wedge between *the object* and *my representation of it*. It is a conceptual distinction. A distinction is also an implicit relation. Imagine dividing a region into two; the dividing boundary performs three functions at once: it defines, distinguishes, and relates or unites the two parts. It becomes the defining characteristic of the parts; whenever we talk about one part, we implicitly invoke the other and the whole through the boundary. The object-experience distinction manifested in the "I think" is such an implicitly-relational boundary. It imparts a minimal conceptual complexity in the integral structure of intelligible experiences. Kant called the resultant structure the *unity of consciousness* or the *synthetic unity of apperception*, which is basic and original to all experiences. "The principle of apperception is the highest principle of the whole sphere of human knowledge." "It is the unity of consciousness that alone constitutes the relation of representations to an object, and therefore their objective validity and the fact that they are modes of knowledge."[122]

The immediate contents of experiences are random and haphazard. The categorical distinction enables us to differentiate two elements, so that regularities are attributed to objects of experiences, and the lack of regularities regarded as subjective. To put it the other way, the object is posited as that which lacks arbitrariness and conforms to some rules. Since regularities and rules are general, they run through the multitude of haphazard representations, uniting them and making them intelligible. Kant said: "An *object* is that in the concept of which the manifold of a given intuition is *united*." "If we enquire what new character *relation to an object* confers upon our representations, what dignity they thereby acquire, we find that it results only in subjecting the representations to a rule, and so in necessitating us to connect them to some one specific manner."[123]

Kant did not say that a proposition is objective if it does not mention subjective representations. He argued that objective judgments require a minimal conceptual structure that incorporates various representations and distinguishes the object from them. He summarized the significance of the structure as one of the two highest principles of objective knowledge: "The conditions of the *possibility of experience* in general are likewise conditions of the *possibility of the objects of experience*."[124] The Transcendental Deduction, of which the principle is the conclusion, caused Kant the most labor. However, he did not invoke the principle when he tried to develop the metaphysical foundation of mechanics. I think he would be pleased to see the version of classical mechanics retrofitted with the symmetry structure.

§ 16. The Illusion of the Observer

In the working understanding of quantum mechanics, eigenvalues are interpreted as the outcomes of measurements; a specific eigenvalue is *observed* or *measured* in an experiment. The explicitly empirical connotation has engendered much controversy in quantum interpretations. Bell said: "The subject–object distinction is indeed at the very root of the unease that many people still feel in connection with quantum mechanics. *Some* such distinction is dictated by the postulate of the theory, but exactly *where* or *when* to make it is not prescribed." He quoted passages from standard textbooks with empirical terms and complained that they "presuppose in addition to the 'system' (or object) a measurer (or subject)."[125]

This section examines the meaning of observation. It refutes the doctrine that the concept of "observed value" implies the inclusion of a substantive observer in physics or the division of the world into the observer and the observed. I argue that the subject is not a part of the world but is expressed in the logical form of quantum mechanics and is exhausted by the form. The illusion of the observer arises from the adoption of a categorical framework too narrow to accommodate the rich structure of objective knowledge.

Some interpreters argue that empirical connotation is anthropocentric and should be purged from physical theories; even the notion of observables must be replaced by that of physical quantities. We should distinguish a *theory* from its *statements about objects*. Objective statements should be free from empirical and anthropocentric notions, but the theories should be allowed to contain whatever means they deem necessary to produce such statements. Physics is an empirical science, and our empirical capability is limited. It is unreasonable to insist that the empirical condition can only be tacitly understood but not explicitly stated. Einstein argued that "it is the theory which first determines what can be observed."[126] If a physical theory is to distinguish what can be observed from what cannot, it must incorporate the criterion of observability with its anthropic connotation. All we demand is that the empirical elements are properly included so as not to contaminate the statements about physical objects. I argue that one of the wrong ways to account for the empirical aspect of physics is the notion of the observer.

The Vain Claim of Metaphysical Completeness

The idea of the observer originated in the Copenhagen interpretation, although it may be fanned by the phenomenalist doctrine of logical positivism, which flourished at the time. One of its roots lies in the Copenhagenist claim that quantum mechanics is absolutely complete and final, even in the face of the quantum measurement problem. Bohr said quantum mechanics "exhausts the possible information about the objects." "A more detailed analysis of atomic phenomena . . . is *in principle* excluded." Einstein objected; he saw clearly the implication of the doctrine: "If one wants to consider the quantum theory as final (in principle), then one must believe that a complete description would be useless because there would be no laws for it."[127]

The present situation, including our inability to tell what happens during quantum measurements, is incompatible with the conjunction of the following three assumptions: (1) A measurement is a physical interaction between physical objects; (2) physical interactions are explainable by physical laws; (3) quantum mechanics is the absolutely complete and final theory. Copenhagenists insist on retaining the third assumption; hence they must somehow revise the first two. In the revision, measurement becomes observation. The most handy candidate for something not covered by physical laws, standing in relation to objects, and likely to fill up gaps in an empirical science, seems to be the observer. The peculiarity of the Cophenhagen "agencies of observation" is that they do not obey the laws of quantum mechanics. Bohr and Heisenberg were careful to talk only about the agents of observation, although Bohr had blurred the line by asserting that it is impossible to delineate the agent clearly. Their followers pushed their doctrine to its logical conclusion and talked about consciousness and the observer-created universe.

When we say that the state vector gives a complete description of the quantum system, we mean that for systems that fall within the scope of quantum mechanics, quantum descriptions are complete. This conditional completeness is acknowledged even by Einstein.[127] Scientific theories generally do not claim to be absolutely complete and exhaustive. The metaphysical rouge does not add one bit to the glory of quantum mechanics. Therefore we reject the claim to absolute completeness and regard the process of quantum measurements as an open physical question.

The rejection of the finality claim mitigates the requirement for the observer or his agent in interpretations. However, we still have to clarify the meaning of the empirical terms in quantum mechanics. Do these terms imply the presence of the observer? Does "the eigenvalue is observed" imply "someone observes the eigenvalue" within quantum mechanics? Does it mean that the world is divided into two parts, the observer and the observed? I argue the answers are no. Generally, the notion of a nonphysical observing substance that stands in explicit relation to physical entities is illusory. Specifically, quantum mechanics has stubbornly resisted the introduction of such a notion. The observer has no place in the working understanding of quantum mechanics. However, it hides in some sophisticated interpretive theories, as discussed in the next section.

A conceptual gap similar to that due to the lack of a quantum measurement theory exists in our everyday understanding of perceptual experiences. We have no satisfactory theory of perception. Even if we succeed in correlating the characteristics of photons impinging on our retina with the firing patterns of neurons, we still do not know how the patterns give rise to clear ideas such as "a tree." Nevertheless, we understand "a tree" despite the problem of perception. The categorical framework that underlies the common-sense understanding is complicated. It suggests that we employ a similarly broad framework in interpreting quantum mechanics. A richer categorical framework enables us to make less informative statements, for it is purely logical and has no factual content by itself. An interpreter is like an artist; he must know when to stop. The lack of a satisfactory theory of perception or quantum measurement means that we do not know how intelligible experiences and physical objects are substantively related or what happens in measurements. The interpretation should reflect the lack of knowledge.

Explicit, Implicit, and Reduced Implicit Relations

Let's see how careless reasoning can lead to the notion of the observer. We know there is no unowned experience, no pain floating around in the room, as the dying Mrs. Gradgrind hallucinated in Charles Dickens' *Hard Times*. Thus we assume an experience implies a subject, the observed implies an observer, and the two are connected by some kind of relations, probably causal relations. The causal link leads us to think of the observer as something substantial. But a corpse is no observer. Thus we conclude the observer is something not quite physical; it is a kind of thinking substance often called consciousness, the mind, or the Cartesian subject.

In most arguments for the observer, it is tacitly assumed that since experiences are relational, an explicit relation must obtain between an experience and a subject. Thus someone having an experience is logically the same as someone having a pen. However, explicit relation is not the only kind of relation. Consider a dance and its dancer. There is no dance without a dancer, but is there an explicit relation? Does a dancer wear a dance as she wears a skirt? Is the dancer meaningful apart from the dance? Or is the meaning of the dancer exhausted by the dance? The concept of dance has certain complexity incorporating the sense that it is danc*ed*, and the dancer has no more meaning than that the dance is danc*ed*. If we say a relation obtains between the dance and the dancer, the relation can only be *implicit* in the concept of dances.

Leibniz distinguished two kinds of well-founded relations, those "in the subject" and those "out of the subject." I call them implicit and explicit relations. *Explicit relations* between *a* and *b* are logically expressed in the form *Rab*, for instance, *a* loves *b*. *Implicit relations* are expressed in the form *Fa*, where *F* is a relational property of *a* such as being in love with *b* or simply being in love. Implicit relations are also called relational properties. To cite Leibniz's example, the paternity in David is an implicit relation, the sonship in Solomon another. The father-and-son relation common to David and Solomon is an explicit relation that is well founded, as the first born makes the man a father.[128]

One difference between implicit and explicit relations is that *b*, the object *a* relates to, is always specified in explicit relations but not necessarily in implicit relations. The paternity of David does not specify Solomon; it does not even specify the number of David's progenies. I call those implicit relations in which one of the relata is not specified *reduced implicit relations*. They are found in reduced passive sentences such as "the town is destroyed," "he is elected president," or "the bed has not been slept in." Suppose we try to translate them into explicit relations by introducing an indefinite subject and writing "something destroys the town" and "someone elects him president." The translations are not satisfactory. For someone or something is singular, whereas the reduced passive may demand plural or collective subjects. Many factors and processes, some not well defined, may have contributed to the town's destruction. A citizen votes for a candiate but does not elect him, and if we say "the citizenry elects him," we face the ontological difficulties regarding collective subjects. Thus even the indefinite subject in the explicit relation may be too specific and too restrictive. For the negative case, if we want to avoid inventing a nobody who has slept in the bed, we must account for every one in the world and assert that each has not slept in it, which is a mess.

Reduced passive sentences are less informative than explicit relational sentences. However, saying too much is as undesirable as saying too little. If "he is in love" amply explains his out of the ordinary behavior, then "there is someone whom he loves" may be saying too much. Reduced passive sentences make perfect sense, and their reduced implicit relations allow us to say just what we want and to be silent on the rest. The right to remain silent is available only in a framework more liberal than one that allows only explicit relations.

"The eigenvalue is observed" is a reduced passive sentence. Additional information is required to turn it into the explicit relational "someone observes the eigenvalue." No such information is provided by quantum mechanics, and no justification for it is provided by any interpretation. Subjectivist quantum interpretations arise from making the relation of the eigenvalue to the observer explicit and jamming the explicit relation into quantum mechanics. In this process they have introduced the illusory notion of the observer.

The content of an observation is the object, and I argue that *the logical relation of the content to the observer is reduced implicit in genuine observations.* The content of an observation has an irreducible conceptual complexity that signifies it is a content and that implicitly relates the object to some unspecified subject falling outside the content.

The Origin and Fallacy of the Observer and Its Instrument

The dominant kind of relations in physical theories are *physical interactions*, expressed as "*a* interacts with *b*." These are explicit relations. Examples are "the earth and the sun gravitationally attract each other," "the photomultiplier detects the photon," or "Tom sees Mary." The last example may be a surprise, but remember we are only concerned with the form of the relation and should not automatically read more meaning into "Tom" and "see." Formally, "Tom sees Mary" is not different from "the camera captures Mary's image;" neither conveys any sense of recognition that marks intelligible observations. This can be seen by the possibility of qualifications such as "Tom sees Mary, but thinks that she is Jane." In quantum mechanics, the coupling of two quantum states $|\phi> \otimes |\psi\rangle$ as in Eq. (2.5) is a physical interaction.

In many philosophical theories, observation is an explicit relation *Rso* between a subject *s* and a thing *o*,

$$s \text{ observes } o. \tag{5.6}$$

I will call this a *detection statement*. It is the source of the saying that the world is divided into the observer and the observed, which is connected by the relation of observation.

Formally, (5.6) is the same as a statement of physical interactions. The passive receptor shows no sign of awareness. The result is not a subject–object relation but a thing–thing confrontation. The meaning of observation is supplied by the anthropomorphic subject or verb. If the verbal nuance is lost in abstract expressions, it is supplied by making substantive qualifications of the subject,

$$s, \text{ which is such and such, observes } o, \tag{5.7}$$

where "such and such" specifies some human conditions. This is a common move of logical positivists and empiricists, whose observation–theoretical distinction depends on substantive qualifications. What counts as qualifications for observers has always been ambiguous and controversial. It is not unnatural to go to the extreme and write "*s*, who is conscious, observes *o*." Since the explicit relation puts the subject on equal footing with the object, the last qualification ushers in the illusion of some kind of not-quite-physical substance.

Insisting on explicit relations is the line of Copenhagenist reasoning. For example, von Neumann said, referring to Bohr: "Experience only makes statements of this type: an observer has made a certain (subjective) observation; and never any like this: a physical quantity has a certain value."[129] His experiential statement is in the form (5.7), which is why he has to insert the qualifier "subjective." His subsequent argument shows that once this sort of experiential statement is adopted, we are in an infinite regress, for the "subjective" can be inserted anywhere. This is not surprising, for the qualification is logically arbitrary.

Invoking the mind or consciousness does no help one bit if they are put in explicit relation with the object. Logically, they are all substantive descriptions of the subject and play the same role as von Neumann's "subjective." Wigner pointed out that a friend's mind can be assigned a state vector and turned into a quantum object. Nor does it help to appeal to "my consciousness," for consciousness is reflexive. Bohr had described a situation in which I divide myself into an infinite retrogressive sequence of "I"'s who consider each other. As long as the Egos are represented by state vectors standing in explicit relation with objects, they all become quantum objects, and no observation obtains. All we gain is an infinite regress. Von Neumann's chain of quantum measurements never ends, because it does not even touch the genuine meaning of observation.[130]

The same logical mistake occurs in various guises in the discussion of quantum measurements. Interpreters essentially try to push "*s*, which is a classical instrument, measures *o*, a quantum object" into quantum mechanics. They are unaware that substantial qualification falls within the jurisdiction of physical theories, and being classical is a predicate quantum mechanics disallows.

The Logical Form of Observation

A common error in philosophy is the introduction of extraneous substantive concepts when one tries to say something in a categorical framework too impoverished to carry the statement. The fallacies of the observer and the classical instrument are examples. Various attempts to introduce the notions of observation and observer into quantum theories fail because they all try to overcome a *formal difference* by a *substantive qualification*. The agent, the instrument, and the division of the world into the observer and the observed are all substantive notions. In contrast, the concept of "observed" in the interpretation of eigenvalues is an abstract qualification that signifies no more that it is the *content* of knowledge. There is a logical chasm that cannot be bridged by substantive means. It can only be understood by adopting a richer categorical framework.

Consider the "camera sees the theft" and "the police officer sees *that* a theft is being committed." The first "see" is a literary license, the second expresses a genuine observation. The difference is made by the "that," which ushers in a totally different logical form. *Observation statements* have the general form:

$$s \text{ observes that } E(o). \tag{5.8}$$

The that-clause conveys the *experiential content* $E(o)$, which is a judgment or proposition E about an object o made by the subject s. The content may contain indexicals such as "this," which reflects on the subject that points. However, the object does not stand in explicit relation with the subject in (5.8). A human observation is always an observing that. In daily speech, s is usually a person or a personified animal or a smart robot, for observation statements express the *intelligibility of experiences*. Unlike "Tom sees Mary," "Tom sees that there's Mary" conveys the sense of recognition; it does not admit qualification such as "Tom sees that there's Mary, but thinks that she is Jane."

Observation statements are so common in daily conversations that they are often abbreviated, especially in first person singular cases. The abbreviation may be confusing. "I see a tree" has the grammatical form of a detection statement, but its logical meaning is observational. I am not asserting there is something that happens to be me and that is detecting a tree, I am reporting my experience of seeing a tree. Thus even when they are *grammatically* similar, detection and observation statements have totally different *logical* forms. The logical difference is all-important for us. For the suggestive nuance of anthropomorphic words is lost when the propositions are written in abstract symbols. The logical form alone remains to signify observation.

Logically, detection statements are extensional, and observation statements are intensional. In *extensional* propositions, terms with the same referent can be unconditionally interchanged without changing the truth value of the proposition. In *intensional* propositions, they generally cannot.[131] For example, suppose Smith, who is known to the security guard of an office building, walked in one day, nodded to the guard, and went on to steal some documents. In talking about the theft, we can equally say "the guard saw Smith" and "the guard saw the thief." The substitution is legitimate because Smith is the thief and extensional statements do not signify recognition. However, if we carelessly substitute "Smith" by "the thief" in intensional statements, we might get a letter from the guard's lawyer. For "the guard saw that Smith entered" is true but "the guard saw that the thief entered" is false, unless the guard was an accomplice. One need not go to law school to explain why the substitution is not warranted: The guard did not know.

Roughly speaking, extensional statements tacitly assume omniscience, which underlies unconditional substitutivity of co-referential terms. By restricting substitutivity, intensional constructions introduce the notions of finitude and partiality. The "that" introduces a "frame" or "horizon" that opens a finite context of intelligibility, which brings the notion of experiential content. "That" signifies $E(o)$ is *observed*, but it is not anything substantive (Fig. 5.2). *Subjectivity is conveyed not by a predicate but by the logical form of observation statements. It is exhausted by the nonsubstitutivity of co-referential terms in experiential contents.*

Intensional logic is much more complicated than extensional logic. The omniscient view is conceptually less complicated; life would be so much easier

Figure 5.2 Left: In extensional constructions, an observer is just another thing on the same footing as the object. Right: In intensional constructions, the content of an observation includes the empirical object but neither the observer nor any explicit relation between them. The formal structure of the content includes a distinction between the physical object and the experience of it. This distinction, like a horizon, signifies the self-awareness that accompany intelligible experiences, and the self-awareness exhausts the meaning of the observer.

if we were God. There are a variety of intensional constructions, which are indispensable for the description of natural languages (see note 131). The ones closest to our topic are associated with propositional attitudes involving verbs such as say, think, know, or believe. As in the case of observe, the that-clauses following these verbs represent judgments, which are finite and partial. Different judgments $E(o)$ and $E'(o)$ can be made about the same object o; the additional variable accounts for the notion of subjectivity or the subject's intellectual contributions to the experiential content. The richer conceptual structure of intensional statements makes room for the notions of possibilities, errors, and the theoretical aspect of observations. These notions find no place in the logical structure of extensional statements.

As far as the content of observation is concerned, the specific observer is inconsequential and can be suppressed. $E(o)$ retains its nature as a content even when (5.8) is put into the reduced passive form, "it is observed that $E(o)$" or

$$E(o) \text{ is observed.} \tag{5.9}$$

The intensional characteristic of the that-clause remains intact. Some information is lost in the transition from (5.8) to (5.9). Demonstratives and indexicals that reflect on s must be properly treated, for instance, by invoking certain cartographic and dating systems. However, to have empirical significance, these systems must be ultimately based on some demonstratives such as "here is Greenwich." Thus the subjectivity of the content persists, although it is pushed one step back. Statements in the form (5.9) are employed in quantum theories: "the eigenvalue a_i is observed" or "the observed eigenvalue is a_i."

The Content of Observations and the Concept of Objects

The content of observation is the empirical object; thus the logic of observation should agree with the general concept of objects discussed in the preceding section. It does agree; what marks observation statements from detection statements is the intelligibility of experiences, which is also what marks the categorical framework of empirical objects from that of the given. The variety of observational contexts, $E(o)$ and $E'(o)$, is analogous to the various representations $f_\alpha(M)$ and $f_\beta(M)$ in the concept of objects. In the context of reference, $\{x^\mu\}_\alpha$ and $\{x^\mu\}_\beta$ are co-referential terms, both designating the external object x (§ 21). The intensionality of the designations is clear. We cannot simply substitute the value of $\{x^\mu\}_\alpha$ for that of $\{x^\mu\}_\beta$ in $f_\beta(M)$, which would almost surely yield wrong answers. We must transform the entire representation, analogous to changing the perspective of perception from $E(o)$ to $E'(o)$.

The content of physical theories in the representational form is the logical equivalence of the content of perceptual experiences. Quantum mechanics contains statements such as "the system in state $|\phi\rangle$ has amplitude $\{a_i c_i\}$." It has the same logical form as the that-clause in an observation statement, because it has invoked a specific representation. Unitary transformations insure that we can also say "the system in $|\phi\rangle$ has amplitude $\{b_i d_i\}$." Again we cannot simply substitute $\{a_i c_i\}$ in the A representation by the value $\{b_i d_i\}$.

The finitude and partiality of human experiences lie as much in the limitation of intellectual representation as in the limitation of sensual capability. The separation of the amplitudes and eigenvalues in quantum mechanics is an additional complication that involves the general limit on the sensual aspect. It is distinct from the logic of observation, which is a purely conceptual matter. The intelligibility of quantum mechanics in the absence of a satisfactory theory of measurement highlights the relative unimportance of perceptual mechanisms in our way of understanding. The mechanism of perception belongs to the content of some empirical theory and not the logic of observation. The quantum measurement problem is frustrating, but it does not imply a different logic that calls for a substantive observer.

The Illusions of the Self and the God's Eye View

The unity of consciousness, which distinguishes objects from experiences of them, is central to Kant's philosophy. Yet Kant emphasized that the unity is only a *formal* unity that calls for a complicated conceptual framework. He said: "I am conscious of myself, not as I appear to myself, nor as I am in myself, but only that I am." I-think "can have no special designation, because it serves only to introduce all our thought."[132] Since there is no unthought thought, the meaning of I-think is exhausted by the thought as such. I am aware of myself only by being aware of the contents of my experiences as all *mine* and being aware of the object as distinct from my experience of it. It is important for theoretical reason that I-think does not convey the full significance of a person; the assumption that it does leads to illusions. Conversely, theoretical reason is incapable of fleshing out the full concept of a person, leaving room for freedom, autonomy, and practical reason.

Consciousness signifies our openness to objects, not a binary relation between a Cartesian subject and an external object. It is like the horizon that opens a field of intelligibility but is itself not an item in the field. What appear together with the horizon, fall within it, and serve as the locus in which various representations are combined are only objects. The field of sight contains neither the eye that sees nor the relation of the eye to the object; the significance of the eye is exhausted by the horizon of the visual field. It is the abstractness of consciousness that demands the intensional construction of observational statements "I observe that $E(o)$." Since "I" does not occur in $E(o)$, which is the content of empirical knowledge, it has no empirical evidence, and we cannot apply the concept of object to it. Thus the world is not divided into the subject and the object. The world contains only objects, which can include physical bodies of human beings; the subject is no more than the abstract bounds of experience and knowledge. Kant said in his criticism of the Cartesian subject: "Rational psychology owes its origin simply to misunderstanding. The unity of consciousness, which underlies the categories, is here mistaken for an intuition of the subject as object, and the category of substance applied to it. But this unity is only unity *in thought*, by which alone no object is given."[133] The illusion of a thinking subject arises from confusing the unity of consciousness with the consciousness of a unity.

Kant was strongly influenced by Hume, who similarly argued that the self, regarded as a substantial subject, is illusory. Hume said: "When I enter most intimately into what I call *myself*, I always stumble on some particular perception or other, of heat or cold, light or shade, love or hatred, pain or pleasure. I never can catch *myself* at any time without a perception, and never can observe anything but the perception. When my perceptions are remov'd for any time, as by sound sleep; so long am I insensible of *myself*, and may truly be said not to exist."[134] *Myself*, the observer, can never be found in the content of my observations, and it vanishes if all contents are removed.

The arguments of Hume and Kant have been confirmed by quantum mechanics. The observer or instrument *qua* observer is never found in the content of quantum mechanics. It is not because physicists have not tried to find it; they have tried very hard. Many attempts to pull the observer into quantum mechanics failed. Whenever an instrument is put in the content, quantum mechanics faithfully treats it as an object. Consequently the instrument ceases to be an instrument *qua* instrument that returns definite numerical results. The content of knowledge comprises only objects; there is no room for the observer, even if it is called an instrument.

The experiential content can be complicated. It can include many objects and their physical interactions. The that-clause in an intensional statement can contain explicit relations, and we can write it as $E(o_1, o_2, \ldots R_1 o_1 o_2, \ldots)$. We have large areas of knowledge that are relatively thorough so that each area can be regarded as a large content. We can envision the development of the disciplinary sciences as attaining higher grounds and gaining wider views. However, no matter how high we climb and how sweeping the vista is, the view is always bounded by a horizon. The observ*ed* is a formal distinction that signifies the horizon of knowledge, which can be realized in any way physical theories see fit. As "I think" accompanies all my experiences, "we think" formally accompanies all our knowledge. Finitude is a human condition that cannot be obliterated by science. Perhaps we can broaden our horizon to include the physical interaction between quantum and classical objects, but we can never step out of our horizon to attain God's position.

§ 17. The mentalism of Yes–No Experiments

Mentalism is profuse in quantum interpretations. Consciousness and other minds take their bow, and the participatory universe that could not exist without our acts of observation clamors for the spotlight. These ideas are popular, but they are poorly developed and are sometimes referred to by philosophers as "the philosophical sayings of physicists." The theories of philosophers are usually more sophisticatedly argued, but they are not free from mentalism. Their mental element is glossed with highly technical terminologies and becomes unobtrusive. A notable example is the yes–no experiment, not in the abstract sense as it occurs in mathematical axiomatizations, but in the substantive sense as it occurs in philosophical interpretations. The yes–no experiment is also called test, event, filter, question, experimental question, or experimentally verifiable proposition. Whatever it is called, it squarely embeds the act of observation into the primitive concept of quantum mechanics.[135]

We should distinguish a physical theory's mathematical foundation from its conceptual foundation, and its axiomatizations from its interpretations. An axiomatization involves certain interpretation in its choice of primitive concepts. The meanings of the primitives are often prima facie clear but are seldom scrutinized by mathematicians or mathematical physicists, who are mainly concerned with mathematical foundations. They concentrate on the internal coherence of axiomatic structures, and seek to set forth the most general and powerful mathematical concepts. Analysis of meaning is the job of philosophical interpreters concerned with conceptual foundations. Their task is to clarify the connection between the mathematical concepts and the objective world.

Philosophers of physics deal with four sets of vocabularies: technical terms in mathematics, physics, and philosophy, and words of everyday speech. The vocabularies share many words but not necessarily their meaning. When mathematicians and physicists choose names for newly introduced technical concepts, they often pick suggestive words. However, a word thus used becomes technical and drops its ordinary meaning. To restore its ordinary meaning involves an extra interpretive step that may be erroneous. Philosophers, whose main task is the clarification of meaning, should be doubly careful not to confuse the suggestive nuance of the technical names with significant interpretations.

The Lattice-Theoretic Axiomatization

Von Neumann noted in his classic treatise that projection operators can have only two eigenvalues, 1 and 0, and suggested that they can be regarded as a kind of *proposition*. Based on the idea, he and Garrett Birkhoff developed quantum logic, which has attracted much attention among mathematicians and philosophers.[136] I think the logic inspired by quantum mechanics has great potential applications, but I do not discuss it here, for I cannot see any clear relation it has with quantum *physics*. Mackey's axiomatization of quantum mechanics contains a primitive concept called *questions*, which admit two answers, 1 and 0, or yes and no. The projection operators on a Hilbert space form a structure called a *lattice*. The lattice-theoretic axiomatization abstracts from the Hilbert space and focuses on the features of the lattice, whose elements are called *propositions*. A proposition has two values, called *true* and *false*, and the combination rules among propositions strictly resemble the rules of logic. Hence the entire structure is often called a *logic*.[137]

The Interpretation of the Axiomatization

The mathematics of the lattice-theoretic axiomatization is impeccable. The names "proposition," etc., are natural in view of the mathematical characteristics of the concepts. However, they are nevertheless names for abstract concepts. When physicists look at the axiomatization, they would consider "proposition" or "logic" as mathematical jargon, whose meaning is to be assigned afresh by physical considerations.

The abstract concept of proposition or question has no natural interpretation in physics. In many interpretations based on the lattice-theoretic axiomatization, it is given an operational meaning as yes–no experiment. A *yes–no experiment* is one whose outcome is the answer to the question: "Is a_i the outcome of the measurement of the observable A?" Here "question," "yes," and "experiment" are no longer technical terms but take on their ordinary sense. As the primitive and fundamental concept, the yes–no experiment is *the* basic observable upon which all other observables are constructed. It differs from ordinary observables in being an observable of an observable. It demands the observation of the value "yes" or "no" *in addition* to the eigenvalue a_i. Physicists such as H. Margenau and L. Cohn had criticized the postulation of such universal supermeasurements on top of specific measurements such as that of spin and momentum, which have clear meaning in physics.[138]

The Object Language and the Metalanguage

The yes–no experiment in interpretations hides the confusion between physical and epistemological concepts, or concepts internal and external to a physical theory. Physics is about the objective world. It employs physical concepts such as particle and field, spin and momentum, terms that belong to the object language. Epistemology is about knowledge and theories. It employs a completely different set of concepts, which include proposition and question, syntax and semantics, law and explanation, truth and falsity. These epistemological terms have no place in the content of physical theories. They belong to the metalanguage with which we talk about the theories.

Consider Heisenberg's suggestion that the wavefunction represents not a microscopic system but our knowledge of it.[139] The proposition sounds both indisputable and absurd. Why? Because we variously look at quantum mechanics externally so that "knowledge" is understood in the meta-language, and internally so that "knowledge" is understood in the object language. The former is obvious; the latter senseless. Externally, wavefunctions are our knowledge, so are Newton's laws and government statistics. Internally, wavefunctions characterize microscopic systems, Newton's laws govern planetary orbits, and statistics describe populations. All sciences are our knowledge, but the contents of the sciences are features of the objective world. The content of knowledge is not knowledge but objects and their characteristics. Thus "wavefunctions are our knowledge" is either trivial or wrong. It is wrong in interpretations, for interpretations are confined to contents.

Similar confusion occurs in the yes–no experiment. Often we say "it is red" and "yes, it is red" are the same. In so saying we regard the "yes" as redundant and can be eliminated. However, the "yes" in yes–no experiments is anything but redundant; it is the outcome of an experiment, the measured quantity in addition to redness. It is represented by a distinct eigenvalue associated to a special observable. When "yes" is thus significant, we easily slide from "it is red" to "yes, it is red" without realizing that we have shifted from the object language to the metalanguage. Of course, if it is red, then "it is red" is true. However, the logical difference is great. "Red" describes an object. "Yes" describes the proposition "it is red" by affirming its truth. The truth of a proposition nominally means its correspondence with facts. Yes and no, truth and falsity, agreement and disagreement with facts, are not physical but epistemological concepts that belong to the metalanguage. Being true or false is the feature of the content of a physical theory, not a part of the content. There should not be a term in a physical theory that stands for truth. However, truth is precisely what the concept of yes–no experiments assumes. It interprets a specific observable in quantum mechanics as a correspondence between a proposition and a fact. Thus it brings the intellectual act of comparing the proposition and the fact into physics proper.

Proponents say yes–no experiments are routine. I counter they are never performed, if the cognition of an experimentalist is not included in the account of the experiment. Consider the simple experiment of throwing a die, the outcome is Ace or Two or . . . , with no inkling of yes or no. Or consider the measurement of a spin-$\frac{1}{2}$ system, the outcome is spin up or down, not yes or no. Yes and no are not results of any physical experiment. They are the outcomes of anthropic experiments that consist of someone looking at the physical outcome and asking: "is the measured velocity 60 cm/sec?" The answer to this question, yes or no, is given by a person, not by nature. Nature's answer is simply 60 cm/sec or some other value. Experiments are always performed by persons. That is why yes–no experiments acquire the deceptive air of a commonplace. *The yes–no experiment depends on the act of cognition.* To include the cognitive act of some person in *the* basic observable is fundamentally mentalistic. Calling the yes–no experiments by other names changes the packaging but not the merchandise.

Mental Selection

Technically, a projection operator is one term in the spectral expansion of a physical observable such as momentum or spin. The question is whether it should be accorded individual significance and made into a superobservable in a physical theory. In the bulk of physics literature, ordinary observables are treated as integral unities. An observable is expanded in terms of its complete spectrum and a quantum state in terms of the complete sets of eigenstates of an observable. The individual terms in the expansion have no independent meaning until the last step, when the absolute square of the state is taken and the

Born postulate invoked. In contrast, the yes–no experiment puts meaning on the individual terms of expansion from the very beginning.

The expansion of a state in a complete sum as in (5.2) can be called an *analysis*, and the picking of a specific term in the sum a *selection*. Analysis of a state is part of quantum physics and does not irreversibly damage quantum properties. This is most apparent in various "haunted experiments," where quantum systems undergo various distortions represented by various analyses, and yet finally recover their original properties. Consider the Stern–Gerlach filtering of spin-$\frac{1}{2}$ particles in the same state $|\phi\rangle$. Analysis consists of separating the up and down spin constituents by an inhomogeneous magnetic field. The separated beams, even if they had been miles apart, can be brought together and properly recombined to restore the original state $|\phi\rangle$. In contradistinction to analysis, selection involves the extra step of blocking one of the constituent beams and isolating the other. This extra step destroys the original state.[140]

The selection step belongs marginally to the logical structure of quantum mechanics, which says nothing about the selection process but mentions its result by curtly stipulating $|c_i|^2$ to be the probability that a_i is the measured value. Redhead remarked that the selection of parts of a system is not a physical process, "it is a 'mental' operation that we perform when we decide to focus our attention on the subensemble" that shows a specific value. This does not mean we cannot perform physical manipulations to realize our mental selection; it means the selection is not part of quantum physics.[141] In physicists' calculations, the selection signifies the "measurement" that marks the *end* of quantum physics. In contrast, the yes–no experiment puts the selection process at the *beginning* and makes mental operations and human involvement into the primitive of quantum mechanics.

One way to preserve the yes–no experiment is to interpret quantum mechanics in a strict instrumentalist sense. Quantum mechanics is totally meaningless; it is merely an algorithm for manipulating the sequences of 0s and 1s obtained in experiments. This interpretation is irrelevant to physics.

"Operational meaning" or "instrumental meaning" is the mainstay of logical positivism. It appears to be materialistic and objective. It relies on many common-sense notions. This poses no objection if the notions are critically examined. The trouble is that often some superficially obvious notions are unquestioningly codified in the operational definition and shielded from criticism, leading to adverse results. The yes–no experiment is an example.

Undoubtedly, lattice theory and quantum logic are very powerful and general structures. However, in philosophical analysis, the most powerful concepts in the abstract may not be the best choice as primitives. The subject matter is more important. Suppose we want to understand a person. We have ample techniques to dissect him, reduce him to organs, tissues, cells, molecules, and atoms. If we did these, we could publish papers that may be valuable in their own right, but we would end up knowing very little about the person. He was killed by the first cut at the throat.

There is a question physics professors often throw at students who have covered the blackboard with Green's functions: What is the *physics* behind all the calculation? It is also sobering to ask in interpretations: What is the *philosophy* behind the logic? Technical prowess is desirable, but it cannot replace thinking. Saul Kripke reflected at the end of a highly technical philosophical paper: "Logical investigation can obviously be a useful tool for philosophy. They must, however, be informed by a sensitivity to the philosophical significance of the formalism and by a generous admixture of common sense, as well as a thorough understanding both of the basic concepts and of the technical details of the formal material used . . . . There is no mathematical substitution for philosophy."[142]

6

The Event Structure
and the Spatio-temporal Structure:
Local Fields

If we imagine the gravitational field, i.e., the function g_{ik}, to be removed, there does not remain a space of the type (I) [Minkowski space], but absolutely *nothing*, and also no "topological space". . . . There is no such thing as an empty space, i.e., a space without field. Space–time does not claim existence on its own, but only as a structural quality of the field.

ALBERT EINSTEIN
Relativity, The Special and General Theory, Appendix V

The inhabitants of the universe were conceived to be a set of fields—an electron field, a proton field, an electromagnetic field—and particles were reduced to mere epiphenomena. In its essentials, this point of view has survived to the present day, and forms the central dogma of quantum field theory: *the essential reality is a set of fields* subject to the rules of special relativity and quantum mechanics; all else is derived as a consequence of the quantum dynamics of these fields.

STEVEN WEINBERG
"The Search for Unity:
Notes for a History of Quantum Field Theory"

§ 18. The Whole and the Individual: Field and Event

Going from classical mechanics to quantum field theory, the focus of physics changes from locomotion to dynamical interaction. The transformation brings interrelated changes in the basic material, dynamical, and spatial concepts. The primary form of matter changes from discrete mass points in empty space to continuous fields comprising discrete events. The primary dynamical concepts change from action-at-a-distance to coupling-on-the-spot, from external forces to interactions generated by the interactants themselves. Unlike classical and nonrelativistic quantum mechanics, where spatial position is a dynamical variable and time is the lone parameter, position is now a parameter on equal

119

footing with time. Some, although not all, of these changes also occur in other field theories, classical electromagnetism, and general relativity.

Field theories present a full world. The idea of a full world is not new; to the ancients, fullness was intuitive, and emptiness was not. Aristotle said nature abhorred a vacuum, and the void posited by the Greek Atomists was ignored for a long time. The ontology of ponderous bodies moving in empty space gained currency only with the triumph of Newtonian mechanics. It was rejected by many of Newton's contemporaries, including Descartes and Leibniz. The disparate views on the form of matter played no small role in the Newton–Leibniz dispute over the nature of space; a full world with no empty space makes the ideas of container and occupancy otiose. Voltaire remarked that in crossing the English Channel, he had left a plenum and landed in a vacuum. Field theories put us back in the plenum.

The vacuum and the plenum each has its difficulties. Each is involved in its own way in the perennial philosophical problem of the tension between individuals and the community to which they belong. Most philosophical theories have difficulties accounting for the interdependence of the entity and the whole, resulting in the dominance of one at the sacrifice of the other. The tension first appeared in the debate between the Greek Eleatics and Atomists, and did not pass with them. Those who posit a set of entities have trouble uniting them into a whole; the cementing elements are extra and easily give way under skeptical attacks, resulting in solipsism. Those who posit a whole have trouble differentiating determinable parts and admitting entities, without which little can be articulated. Leibniz's Monadology is outstanding for its attempt to formulate a genuine whole with genuine individuals. Yet Kant remarked that without the appropriate concepts of space and time, Leibniz's system succeeds only with the help of God.

Atomism holds sway; the difficulty of the holist seems to be more basic. However, holists do have an important point. They remind us not to forget the ideal elements hidden in the concept of discrete individuals and the presuppositions of postulating sets of entities as the semantic domain of discourse. Epistemologically, the entity that announces itself by stimulating our senses is a myth. Cognitive scientists tell us how much is required to individuate entities in our visual field, which is usually a continuum of most complicated features.[143] That is why computers have such a hard time recognizing figures and patterns. The sound waves impinging on our ears are no less continuous; computers have no easier time in parsing speech. It is more likely that we recognize entities by their response to our manipulation than our response to their stimulation. Physically, there is no entity so isolated that it is totally decoupled from others. Apples and stones and rivers are things. Yet apples hang on branches, rocks bury deep in earth, and rivers constantly exchange their contents with the ground and the atmosphere. Things support and resist each other; they reflect and scatter light. In physics, too, isolated systems are idealizations. The world is an interactive system. We ideally carve out discrete entities by neglecting weak couplings. In most cases, the neglected coupling is weak enough so that the approximation of discrete entities is good. In some cases, as in the strong nuclear interaction or the electron cloud in an atom, no good approximation is available. Good or bad, they share the general idea of approximations and idealizations.

Having acknowledged the holist's point, we still think that the basic units of our experiences and discourses are entities that can be individually referred to and made into subjects of propositions. Our task is to articulate the concept of entities without neglecting the holist's point. Imagine a picture with a rich nuance of colors but no clearcut contour or conspicuous point on which we can put our finger and say *it* is red. Imagine that the picture is four dimensional and extends without bound, and that it represents the material universe as some kind of undifferentiated stuff. The stuff contains all subtleties and complications of the world. Perhaps God or some superintelligence can comprehend all details of the four-dimensional whole in a single glance. We mortals cannot. Without differentiation we cannot say much beyond Parmenides' poem:

> Nor was it ever, nor will it be, since it now is,
> All together, one, continuous.[144]

We must analyze and differentiate to gain definite knowledge. Imagine a blank domain such as a sheet of white paper, then divide it into two parts by drawing a curve across it. The boundary becomes the defining characteristic of each part and the intrinsic relation between them. Now imagine that we mentally draw an infinitely dense grid of curves over the four-dimensional colorful picture. The conceptual scheme partitions the picture into a world of parts implicitly related to each other and to the whole, and yet each can be singly referred to. We intellectually crystallize the amorphous continuum into a continuous world comprising determinable entities. These entities are not isolated; being the result of a systematic partition, they are radically situated and implicitly related from the beginning.

The preceding simplistic and one-sided metaphor makes two points. The first is the idea of systematic partitions; the second concerns space–time. If fields have made a substantival space obsolete, they have not endorsed the positivist theories that assert space is a set of relations among nonspatial entities. Field theories bring to relief the Parmenidean and Kantian insight that entities are not individuated without the help of spatial and temporal concepts. However, spatio-temporal concepts alone are insufficient for individuation. As described in the metaphor, the resultant entities are phonies; they have no individuality at all. The purely geometric scheme is arbitrary because it completely ignores the qualities of the resultant entities. It is like some political arrangements of former colonies based on territorial consideration alone; the result is endless turmoil. A satisfactory scheme of individuation must account for the resultant properties of the entities and the whole.

How are entities defined in the plenum? How do entities maintain their individuality in interactions? No one can answer better than field theories, which address these questions with a rigor unmatched in philosophy. A field is a whole, but contrary to its popular image, it is not amorphous. *A field is a genuine whole comprising genuine individuals, a continuous world with discrete and concrete entities, technically called events.* The discreteness of the events and their mutual interaction are both clearly articulated. The articulation demonstrates that the concepts of entities and space–time are thoroughly integrated. Together they constitute the conceptual framework for *a plurality of individual objects*, which presupposes the primitive structure of *objects* discussed in § 15. The plurality is not yet a community. Although the objects are situated within a spatio-temporal structure, they are unrelated and solipsistic, and their properties cannot be compared. In the next chapter, the general concept of objects is enriched with the notions of *relational properties*, both causal and spatio-temporal. The result is the minimal conceptual structure underlying our knowledge of the *objective world* or a world of interacting objects.

Field theories with local symmetries, which include general relativity and quantum field theory, individuate events by their two symmetry structures. A symmetry structure characterizes a unity (§ 6). The local symmetry of a field theory secures the unity of each event; the spatio-temporal symmetry secures the integrity of the field system as a whole. The general idea is similar to the novelty of Leibniz's system of monads. Each event in an interacting field system is partless and spatio-temporally indivisible. Yet it is endowed with its own internal structure that incorporates all its possible qualities, including relational properties that underlie interaction. All events in a field are of the same kind; their possibility structures are alike, being all mirror images of the *local symmetry group.* The local symmetry structure affirms the individuality of an event by endowing it with its own choice of representation for definite description. The distinctiveness of the events is affirmed by the separate and systematic introduction of their numerical identities. The system of identities is the primitive spatio-temporal structure of the field system. These ideas are schematically summarized in Fig. 6.2. Reference to the entities involves the structure of the *spatio-temporal group*, which takes account of concepts such as singular terms and coreferentiality (Fig. 6.4). Mathematically, all these structures can be summarized by fiber bundles. The following three sections investigate the meanings of the elements in these categorical structures, with special attention to how the concepts of individuals and space–time are integrated.

§ 19. This-Something: Events or Local Fields

"Entity" is the Latin equivalent of "thing." Ordinarily, things are enduring. I retain the usage and use "entities" more generally to include momentary things, and "individual" even more generally. An entity is an individual that can be singly picked out, referred to, and made into the subject of propositions. It is an instance of some general concept. It has properties, but is itself not the property of something else. It is a particular individual and has its own identity, which distinguishes it from all other entities, including members of its own kind, with which it shares many or all qualities.

The requirement of identifiable reference suggests that what is usually called "particles" in quantum field theory are strictly speaking not entities. As quanta of field excitation, particles are like modes of waves and lack individual identity (§ 9). *What then is an entity in a field? How do we unambiguously refer to it?*

Peter Strawson investigated the conceptual scheme in terms of which we think about individuals. He concluded that identifying reference to individuals requires a spatio-temporal framework.[145] The trouble with most contemporary philosophical analyses of the concept of individuals is that they take the notion of space–time for granted. Often they assume a substantival space–time that the individuals occupy. The substantival space–time has been disputed since the advent of the theories of relativity. The philosophy of space–time boasts a large literature, but almost all of it takes the notion of individual entities for granted, assuming that the entities can be unambiguously referred to without spatio-temporal concepts.[146] There is little cross reference between the two bodies of philosophy, although each is finding problematic what the other uncritically depends on.

Field theories show that *the concepts of entities and space–time are inseparable*; entities are individuated within a world that is spatio-temporal. Since we must arbitrarily pick an entry point for our investigation of this structure of thoroughly interrelated concepts, our initial position must do some violence, which we hope to repair as the discussion progresses. In this section, I consider entities as far as possible without mentioning space–time. Their inalienable spatio-temporal significance is revealed in the subsequent sections.

There are several outstanding philosophical problems regarding the concept of entities. One is the identification and reidentification of a thing through time and changes. Another is the identification of a composite system as a unitary entity. These important problems can get very complicated. Fortunately, we need not worry about them in our reflection on fundamental physics. The entities that concern us, events in an interacting field system, are simpler than the simple things in traditional philosophy. They do not endure and they have neither spatial nor temporal parts. We can regard them as the basic entities or basic individuals of the physical world.

Individuality, Kind, and Qualities

Aristotle staked out various sciences. He also contemplated a science of being-*qua*-being or things in general, as distinct from the disciplinary sciences whose topics are certain kinds of things. He was specifically interested in primary beings, which are those subjects of propositions to which all our descriptions of the world ultimately refer. Aristotle found that the general concept of a primary being incorporates two elements: it is both a *this* and a *what-it-is*. Examples are this man or that horse. An entity is not simply singled out to be the subject of discourse; it is always singled out as something. When I say "look at the thing there," what I point out is a thing, say a sculpture, not a bare particular or a space region, which I describe as a sculpture. An "I know not what it is" is an unfamiliar thing, not a bare thing that has all its qualities stripped away. The entity, *this-something*, is the irreducible unit of discourse. Only after we have grasped the *this-something* can we go on to describe it as *such and so*, as "the horse is white."[147] The distinction between a synoptic "something" and specific "such and so" finds a parallel in the distinction between a broad type represented by a class of state spaces and the values of properties types represented by functions or operators in physical theories (§ 11).

The notion of an entity as this-something is common-sensical; Aristotle arrived at it by analyzing common discourse. Yet it has been much challenged. It is not a simple and unanalyzable concept; it has at least two irreducible elements, one for an entity's numerical identity, the other its kind. The conceptual complexity is a major point of contention.

Some philosophers argued that the complexity can be eliminated by adopting the *cluster-of-qualities* notion of things, which says that the concept of entities is exhausted by that of qualities. An entity is no more than the sum of its qualities. No two entities can be alike in all qualities, hence the notion of numerical identity can be discarded. The doctrine that descriptions alone suffice for the concept of entities was championed by Leibniz, and was espoused by Russell and A. J. Ayer in our times.[148] It is related to the doctrine that general terms alone suffice for discourse and all singular terms can be eliminated, which is promoted by W. V. O. Quine. As Strawson pointed out, shorn of the support of expressions such as Fa, $\exists x(Fx)$ reads "F instantiates" and not "there is something that is F." When "this is a horse" is logically outlawed, we can only say "being-horse instantiates" and not "there is a horse."[149]

There is the statistical mode of discourse where the independent variable is the value of a quality type (Appendix A3). For instance, in "those earning less than $10,000 a year constitutes 10% of the population," we pick out a group of households by specifying a level of annual income. However, there is generally no guarantee that each value, no matter how finely defined, picks up only one individual.[150] It is a familiar complaint that statistics overlooks the particularity of individuals. Thus although superficially statistical statements appear to support the cluster-of-qualities view, upon a closer look they refute it by showing that the notion of entities is obscured when only qualities are invoked.

The cluster-of-qualities notion of entities has great difficulty in accounting for our daily understanding. Identifying descriptions using exclusively general terms are difficult to find. Most philosophical papers concentrate on celebrities, but not many people have such distinction as the discoverer of the equivalence of mass and energy; even fewer things have such distinction as the inspiration for the inverse square law of gravity. We often talk about persons or things of which we know little and cannot provide any identifying description. Far from being nonsense, such talks are the crucial means to gather information. We understand that even a faceless "nobody" has his identity, often represented by his name. As we repeatedly hear his name mentioned, we pick up information on him. The dearth of substantive content exemplifies the complexity of the general concept of entities.

The difficulty of identifying descriptions is more severe in scientific discourse, where we are faced with countless entities that are more or less the same qualitatively. Quine suggested inventing *ad hoc* unique predicates such as "pegasizes" for Pegasus. However, he did not consider the way to safeguard the uniqueness of the predicates, which is paramount in scientific theories. Leibniz was more cautious and systematic. He introduced a peculiar quality type, called the "point of view," which has different values for different simple entities. The uniqueness of each value and their mutual compatibility are guaranteed by the pre-established harmony. Unfortunately, the harmony must be underwritten by God; as Kant remarked, Leibniz was pushed to it because he lacked the proper concept of space. Most other proponents of the cluster view appealed to the spatial and temporal positions of the entities as their unique qualities. This move is also fraught with difficulties. For as Leibniz argued, space points are themselves qualitatively alike and contribute nothing to differentiation. Spatial relations do no help either because the world can be symmetric so that the relations are alike. If we treat space points as distinct despite their identical qualities, then we have already gone beyond the concept of qualities. The notion of numerical identity has been smuggled in.

The appeal to space–time or an additional peculiar quality exposes the inadequacy of the cluster-of-quality view. Ordinary qualities are insufficient for the notion of entities. If we abstract from the substantiality of space–time points, what the spatio-temporal augmentation amounts to is that an extra general concept besides qualities is required for the notion of individuals. Let us call the extra concept the numerical identities of entities.

The *this* of this-something expresses the numerical identity of an entity. Philosophers have proposed several identity criteria for various composite things, but all criteria seem to lose weight for partless and momentary entities.[151] The simple case suggests that the concept of identity is distinct from and more basic than the criterion. Ludwig Wittgenstein said, "identity is not a relation between objects . . . . To say of *two* things that they are identical is nonsense, and to say of *one* thing that it is identical with itself is to say nothing."[152] Identity does not say anything beyond *one thing*; rather, it discloses the meaning of *being an entity,* and the disclosure signifies our primordial understanding. Heidegger said identity means: "To every being as such there belongs identity, the unity with itself."[153] "*A* is *A*" is tautological because identity is constitutive of the general concept of the individual that *A is*. Identity ensures that an individual is one and cannot be two, and that two individuals cannot be one.

Identity is a formal concept, not to be confused with substance or some substratum that supports qualities. It is the condition for the possibility of direct reference to things. Figuratively, if names are "tags," then the concept of things has to provide "bulletin boards" for the names to be tagged. The bulletin board is furnished by the formal concept of identity, which makes naming possible.

The dictionary says that besides *oneness* and *sameness*, identity also means *individuality*. People often complain that they lose their identities, meaning they lose their peculiarities and are assimilated by the crowd. To have identity is to be distinctive, which implies others from which one differs. Thus identity implies the general notions of *difference* and *diversity*. Entities, as individuals separated from each other, presuppose the general concept of plurality. There may be only one entity, the "only" signifies the lack of plurality. The posit of an entity is simultaneously the posit of the possibility of a diversity. That is why substance terms such as apple and orange are also called count nouns, because the notions of unit and multiplicity are built into the concept of entities.

Kinds and Possibilities

The "something" of this-something signifies only the broadest classification of entities into *kinds*. The distinction between what-it-is (the kind a thing belongs) and such-and-so (the thing's definite descriptions) is not apparent in predicate logic, but it is not absent. Suppose F represents the property "flies," then $(\exists x)Fx$ reads "some fly." This makes no sense because "some" and "all" are syncategorematic terms that have no references of their own. We immediately ask, some *what* fly? This question is usually accounted for by one of two ways. We can specify in advance the domain of quantification, say, x ranges over all birds. Alternatively, we can restrict a broader domain by using another predicate letter, say, B for being bird, then writing $(\exists x)(Bx \wedge Fx)$ for "some birds fly," and $(x)(Bx \rightarrow Fx)$ for "all birds fly." Unlike $(\exists x)Fx$, $(\exists x)Bx$ reading "there are some birds" makes sense. The fact that some predicates such as being white cannot be invoked alone confers special status to those predicates that can be invoked alone. These special predicates signify the kinds of entities. Many names have their kinds explicitly stated, for example, "Mount Everest."

The what-it-is, under which an entity is singled out, is called a *sortal concept*.[154] A sortal concept includes the stipulation of what counts as an instance under it, and is linguistically embodied in a substance term. Substance terms are mainly common names of *kinds of entities*. They have the peculiarity of carrying intrinsic and definite principles of individuation. "Apple" and "planet" and "molecule" are substance terms, one does not understand them unless he knows what counts as an apple or a planet or a molecule. Substance terms stand in contrast to mass terms; which represent kinds of stuff or material, for instance "air" and "gold." Mass terms have no intrinsic principle of individuation. They accept supplementary individuation such as a cup of water or a ring of gold, but the supplements are extrinsic and arbitrary.[155]

What is a *kind of entity*? A kind cannot be defined by its extension, for without the concept of apples we do not know what to gather as the extension of "apple." Thus the definition of kinds must appeal to qualities. Traditionally, philosophers try to stipulate a set of common or essential traits or to hold up a paradigm member. Their prime examples are atomic and molecular structures. For instance, gold as a natural kind is defined by its atomic structure. Such stipulations meet with great difficulties in many applications where the descriptions are not so simple and clear-cut. There may be deviants that neither resemble the paradigm nor exhibit the common traits and yet are clearly accepted as kind members. Wittgenstein, for example, argued that often there is simply no single feature common to the members of a kind, which exhibit at most some "family resemblances."[156]

We can distinguish *two concepts of kinds*, one for phenomenological classification and another as an active structural element of the concept of entities. We are more interested in the latter. Physical kinds are sometimes designated by some parameters, for instance, the atomic numbers. The parameters are useful for phenomenology. However, when we examine the structure of physical theories more closely, we find the parameters are abbreviations of some complicated structures. Six each of neutrons, protons, and electrons thrown together does not make carbon, what makes carbon is the peculiar structure that binds these elements together. The atomic numbers are more like shorthand notations for atomic structures than definitions of essential traits.

In modern physics, a kind is often represented by a class of *state spaces*. Sometimes the state space is the group space of a symmetry group. As discussed in §§ 3 and 11, a state space includes all *possible* states of a system. Similarly, a group circumscribes the range of *possible* properties that a kind of entity can assume. For instance, the symmetry group $U(1)$ specifies the kind of entity participating in electromagnetism by systematically accounting for all possible phase factors of the entities. *In physical theories, a kind of entity is defined by a systematic circumscription of admissible properties; things of the same kind share the same possibilities.* Being a member of a kind, an entity can be in any of the possible states, although it actually is in a particular one. Thus the concept of kinds incorporates the distinction between the possible and the actual. It naturally allows counterfactual statements.

The systematic definition of possible states means they are not arbitrary but correlated. It leads to a coherent structure of possibilities delimiting the range of the actual traits of the members. The coherence of the structure determines its membership in cases where two possible ranges overlap. There may be kinds that admit only one or a few possible states; in such cases one can recognize common traits. However, generally possible states are not common features, for possibilities are abstract. An entity actually realizes only one of them.

In mathematical physics, the possibility space or the state space of a kind is exactly defined. In more complicated situations the definition may not be exact or even feasible; hence we have notion of norms and essential traits. However, even here a vague notion of possibilities is present. To make the concept of possibilities basic to the general concept of objective properties has a venerable tradition. Aristotle considered modal syllogisms. Leibniz argued that possibility precedes actuality. Modality is one of Kant's four groups of categories. Existentialists generally hold that the essence of a being lies in its possibilities. Since Kripke introduced the possible world semantics in the 1950s, modal logic has become a powerful tool to formalize natural language.

Events: Entities in Field Theories with Local Symmetries

The concept of this-something is incorporated in field theories via their two symmetry structures. As discussed in § 15, a symmetry structure incorporates the elements of representations, transformations in a group, and certain invariant features. *The idea of kinds is embodied in the local symmetry group,* which pertains not to spatio-temporal but to qualitative features. The symmetry group circumscribes a set of possible states and defines a natural kind. The domain of its transformations is the state space of a single object. With the localization of symmetry transformations, a state space is set up for each point in a field, turning the point into a distinct object, called an *event* (Fig. 6.1). The autonomy of an event is expressed by its freedom to pick its representations independently of its neighbors. Localization and identification are meaningful only within a global whole, which is represented by the spatio-temporal group. *The identities of the events are the invariants in the spatio-temporal symmetry structure.* The "this," which goes beyond an abstract identity and includes definite indexicals, involves the invariant and other elements of the structure (§ 21). Thus the two symmetry structures jointly characterize a plurality of entities.

Before I go on let me clarify some terminologies and distinguish the meaning of matter, field, particle, and field quantum. When physicists say they are investigating the structure of matter, they mean more than what are called matter fields. I use "matter" and "material" in the broad sense that includes both matter fields and interaction fields. Both types of fields are energetic, and energy can be easily converted from one field to the other. For instance, in pair creation, a quantum of the electromagnetic field converts into a quantum of the electron field and a quantum of the positron field.

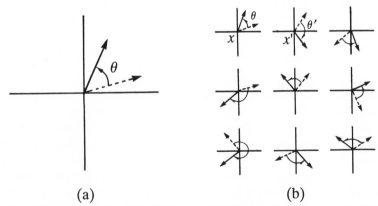

(a)   (b)

Figure 6.1 (a) The domain of symmetry transformations is the state space of a single entity. (b) Symmetry transformations are characterized by certain parameters, say θ. When θ is allowed to vary from point to point, the synmmetry is localized. Since the transformation at x is independent of that at x', a state space is effectively set up at each point. Each state space belongs to a discrete event $\psi(x)$ identified by x.

I use "particle" in its ordinary sense. A *particle* is an individual entity. Particles are usually well defined spatially and temporally, but we can hide the spatio-temporal conditions, as long as we can somehow unambiguously individuate them and refer to them. Classical mass points are particles. In quantum mechanics, an electron as an isolated system represented by a state vector or a wavefunction is a particle. In the quantum case, the isolation is done outside the theory, and can lead to problems when we consider quantum multiple-particle systems (§ 23).

A *field quantum* or a discrete mode of field excitation is usually called a particle. Field quanta have less individuality than what we ordinarily think of as things because they do not admit exact ostensive reference. Also, they are not well defined in the presence of interactions (§ 22). To avoid confusion, I try not to use "particles" for field quanta, but I do not always succeed; "virtual field quanta" sounds too outrageous.

A *field* is a dynamical system with an infinite number of degrees of freedom. A *free field*, say an electron field or an electromagnetic field, is an idealized component of an interacting field system obtained by neglecting interactions. Each degree of freedom $\psi(x)$ of a free field is designated by a value of the continuous parameter x, analogous to a bead in a vibrating string indexed by an integer (Fig. 4.2). When x is fixed at a specific value, $\psi(x)$ is an entity in the field. With x regarded as a variable, $\psi(x)$ is often called the *local field* or the local field operator, which distinguishes it from "field" as in the dynamical system as a whole. The local fields are the focus of field theories.

Consider an interacting field system with N free fields represented by local field operators $\psi_i(x)$, $i = 1, \ldots N$, where x is the four-dimensional spatiotemporal parameter. The field system is characterized by the Lagrangian in which all interaction terms are of the form $\psi_i(x)\psi_j(x)\psi_k(x)$; that is, all interactions among the free fields occur at a point. The interacting system itself must be considered as an integral whole, in which the free fields are approximations, for the charges of the free fields are also the sources of interaction. *All local fields with the same parameter x and their products constitute an event. An event is an entity in an interacting field system.* An event is extensionless in all four dimensions; hence it is spatially and temporally indivisible. However, it is analyzable into idealized free local fields and their interaction. The indivisibility and analyzability are compatible with the nonspatial characteristic of quantum properties (§ 12).

An event is a dynamical quantity; it is the transformation of the state of the field system at a certain point. For example, the event $\psi_i(x)$ may represent the excitation of certain modes of type i, $\psi_i(x)\psi_j(x)\psi_k(x)$ may represent the excitation of certain modes of type i and deexcitation of certain modes of types j and k. Our events are concrete; they are distinct from the events in many general relativity texts, which are merely points of a bare manifold. Our technical "event" differs from the ordinary usage of "event," which means a happening to enduring things. However, the two are related. The concept of enduring things can be constructed from our events. Conversely, if we regard things as the basic concept, then our events become incidents. Finally, our events are not to be confused with "temporal stages" or "time slices" of things posited by some philosophers. Such stages and slices are like the phonies in the geometric partition of § 18, they lack the coherent internal structure of entities as will be discussed.

In this section, we neglect the spatio-temporal connotation of the parameter x and consider only its *uniqueness* and *integrity*; x is unique to the event $\psi_i(x)\psi_j(x)\psi_k(x)$ and x has no parts. The *numerical identity* of the event is embodied in the parameter x. The *kind* of events is embodied in the local symmetry group, which prescribes the internal structure or state space for each event in the field system. Together they realize the formal structure of *this-something*. To illustrate the interpretation, let us first consider the idealized case of a single free field $\psi(x)$. Interactions, considered in the next chapter, will modify the concept of events developed here by adding explicit relations and relational properties.

Partition of Unity and the Quotient Space

To see exactly how an event is conceptually analyzed into identity and quality, we have to study the structure of local symmetries. Instead of repeating the usual localization procedure as presented in § 10, I will here illustrate the formation of concrete individuals under sortal concepts by the *fiber bundle formulation* of gauge field theories. The fiber bundle formulation is general. Fiber bundles with local symmetry groups apply to general relativity as well, and other less elaborate bundles apply to classical mechanics. Thus it provides a kind of unified framework for us to compare the conceptual structures of various major physical theories.

The fiber bundle, with its topological and geometric trappings, is complicated (Appendix B). However, its primary ideas of fibrillation and bundling can be illustrated by the idea of the *partition of unity* familiar in set theory. Here a local symmetry group divides up all possible qualities and the quotients are introduced as the identities of the parts.

Let us start with a set D of abstract qualities $\theta, \theta' \ldots$ and a *symmetry group* G. In field theories, both D and G are continua, but this specific feature plays little role in the following. G may be the unitary group $U(1)$, $SU(2) \times U(1)$, $SU(3)$, or the Poincaré group, which, respectively, leads to the field theory for electromagnetic, electroweak, strong, or gravitational interaction. At this starting point, both D and G are abstract and meaningless. Our aim is to find the minimal conceptual structure in which we recognize events as individuals. The elements θ and the group G will acquire meaning as the events emerge.

The group G is constitutive of a sortal concept. It acts on the set D and collects a subset of elements; the subset is technically called a G orbit. Group actions constitute an equivalence relation or relation that is reflexive, symmetric, and transitive. Let $\sim$ denote the equivalence relation corresponding to G. We pick an element θ and gather all elements θ' that stand in relation $\sim$ to θ. The result is a G-orbit

$$\theta \sim \theta' \Rightarrow \theta \text{ and } \theta' \text{ belongs to the } same \text{ } G\text{-orbit.} \tag{6.1}$$

The G-orbit containing θ is denoted by $\{\theta': \theta \sim \theta'\}$. We pick another element ζ and repeat the process. We do so until all elements in D are accounted for. Since $\sim$ is an equivalence relation, the resultant G-orbits do not share common elements. Two G-orbits either have no element in common or share every element, in which case they are identical. Thus the action of G sorts the elements of D into pairwise disjoint G-orbits,

$$D = \{\{\theta' : \theta \sim \theta'\}, \{\zeta' : \zeta \sim \zeta'\}, \ldots\} \tag{6.2}$$

Each G-orbit is isomorphic to G and to each other. The local symmetry group creates the orbits in its own image, so to speak (Fig. 6.2).

We next confer individual identity to each G-orbit by explicitly saying that it is a unity distinct from other G-orbits. There is a most natural way to do this since D is already neatly partitioned. We introduce a map π that sends all elements θ in a G-orbit onto a single point x in a separate set M. Conversely, the inverse map π^{-1} canonically assigns a unique element x in M to each G-orbit, which we now denote by $\psi(x)$,

$$\pi(\theta) = x; \; \theta \in \psi(x), \; x \in M \tag{6.3}$$

M is not given in advance; it is the quotient of D by the equivalence relation G. There is no point in M that is not associated with a $\psi(x)$ in D. Set theoretically, D becomes a set with an indexing set M:

$$D = \{\psi(x): x \in M\} \tag{6.4}$$

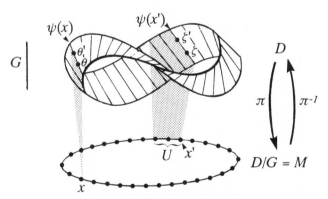

Figure 6.2 The concepts of individuals and space–time. The local symmetry group G collects all possible states $\theta, \theta', \ldots$ admissible for a kind of events into pairwise disjoint G-orbits. Each G-orbit is a mirror image of the group G. The quotient space M is defined by the projection map π, which assigns to each G-orbit a unique identity x. With its numerical distinctiveness explicitly introduced, the G-orbit becomes *an event $\psi(x)$ of the kind G with identity x*. G and π constitute the sortal concept of the events. The identity x of the event $\psi(x)$ is also interpreted as its *absolsute position* in the *primary spatio-temporal structure M*. $\psi(x)$ is an indivisible unit because x is elementary in M. The extension of a finite region U of the field is obtained by integrating over the identities of its constituents.

In field theories, the sets D, M, and $\psi(x)$ all have complicated structures, which are not shown in this simplistic presentation.

Equation (6.4) is not merely a handy way of writing Eq. (6.2); it involves an extra layer of structure. The x in M corresponding to $\psi(x)$ captures the idea that $\psi(x)$ is an entity numerically distinct from all others despite their qualitative similarity. x uniquely designates $\psi(x)$ and marks it out from others; hence it naturally signifies the *identity* of $\psi(x)$. The group G, whose features are typical of all $\psi(x)$, naturally signifies a *kind*. $\psi(x)$, with the conceptual complexity of kind and identity, becomes an *individual entity* or an *event*. The systematic way in which $\psi(x)$ are individuated shows that they are members of a world and implicitly related. M, the set of all x, can be interpreted as the *identifying system for the individuals*. I further interpret the elements θ as *possible properties* for the kind of entities specified by G. The *actual property* of an entity is a specific element among the possibilities. All entities of a kind have the same set of possibilities, but not all are actually similar in all respects.

Mathematically, the quartet (D, M, G, π) constitutes a *fiber bundle*, where D is called the total space, M the base space, π the projection map, and $\psi(x)$ a fiber. D, M, and G are all differentiable manifolds. The entire fiber bundle (D, M, G, π) is required to represent a field theory with local symmetry group G. Each fiber $\psi(x)$ represents an event with identity x and independently transformable phase space patterned after G, the local symmetry group that determines the kind of field it belongs. Technically, M functions as the *parameter space* with parameter x.

The collecting action of the symmetry group does *not* specify which element in the G-orbit corresponds to the identity element of G. The resultant conceptual structure includes no common "baseline" across the individual phase spaces, and each event can choose its own convention of definite description. In general relativity, this is familiarly known as the freedom of the local light cone structures to rotate. The absence of rigid baseline makes the events into autonomous individuals. As it stands now, the individuals are solipsistic because their properties cannot be compared. There is yet no device that enables us to tell if θ for $\psi(x)$ is the same or different from ζ for $\psi(x')$. However, the foundation for the construction of a community is already present, for the solipsistic events are already embedded in a world.

Let us compare the above set-theoretic consideration to the more standard description of field theories. In § 9 we saw that the parameter x for a field $\psi(x)$ is the continuous limit of indices of beads in the discrete case. This agrees with the idea of the field as a set with an indexing set in (6.4). However, there is a big difference. Figure 4.2 illustrates a concrete example; the indices in it are definite. Here we are discussing formal concepts; x and the indexing set M are both abstract concepts; the identity x is neither a definite index nor a set of numbers. In physicists' parlance, x and M are coordinate-free. They are the formal conditions for the possibility of definite indexing, for they present definite entities for us to coordinatize. Definite coordinates and indices will be introduced in § 21 and interpreted as names of events.

In § 10, we saw that the symmetry group of a field is localized to each point. This agrees with the idea of the collecting action of the group in (6.1). The disjointedness of the G-orbits accentuates the independence of the state spaces for different field operators. The physics approach introduces symmetry transformations on each point. The set-theoretic approach reverses the order and introduces the point to seal the unity of a group of transformations.

To summarize, the sortal concept that individuates entities in a world involves two operations. The first is the formation of identical equivalence classes of qualities by the notion of kinds embodied in the local symmetry group G. The second is the introduction of the numerical identity for each equivalence class by a project map π. Together they seal the individuality of the event $\psi(x)$. Their systematic nature ensures that the events are individuated within a whole, the idea of which is made explicit by the quotient set M. It is interesting that the G-orbits $\{\theta: \theta \sim \theta'\}$ adequately represent "clusters of qualities." By requiring the additional structures of projection and quotient, physical theories cast their ballots against the cluster view of things.

§ 20. The Basic Spatio-Temporal Structure of the Physical World

In the previous section, I argued that the general concept of entities comprises the dual concepts of kind and identity, and that individuation is achieved systematically. The general concept is realized in field theories by the localization of symmetries. The conceptual structure of the world as a field can be represented by the fiber bundle (D, M, G, π), where $D = \{\psi(x): x \in M\}$ is a set with indexing set M. The event $\psi(x)$ has a typical internal structure specified by the local symmetry group G and an identity x assigned systematically by a map π. M, the system of identities, is the quotient resulting from the partition of all possible qualities into individual events.

In the physics literature, M is usually called space–time and x the position of $\psi(x)$. Therefore *I further interpret the numerical identity x of the event $\psi(x)$ as its absolute position, and the system of identities M as the primitive spatio-temporal structure of the objective world* (D, M, G, π). Both M and x are coordinate-free. The coordinates $\{x^\mu\}_\alpha$ of x I call the *relative position* of $\psi(x)$, whose meaning is explored in the next section.

The primary spatio-temporal structure M is not a free-standing entity but a component of the larger structure (D, M, G, π) representing the objective world. M systematically abstracts from qualitative features and reveals the objective disposition of events in the world. We can call it the "order of situation," a phrase used by both Newton and Leibniz in association with space.[157] The abstraction means that M is not a construction upon the world of events but is inherent in it, just as the circle is inherent in the Möbius strip in Fig. 6.1. The world of events is as spatio-temporal as it is material. Space–time is like the architectural structure of a building. As an integral aspect of the building, the architecture is as absolute and objective as any other aspects. Yet the architecture is not a substance or substratum underlying the building. We can think of the architecture abstractly, but we cannot physically yank it out of the building to become a free-standing furniture of the universe.

I emphasize that M is the primitive and not the full spatial and temporal structure of the objective world. The spatio-temporal structure is so called because it is four dimensional, but it essentially captures only spatial connotations. This does not mean that the field is a four-dimensional landscape. It means that M incorporates only the first layer of the finely analyzed hierarchy of geometric concepts; it includes only the structure described in D1 of § 5. In its coordinate-free form, M contains no explicit relations. It lacks notions such as contiguity, orientation, distance, extension, and all metrical notions, which involve extra constructions. I am justified to call this skeletal structure spatio-temporal because it alone is shared by all physical theories involving space–time and is presupposed by other geometric concepts. The generality of M accords it fundamental philosophical significance.

Some notions of time and spatio-temporal relations are introduced in the next chapter. However, even there we are a far way from the full spatial and temporal structure of the objective world. Important temporal meanings, such as process and irreversibility, not to mention those that are primal to our life, are left out.

Several Ways of Interpreting M

The complexity of the conceptual structure (D, M, G, π) enables us to examine the meaning of M in various ways. In one approach, we start from the end product and consider the familiar world of entities. Entities do not stand alone, they are members of the world, which exhibits a definite pattern. Since entities couple with each other, the equilibrium arrangement depends on their properties. When we systematically disregard their possible qualitative features and focus on their absolute disposition, we arrive at the system of individual positions. Technically, we abstract the spatio-temporal structure M from the world of entities with the map π.

This approach is close to Leibniz's real theory. Leibniz said: "Space denotes, in terms of possibilities, an order of things which exist at the same time, considered as existing together; without enquiring into their manner of existence."[158] This is totally different from the "relational theory" that is often wrongly attributed to Leibniz. Leibniz's theory is commonly misrepresented in the space–time literature. A full discussion is out of place in this work, but some comment is given in the notes.[159]

The structure (D, M, G, π) can be looked at in another way. We can regard the primary spatio-temporal structure M as the scheme of individuation and identification that we project into the world via the inverse map π^{-1} and by which we present the world to ourselves as comprising distinct entities. The individuating scheme determines the kind of entity that emerges. The logic of this view is more apparent in an alternative axiomatization of fiber bundles, which highlights the idea that what we take as objective features are actually a product of a qualitative "dimension" and an identifying "dimension."[160] "To exist" in the original Greek sense means "to stand out from" or just "to stand out."[161] In this sense no entity exists without the primitive spatio-temporal structure, in which individual events first stand out.

This approach has a Kantian flavor. In Kant's critical philosophy, space is what makes identity and difference possible: "plurality and numerical difference are already given us by space itself." "Difference of locations, without any further condition, makes the plurality and distinction of objects, as appearances, not only possible but also necessary."[162] Along similar lines, Weyl argued that the essence of space "is the *principium individuationis.* It makes the existence of numerically different things possible which are equal in every respect."[163] Note the generality of the arguments; the concepts of space and time that are required for the plurality and distinction of objects do not imply any specific geometric property such as those in Euclidean geometry. The approach agrees with Kant that space is not a substance but the precondition for particulars and individuals. It disagrees with Kant that space is a form of intuition. The function of *M* is purely conceptual and comes closer to the "concept of space," which Kant distinguished from "space."

The second approach highlights the absoluteness of the primitive spatio-temporal structure by emphasizing its indispensability in the concept of entities as such. It does not agree with Newton's notion of a substantival space, but it comes close to some of Newton's idea of absoluteness. Newton said: "Space is a disposition of being *qua* being. No being exists or can exist which is not related to space in some way . . . . And hence it follows that space is an effect arising from the first existence of being, because when any being is postulated, space is postulated."[164]

The two ways of looking at *M* are compatible. Since absolute positions are defined systematically, the events they identify are radically situated and implicitly related. The implicit relation obtains among events, not between events and space–time. There is no relation such as embedding or occupation between events and space–time, for there is no space–time apart from the system of events. Absolute position is that aspect of the general concept of events that signifies its numerical identity. The primitive spatio-temporal structure is that aspect of the objective world by which events are numerically distinct from each other regardless of their qualitative features. *The event structure and the spatio-temporal structure of the objective world emerge together. The meaning of the primitive space–time is exhausted by the systematic numerical distinctiveness of the events.* Furthermore, since the scheme of individuation involves the characteristics of the individuals, their properties are interrelated. Anyone who took a cursory look at quantum field theory would be impressed by how strongly the structure of fields is determined by relativistic invariance. No philosophy of space–time can be satisfactory without accounting for this fact.

A key feature of the world of fields is its fullness. Absolute positions are identities of events. There is no identity without an event; hence the notion of "empty space points" is excluded. We ordinarily say that the physical world is the totality of entities. In this sense the world is irreducibly spatio-temporal but not embedded in space–time. Similarly, physical entities situate in the spatio-temporal world and not in space–time points. The qualitative sense of space has been in currency long before the notion of empty Newtonian space; geometry and geography originally referred to features of the world. All topological and geometrical features are defined on the physical world itself. For example, the orientability of the world can be defined as its intrinsic property without invoking a substantive space in which the world is embedded.[165] There is no separate equation of motion or description for M and x is not a dynamical variable for closed systems. Thus space–time is a structure of the world of events but does not claim existence on its own. Fields, which are spatio-temporally structured matter, exhaust the universe. It is not that the matter fills space–time; rather, the spatio-temporal structure spans the physical universe.

As Einstein remarked, in general relativity, classical particles can only appear as regions where the field strength or the energy density is particularly high.[166] Thus the lumpy world we are familiar with is described in terms of the distribution of high-density regions in the world and not in terms of the distribution of entities in a container. When we concentrate on the qualities of the high-density regions, we approximately ignore the qualities of other regions and regard them as merely "being there," or as a kind of backdrop that comes to be called empty space. This is the same sense as we say there is plenty of empty space in a room; we simply ignore the air we are breathing. This sense is compatible with our theory of space–time as an abstraction from qualities.

Field theories investigate the world in the most primitive level. The above derivation, which involves only one kind of event, is especially simplistic. However, we have seen that to articulate even this primitive world requires a minimal conceptual structure more complicated than that in many philosophies, which regard sets of entities as given. Field theories have not added complications, they have made explicit the implicit assumptions taken for granted.

The meaning of the primitive spatio-temporal structure M is neutral to the notions of infinity and continuity. If the world is infinite, so is space–time; if it is bounded, so is M. Space–time is unitary because the world is unitary. The independence on continuity may raise objections, for people sometimes say that the continuity of motion requires the continuity of space. The argument is not sound. There are *two mathematically distinct concepts of continuity*, which separably capture the intuitive ideas of "no jump" and "no gap." To say motion is continuous and exhibits no jumps is to assert the continuity of a function. To say space–time is continuous and has no gaps is to assert that the domain of the function is a continuum. The first statement does not entail the second, for continuous functions and continua are two distinct concepts. Continuous functions can be defined on domains full of gaps, for instance, on the rational line. Based on similar reasoning, Leibniz had concluded that "continuity of motion is distinct from continuity of place."[167]

Being a continuum is not essential to the concept of the world or space–time. The independence is advantageous. For although so far the world is successfully described as a continuum, physicists suspect that the continuum may break down at the minute distance of around 10^{-33} cm. There are already discrete fundamental physical theories, such as lattice gauge theories. The discreteness will not invalidate our theory of space–time.

The general meaning of the basic spatio-temporal structure M is also valid in the cases of Kalusa–Klein or superstring theories, where the dimensionality of space–time is larger than four. In these theories, the higher-dimension space–time is ultimately compactified to "our" space–time, which corresponds to M. Thus we can regard those features that are "rolled up" in compactification as internal structures of an event. The general idea of a quotient space holds in these theories, although the specific features of the various elements are different. The development shows more forcefully the conceptual analyzability and the substantial inseparability of the internal and spatio-temporal structures.

Absolute and Substantival, Relational and Structural

The philosophy of space–time has been a lively industry since the advent of relativity. Despite technical advancements, philosophical issues are still dominated by the Newton–Leibniz debate. However, the meaning of key words such as "absolute" and "relational" has become a bit blurred. Stein found in his analysis of Newtonian space–time that Newton "is by no mean so far as one might be led to suppose from Leibniz's view that the essence of space and time is in some sense relational." C. D. Broad found in his study of Leibniz's theory of space that "at this deeper level, Leibniz's view is in an important sense a form of the absolute theory."[168]

To clarify the issues, I differentiate between the conceptual criterion of absoluteness and the ontological thesis of substantivalism. Space–time is *substantival* if it exists independently of material entities. The relation between matter and substantival space–time is that of occupancy or embedding. Every material point must occupy a space–time point, and no two material points can occupy the same space–time point. It is not necessary that all space–time points are occupied. In a universe without matter, all space–time points are unoccupied and the empty space–time subsists.

Space–time is *absolute* if its concept is presupposed by the concept of individual entities and things, hence absolute space–time is inherent in the world of things. The concept of absolute space–time is necessary only for the concept of the world of things, not for that of the world as such. One can adopt the Parmenidian view in which the world is not conceptualized into things and hence is not spatial. Absoluteness is neutral to the ontological status of space–time. The absolute space is not substantival if the spatio-temporal aspect of entities cannot be materially separated and exist as an independent substance. The substantival space is not absolute if the concept of material entities is posited independently of space. Newton's space is both absolute and substantival. Leibniz denied substantivalism; his debate with Newton is mainly over ontology.

The concept of absolute space–time concerns how we *think* about the world; substantivalism asserts the *existence* of a certain entity. The distinction between them becomes more important in the modern debate whose locus is not space but space–time. In the traditional debate, substantivalism concentrates on space. On the level of philosophical generality, it does not take much to shift the argument from three-dimensional to four-dimensional concepts. A lot more is involved in shifting the argument from the existence of a three-dimensional to the existence of a four-dimensional container for things of as many dimensions. The existence of a filled space–time implies the coexistence of the past, present, and future, which has strong metaphysical implications.

In everyday and philosophical discourses, "relation" usually means explicit relations. Thus Leibniz is said to reject all relations although implicit relations are crucial to his philosophical system. This exclusive sense has caused much confusion regarding Leibniz's "relational theory of space." Nevertheless, I will follow the custom, so that *"relation" unqualified is either generic or an explicit relation.* I will call a system of relations of all types a *structure* or an *order*, as in Leibniz's "order of situations." A structure can contain relations, but it need not. A structure is not relational if it contains only implicit relations. For example, consider a model (S, f) that interprets a formal language. S is a nonempty set of entities. The function f maps every constant a of the language into an entity in S, every one-place predicate letter F into a monadic property of the entities, every two-place predicate letter R into a binary relation on S, and so on. The whole model (S, f) is a structure; it contains implicit relations through the systematic assignment of predicates and relations. A model for a language with only one-place predicate letters is a structure without relations. Qualities that imply some kind of degree or ordering, for example, mass and temperature, can be implicit relations. The implicit relations are made semiexplicit by a conventional scale. Although "a weighs 5 grams" and "b weighs 10 grams" are monadic statements, a and b are implicitly related because "5 grams" and "10 grams" are relational predicates connected through their common scale. Upon the implicit relations, explicit relations such as "b is heavier than a" can be constructed.

The difference between implicit and explicit relations distinguishes two kinds of relational theories of space. The explicit relationist's outlook is atomistic; the implicit relationist's outlook is holistic. Explicit relations can be either internal or external; internal relations are based on the properties of the relata; external relations are not.[169] Positivist theories of space–time are mainly external-relational. Leibniz rejected external relations, for they are not well founded on the concepts of the relata. His real theory of space is implicit relational, in conformity with his basically holistic philosophy, whose entities are monadic and whose concepts are thoroughly interconnected (see note 159).

External relationists presuppose nonspatial entities; hence they are incompatible with absolutism. However, a structural theory of space–time need not involve external or even explicit relations. It can be compatible with absolutism when it asserts that space–time is also the structure in which entities are individuated. *The primitive space–time M is absolute but not substantival, structural but not relational.* It is a conceptual structure with which we comprehend the world.

The Absolute and Structural Space-Time

Space–time M is structural by virtue of its systematic definition. However, it contains no relations; M is not a set of relations among events. The clearly delineated geometric concepts allow us to distinguish the functions of identifying and relating (§ 5). All relations among events, including those of contiguity, have to be separately introduced with the help of extra concepts, as discussed in the next chapter. However, these relations presuppose the structure M. Perhaps we can say that they make explicit the relations implicit in M.

The nonrelational nature of M does not imply that M is substantival. The relational and substantival views prevalent in current space–time philosophy are not exhaustive. In ancient views, space was metaphorically neither a container nor a relator but a kind of divider. Hesiod's account of the creation begins with: "Verily first of all did *Chaos* come into being." *Chaos* is not emptiness; more than two centuries would have passed before the Atomists drew a clear conceptual distinction between empty space and corporal bodies. *Chaos* is a yawn, a gape. The first event in the formation of a differentiated world was the appearance of a gap separating the sky and earth, and by separating them it freed them for our inspection. For the Pythagoreans, the void was some tenuous matter the cosmos drew in to separate things. Aristotle defined places as the innermost motionless boundaries of bodies. He also defined up and down places as where things of various natures tend.[170] The primitive role of space is to separate things, and in fulfilling this role the concept of space is bound up with the concept of the things it separates.

We often regard the impossibility of two things to be at the same place at the same time as a self-evident truth. That the identity of an entity is defined by its spatio-temporal position is so obvious it is often called *the* principle of individuation. The idea originated in Roman times, was held by Thomas Aquinas regarding material bodies, and underwent subtle conceptual changes in modern times. Compare two versions separated by approximately two centuries. Locke said: "For we never finding, nor conceiving it possible, that two things of the same kind should exist in the same place at the same time, we rightly conclude, that, whatever exists anywhere at any time, excludes all of the same kind, and is there itself alone." Popper said: "Two qualitatively indistinguishable material bodies or bits of matter differ if they occupy at the same time different regions of space."[171] Newton's contemporary used the concept of *exclusion*, a relation among things. A relation among things calls for careful stipulations of the notion of things; thus Locke qualified that they are of the same kind. Our contemporary used the concept of *occupancy*, a relation between things and space.

The idea of occupancy is neither necessary nor sufficient for individuation; it is redundant. For no individuation obtains if two things can occupy the same space–time region. The stipulation that they cannot presupposes that they exclude each other. Impenetrability and the capability to exclude, expel, or displace are features of material things we encounter every day. Once these features are acknowledged, things are distinct, and the occupancy criterion becomes otiose. Identifying tags can be pinned on the things themselves instead on their lodgings. Space–time manifests the exclusiveness of things rather than the accommodativeness of some extramaterial substance. This line of reasoning has a long history. When Melissus argued that a void is required for movement, Aristotle countered that it is not so, for "bodies continuously make room for each other," as water parts for objects without gaps. Furthermore, movement is meaningless in a void because the featureless void is unable to differentiate whether a thing has moved.[172]

Impenetrability is a substantive characteristic prevalent in many familiar things. Substantive criteria are risky in metaphysical considerations, for counterexamples are always possible. Leibniz observed: "We find that two shadows or two rays of light interpenetrate, and we could devise an imaginary world where bodies did the same."[173] We need not fantasize; modern physics abounds with quantities that overlap. Quantum mechanical wave packets interpenetrate, and potential barriers are routinely being tunneled through.

The relation of exclusion has wide applications. For instance, property used to be an exclusive right, as the intellectual property of a patent gives its owner the right to exclude others from using certain ideas. The Pauli exclusion principle applies to qualitative fermion states. In all cases, distinction is conferred on whatever that excludes. I have taken the Kantian turn and place the exclusiveness of entities not in substantive properties such as impenetrability but in formal concepts that structure our understanding of the objective world. A categorical framework is posited that formally distinguishes between the numerical and qualitative aspects of things. Exclusiveness is just the formal numerical distinctiveness of entities.

Since exclusiveness is now formal and independent of the notion of "body" with its overtone of solidity, the concept of individuals is clearly defined even in a world where nothing is solid, as the world of quantum fields. In the philosophical literature, Strawson has constructed a world of pure sounds.[174] My theory of individuals and space–time is also applicable to it.

The traditional debate on space–time centers on considerations of motion, acceleration, and rotation. A major argument for the absolute space is Newton's rotating bucket, which tries to show that space is presupposed in the concept of rotation. In it and all variants, the rotating bodies are finite and composite. Hence they are open to the challenge that the trajectory of the orbiter is a composition of rectilinear motions, and that the fictitious force is a manifestation of the relative motion of the parts.[175] A stronger case can be found in the quantum-mechanical concept of *spin* of point particles.[176] Spin is a genuine angular momentum that couples as such to the orbital angular momentum. It stands with mass and charge as the fundamental kinematic or intrinsic properties of elementary particles, which they characterize in the absence of dynamical interactions. In the intrinsic angular momentum we have a case of an "absolute motion" of a free point particle. The deep correlation of spin to space is most clearly shown in Wigner's analysis of the irreducible representations of the Poincaré group. Spin is related to the invariance under rotations of coordinate systems (§ 6). The concept of spin supports the absoluteness of space but not its substantiality. Spin is an intrinsic property, and cannot be regarded as rotation against some substantive background. Physicists had tried to treat the spin semiclassically as the actual rotation of a tiny rigid body; the effort failed.

The mutual dependence of the structures of matter and space–time is also demonstrated in the relation between the masses of elementary particles and the scale of proper time intervals. The *mass* of a particle is a relativistic invariant, independent of coordinate systems, position, or history of the particle.[177] According to Planck's formula in quantum mechanics, the rest energy E_0 of a particle, which is proportional to its mass m, is also proportional to its Planck frequency v, $E_0 = mc^2 = hv$, where h is the Planck constant. This defines a standard of time $\tau = v^{-1} = mc^2/h$. Thus a particle carries with it an intrinsic unit of proper time interval, which is none other than its mass m. There are many elementary particles, but their masses are definite ratios to each other and the ratios are independent of position or history. Different particle masses can be regarded as different units of measurement, similar to the seconds and minutes we use daily. Since mass and spin, the most basic characteristics of particles, are tied to spatio-temporal concepts, space–time is absolute.

The identification of the formal concepts of absolute positions and numerical identities of events explains why predicates invoking specific places and times are not acceptable in universal laws. Consider Nelson Goodman's "grue," which is green before a certain time t_0 and blue afterwards.[178] "All emeralds are green" is a lawlike statement but "all emeralds are grue" is not. For "grue" violates the criterion that laws cannot involve particular entities; it has invoked a particle time that is the identity of a particular event.

Because of its formality, absolute position has no substantive content and is not informative. It says no more than "it is where it is," which expresses the numerical identity of a thing. The qualitative aspect is expressed as "it is what it is." Together the two become "it is it," a tautology that captures the concept of being *qua* being. The notion that no two numerically distinct events can be in the same spatio-temporal position becomes truly tautological. So is the common saying that real things must be "in space." For when any event is posited, space–time is posited, and when space–time is posited, a system of events is posited.

The information contents of statements about definite locations stem from coordination. As with qualitative properties, to say anything definite we must use some conventions, which are the familiar coordinate systems in the case of spatio-temporal qualities. Coordinates enable us to say specifically "it is there" or "it is ten miles north of Boston." These statements are about relative and not absolute positions. The conventionality of coordinates and the relativity of positions are acknowledged and neutralized by the coordinate transformations. The meaning of the coordinates will be discussed in more detail in the next section.

The space–time M is not open to Leibniz's indiscernibility challenge. Leibniz argued that it makes no difference whatsoever, theoretically or experimentally, if the universe is moved to somewhere else in the substantial space. Hence the substantival space is superfluous and there is no reason to posit it. The argument does not apply in our case because there is no empty space for the universe to move into. More important, the concepts of location and movement in space–time are foreign; absolute space–time is not loosely related to the world of events but inseparably bound with it. Our notion of absolute position is similar to Leibniz's "situation"; it is an implicit relation "in the event."

The absolute but not substantive space–time is compatible with an expanding universe. The fact that the universe is expanding does not imply there is empty space into which it expands. It means every galaxy is receding from every other, like specks of dust on the surface of an inflating balloon moving apart from each other. The *surface* of the inflating balloon is an expanding *two-dimensional* space that is not expanding into anything. For the two-dimensional creatures on the balloon, expansion means no more than the fact that the distances between dust specks on the balloon increase while the dust specks maintain their size and provide the scale to measure the relative distances. The three-dimensional case is similar. Clusters of matter dense enough to form self-gravitating bound systems, for example, stars and galaxies, do not expand. Only the galaxies recede from each other. As Leibniz remarked, relative sizes and distances, not the substantival space, underlie the notion of expansion.

The absoluteness of M is seen in counterfactual situations. The event $\psi(x)$ may take on any of its possible properties, but its identity or absolute position x remains fixed. As discussed in the next section, we can construct possible worlds with different qualitative features but with the same immutable spatio-temporal structure M.

The numerical–qualitative duality and its formality imply a categorical distinction between *divisibility* and *analyzability* of events. Numerical integrity and qualitative complexity can coexist in the concept of things. An event has no parts and hence is indivisible, but it is qualitatively analyzable. The analyzability of events underlies the concept of *point interactions*, in which a partless event is ideally analyzed into matter field, interaction field, and their coupling. It also paves the way for the relation of *coincidence*, which engenders difficulty in most other theories of space–time, for coincidence is neither contiguity nor fusion.[179] By definition, no two events can coincide. For more interesting cases, let us consider enduring or extensive things, which can be ideally represented by parametric curves (§ 25). Two enduring things can coincide in an event in which they cross path. The qualitative complexity of the event enables us to discern the two coincidental things as distinctive in some cases. For instance, consider two low-energy electrons come together, interact, and separate again.[180] If the two incoming electrons have parallel spins and hence identical qualities, we cannot identify which outgoing electron is which incoming one. The two processes in Fig. 6.3 are in principle indistinguishable. Thus the incoming electrons lost their identities in the intersection. We cannot say two electrons coincide; instead, we have to admit that four electrons and a two-electron system are involved. If the two incoming electrons have antiparallel spins and hence different qualities, we can reidentify them after scattering by

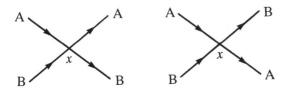

Figure 6.3 In quantum mechanics, the two scattering processes are in principle indistinguishable if *A* and *B* are electrons with parallel spins and hence identical qualities. The two processes are in principle distinguishable if *A* and *B* have antiparallel spins. The difference must be taken into account in all calculations. The scattering cross sections of electrons with parallel and antiparallel spins are different, even if the detectors are insensitive to spin orientations.

their spins and the two processes in Fig. 6.3 are logically distinguishable. In this case we can say that two electrons *coincide* at *x*.

In field theories, the spatio-temporal structure *M* functions as a *parameter space*. Parameter spaces are abstract, and a parameter is an index, which happens to be continuous in our case. *It is the abstractness of M that makes sense of the concept of coincidence.* A substantive space cannot do the job, for two material points cannot occupy one space–time point. We can identify two buildings by one street number, but we cannot stuff two buildings into one number plate.

Against Positivist Relationism

Friedman distinguished Leibnizian relationism, whose concern is ontology, from Reichenbachian relationism, whose concern is ideology.[181] Ontology concerns what there are, ideology stipulates what vocabulary we are allowed to use. Positivist or Reichenbachian relationism maintains that space–time is a set of external relations imposed on a set of pregiven non-spatio-temporal entities, all spatio-temporal concepts and theories are reducible to a set of prescribed relations, and the spatio-temporal relations constitute an isolatable realm of investigation independent of the physical properties of the relata. It was the orthodoxy earlier this century, but has been severely criticized, for example, in Glymour (1977), Friedman (1983), and Earman (1989). I do not repeat their convincing criticism, and am content to add a few brief comments.

Relations convey relevant information concisely and introduce physical properties that are empirically determinable. In the empirical sciences, we often posit a conceptual scheme that includes certain basic spatio-temporal notions. The scheme enables us to define empirically determinable relations such as distance. By measuring these relations, we figure out the specific features of the world. For example, we can determine the specific geometric characteristics of a region of the world by measuring the distances between its points. This is the procedure of physical research. However, philosophy is neither poor-man's physics nor the ideology of physics; it must also examine the presuppositions of the procedure.

To determine the distance between a and b we must be able to unambiguously refer to a and b. Reference to physical objects is more complicated than to abstract points; mathematical entities wear their identities on their sleeves; physical things do not. We have to supply the identities ourselves in the concept of concrete particulars, which involves basic spatio-temporal notions. Also, an abstract symmetric and nondegenerate relation $d(a, b)$ is not empirically determinable. The general definition of distance is the greatest lower bound of all curve segments joining a and b. To determine it we have to use geometric concepts to construct physical paths for the deployment of meter sticks. For the random abutment of sticks yields no distance. Let me refer to the hierarchy of geometric concepts in § 5. Determinate finite distances involve the concept in D7. Yet it is taken to be the primitive by positivists, as if the preconditions D1–D6 do not exist. Positivists confuse *spatio-temporal* concepts with the more restricted *metrical* concepts and forget the general spatio-temporal structure they were using all along. Consequently, they miss the basic significance of space–time and erroneously assert that not only specific metrical features but also the general spatio-temporal concepts are constructions upon relations.

There are more than one way to axiomatize a group of concepts. Some positivists argue that instead of taking points as primitives and defining relations such as distance among them, we can take "distance" (not "distance between a and b") as primitive and define the points in terms of it. Even if the alternative axiomatization is possible, it does not alleviate the difficulties of positivist relationism. The myriad "distances" need to be systematically differentiated and identified. Differentiation and identification are crucial roles for spatio-temporal concepts; they can be called by other names but cannot be ignored. Furthermore, the abstract "distance" by itself has no physical or empirical significance; it gains the usual metrical meaning only in conjunction with the points to be introduced and identified with physical objects. These constructions may again involve spatio-temporal concepts. The metrical structure of the physical world is a high-level construction whose complexity cannot simply be defined away.

Positivism demands all physical meaning be grounded on quantities directly visible to and manually manipulable by us. These quantities are important, but they have many hidden assumptions. To enshrine them as self-evident essentially puts a ban on the critical examination of their presuppositions. Positivism is prevented by its narrow verification theory of meaning from accepting the broader integral conceptual structure of modern physics. For instance, it rules out the possibility that the masses of elementary particles can be the units of proper time intervals. Its insistence on the independent verification of geometry brought it in conflict with many physicists, including Poincaré, Einstein, Eddington, and Weyl. The physicists argued that geometry, mechanics, and physics form an inseparable theoretical whole, and the geometrical properties of the physical world cannot be independently verified apart from other physical laws as positivists claimed.[182] Physics has proved the physicists correct.

Against Substantivalism

Substantivalism claims independent existence for space–time. Having independent existence is much weightier than being objective. We use many structural and relational concepts in our understanding of the objective world. We regard these concepts as objective because they have cash value in physical theories. However, to assert their separate existence involves the additional condition of independence, which is not generally satisfied. That I am conscious does not imply the existence of a substantive consciousness. That the world is spatio-temporal does not imply the existence of a substantive space–time. Existence has to be separately argued and established.

Most recent arguments for substantivalism use highly technical geometric formalism. Often a physical theory casted in the geometric formalism is said to be a "space–time theory," its quantities "geometrized away" into parts of the "the geometry of space–time." These sayings are at best misleading. We should distinguish the formalism of a physical theory from its physical and philosophical significance. A geometric formalism does not imply a theory of space–time.

It is incorrect to consider geometry as a science of physical space. Geometry originally meant the measure of earth, not of space. It had been a science of figures, magnitudes, and extended objects. With higher abstraction, the topic of geometry shifted to pure extension or continuous magnitudes. Extension is the concept abstracted from all things extended. Like other abstract concepts in mathematics, it does not refer to anything physical. Mathematical concepts may bear names such as "field," "curvature," or "metric," which may be descriptive of the original topic from which the abstract theory developed. However, it is characteristic of pure mathematics that its concepts are emancipated from the soil of their genesis. Hence we should not be confused by the nuance of the names. The pure mathematics of geometry finds applications in areas not remotely associated with space, such as the theory of colors. The meaning of the concepts they designate have to be determined individually by the theory that employs them. The metric tensor represents kinetic energy in the modern formulation of classical mechanics. Curvatures can represent field strengths; connections can represent field potentials. The meanings are assigned by physical theories.

We cannot assert the existence of space–time based on the saying that space–time has certain properties. Such arguments, which tacitly assume the existence of the subjects of propositions, are generally invalid. References can fail, as in "the present king of France is bald." Since Gottlob Frege and Russell, analytic philosophers commonly distinguish three senses of the verb to be: the existential use as in "there is x," the identifying use as in "x is the king of France," and the predicative or copulative use as in "x is bald."[183] The analysis clearly exhibits the peculiar status of existential claims. Earlier, Kant argued in his critique of ontological proofs that existence is not a real predicate. Real predicates add to the concept of the subject and enlarge it, but we do not make the least addition to the thing when we further declare that it *exists* or *is*. The "is" in "God is omnipotent" is a copula relating two concepts but making no existential claim for God.[184] Similarly, statements such as "space–time is such and so" or "there are such and such geometrical properties" do not immediately imply the existence of an entity called space–time.

Empty space is closely associated with action-at-a-distance, on which Newton himself frowned. If masses interact via some physical agent, then the interstitial space would not be empty because of the agent's presence. Anyone who has tried to push two iron magnets of the same pole together can almost swear there is something between them besides air and space. This something, the magnetic field, is not removed by the most powerful vacuum pump, an empirical evidence cited by Leibniz against the vacuum.[185] We now know that the electromagnetic field suffuses the universe as the three degree microwave background radiation. Philosophically this fact is not as important as the concept of *interaction fields* that transmit physical effects continuously from point to point and eliminates spooky actions-at-a-distance.

Substantivalism in modern physics has been defended by Michael Friedman, Hartry Field, and John Earman, among others. I agree with their arguments against relationism and conventionalism. However, I do not agree that the only alternative is substantivalism. I note that they use "absolute" in a sense different from mine.[186] Most of them have assumed the concept of entities independently of space–time. Hence their substantive space is not absolute in my sense.

In my sense of absoluteness, the concept of space–time is fused with that of physical entities. The primitive spatio-temporal structure makes events distinct, and its meaning is exhausted by the distinctiveness of events. Substantivalists separate space–time from entities and institute some relation between them. They are not thereby exempted from the task of individuation; the task doubles, for the distinctiveness of entities and the distinctiveness of space–time points have to be separately asserted. Hence a substantive space–time has more elements than my absolute space–time. The extra elements account for its substantiality and independent existence. Is the extra meaning warranted?

Friedman tried to give a clear formulation for "the system of concrete physical bodies as literally embedded in space–time as in some large 'container.'" He posited two sets of entities, a set of space–time points and a smaller set of actual events. He argued that the substantival–relational debate centers on whether the set of events is embedded in or a submodel of the set of space–time points.[187] He called the relationist "Leibniz" but admitted it is only "Leibniz of the positivists." I think a more authentic Leibnizian would not submit that the issue is merely the difference between how the two sets are related. He would argue there is insufficient reason to distinguish two sets of entities in the first place, for the notion of a container space is already assumed in the distinction.

I contend that the distinction between two sets of entities, or between occupied and unoccupied sites, is not meaningful or fruitful in modern physical theories. Friedman mainly considered the ponderous mass ontology. He acknowledged that field physics would render the distinction and hence the debate on the ontology of space–time meaningless. Yet he said: "We should dig in our heels and use the nonvanishing of T [the stress–energy–momentum tensor], not the nonvanishing of g [the metric tensor], as our criterion for an occupied space–time point."[188] Setting $T = 0$ may merely hide the material source in the boundary conditions. In the analogous case of electromagnetism, the Maxwell equations admit a free-field solution, but this does not imply that the plane wave is sourceless. Furthermore, the metric tensor carries energy density, so that $\nabla T = 0$ does not lead to global energy conservation. It is in principle impossible to separate its background and dynamical parts, but energy is energy. In general relativity texts, the metric tensor g is called both space–time and the gravitational field, and most authors shift from one name to the other. The vacillation is not permissible in philosophical categorization. Conceptually, space–time and the interaction field are distinct. The relation between matter and space–time is occupancy, between matter field and inter-action field dynamical coupling, which is much more complicated. Is it better to interpret g as a substantive space–time or as the gravitation field on the same footing as the other three fundamental interactions of the physical world?

Friedman's major argument for the substantival space–time is its "theoretical unifying power." Unifying power lies not in mathematical formalisms but in the physical concept that summons the formalism. Powerful formalism can lead us to see the physics, but the physics alone has the power to determine ontology. We can write Newtonian mechanics in the form of general relativity with the stipulation that the connections and metric tensors vanish. However, the formalism has no physical significance. As Anderson pointed out, it is merely "a trivial extension" that introduces no new physics.[189] The development of relativity theories is clear about the physics that calls forth the significance of the differential geometric formalism. Einstein said: "The theory of relativity may indeed be said to have put a sort of finishing touch to the mighty intellectual edifice of Maxwell and Lorentz, inasmuch as it seeks to extend field physics to all phenomena, gravitation included."[190] The concept of fields integrates the concepts of matter, space–time, and dynamical interaction. Its unifying power exceeds that of substantival space–time. Furthermore, its has full physical significance. A space–time philosophy in the light of modern physics cannot evade weighing the claims of interaction field and the substantival space–time.

I maintain that the interpretation of g as substantival space–time does not unite but fragment. We can write classical mechanics, general relativity, and quantum field theory all in the unified formalism of fiber bundles, in which the spatio-temporal manifold is an element (§ 10, Appendix B). General relativity clearly falls in the company of quantum field theory. They both invoke the concepts of local symmetry group, connection, and curvature, which do not occur in classical mechanics. Furthermore, the curvatures can all be physically interpreted as interaction field strengths. The physical significance of interaction fields facilitates the unification of fundamental physics, toward which physicists are striving. Separating gravity and interpreting it as substantive space–time has, as Weinberg said, "driven a wedge between General Relativity and the theory of elementary particles."[191] Of course, the interpretation of g as the gravitation field does not prevent g from having the specific characteristic of being responsible for most of the metrical properties of the universe. Since the gravitational field acts on all matter, it may be useful to give it a special name such as "curved space–time," but that would be a matter of nomenclature, not philosophy.

Finally, it is dangerous to use the notion of unification for ontological arguments. Unifying power describes the coherent relation among concepts, but ontology requires the further relation between theoretical concepts and the world. The idea of an almighty being has great unifying explanatory power, but that does not imply that God exists. The danger of overvaluing unification is amply exposed by Kant. Kant fully appreciated the role of systematization and unification in science: "Multiplicity of rules and the unity of principles is a demand of reason." Reason posits the unity of knowledge as "the *criterion of the truth* of its rules." However, important as it is, the unity is only a projection and a problem. Reason's unifying role is hypothetical and regulatory. If it gets carried away and mistakes the projected unity as real and an object of knowledge, it generates all kinds of metaphysical illusions.[192] The substantival space–time can be regarded as one of such illusions.

Field and Earman both asserted that field theories support substantivalism. Field said: "According to the substantivalist view, which I accept, space–time points (and/or space–time regions) are entities that exist in their own right. . . . A field is usually described as an assignment of some property, or some number or vector or tensor, to each point of space–time; obviously this assumes that there are space–time points." Field gave little supporting argument. Earman called Field's thesis "manifold substantivalism" and fleshed it out by rehearsing definitions in differential geometry, concluding: "It is clear that the standard characterization of fields uses the full manifold structure." He said: "When relativity theory banished the ether, the space–time manifold M began to function as a kind of dematerialized ether needed to support the fields. . . . In modern, pure field-theoretic physics, M functions as the basic substance, that is, the basic object of predication."[193]

There are two categorical confusions in the arguments of Field and Earman, that between pure mathematics and physics, and that between conceptual and ontological analysis. We must distinguish "field" in mathematics and "field" in physics. In algebra, a field is a set with two operations, the number system is a field. In geometry, a field, for example, a vector or a tensor field, is an assignment of some quantity to each point in a manifold. These mathematical fields should not be confused with physical fields, even if a physical field theory uses some mathematical field concepts.

In the writings of Field and Earman, "field" in the arguments is a mathematical field, but "field" in the conclusions quietly shifts to a physical field. Since the meanings of "field" are categorically different in the two cases, the conclusions do not follow at all. Let us grant that the characterization of physical fields uses the manifold. However, this is just the beginning and not the end of physical and philosophical argument. For at issue is not abstract conceptual analysis but the ontological status of physical entities. How does it follow that what the manifold represents exists in its own right? Why can't the manifold represent a structural quality of physical fields? The manifold and the constructions upon it are used in more than one aspect in field theories. In the structure (D, M, G, π) that represents field theories, not only M but D, G, and the local fields $\psi(x)$, are differentiable manifolds. Many vector fields are defined not on M but on D. The symmetry group G and its copies represent the internal state spaces of events, which are decidedly not spatio-temporal. Does that imply that the state spaces too are mysterious ether? These questions are totally ignored by Field and Earman.

It is a common error in philosophy to assume that if we have a word or a concept, then there must be some independent entity in nature that answers to it. We can analyze a complex concept into various elemental concepts, but that does not imply the thing represented by the complex concept can be decomposed into various parts, each corresponding to an elemental concept. For instance, the concept of a person is complicated, and we analyze it into memory, attention, emotion, and many other aspects. However, this does not imply that each of these is a entity that exists on its own. Ontology demands its own notion of integrity.

Consider three employments of similar mathematics in physics. (1) There is gravitation field, which is mathematically represented by the metric-tensor-field-over-a-manifold. (2) There is dematerialized ether, mathematically represented by a differentiable manifold, and there is gravitation field, represented by a tensor field over the manifold, and the ether supports the gravitation field. (3) There is dematerialized ether, represented by a differentiable manifold, and there are various qualities, known as physical fields and represented by tensor fields, and the qualities are assigned to each point in the ether. The substantivalist interpretations (2) and (3) both assert that the physical world is separable into a substratum plus something else, and that the substratum, represented by a bare manifold, exists by its own right.

Once space–time is ontologically separated from the rest of the world, some relations between them have to be established. According to Field and Earman, the relations are "support," "underlie," and "assign." As cohesive relations between two substances, they must be physical, for our concerns are ontologies, not ideas. Usually, to support is to counteract some forces, as a shelf supports a book against gravity. I can find no explanation, in Earman's writing or in the physics literature, of how the substratum physically supports or underlies the field. The meaning of assignment is even more obscure. In pure mathematics, a function assigns a number to each element in a set. The assignment is conceptual, representing the way we think. In the ontological context, what assigns? Is all the extra ontology justified in any way?

The antisubstantivalist interpretation (1) upholds the ontological integrity and the conceptual complexity of physical fields. We analyze the concept of an existent but do not physically reduce the existent into subentities. The physical topic, supported by experiments, decides what are primitive physical units. Field theorists have made their point. As Einstein said, the field strengths are "the ultimate entities, not to be reduced to anything else." "According to general relativity, the concept of space detached from any physical content does not exist."[194]

Earman's arguments are all directed at relationism, but he acknowledged there are other views besides substantivalism and relationism. Paul Teller proposed the alternative that space–time is a physical quantity or property. Earman dismissed the thesis, saying modern field theories support the view that space points are "concrete things," and that "fields are attributes of the space–time manifold."[195] However, his presentation of field theories fails to address Teller's point. I argue that field theories are closer to Teller's view than Earman's.

As the passages quoted in the beginning of this chapter show, to field physicists, the fields constitute the fundamental ontology. Space–time is a structural property of the fields, not the other way around. This agrees with the common sense notion that the basic ontology is a thing, not a spatial region. Consider a piece of graph paper: The grid is a property of the paper; the paper is neither a property of the grid nor supported by it. Take any other object: We say a book has a rectangular shape; the basic object of reference is the book. We do identify the book by its position, as we refer to a local field $\psi(x)$ via its absolute position x. However, the basic object of predication is not the position but the local field $\psi(x)$. $\psi(x)$ excites, deexcites, and changes kinds in interactions; x does not. We say "the local field $\psi(x)$ designated by the coordinates $\{x^\mu\}_\alpha$ has such and such phase," as we say "the egg there is rotten." We do not say "the space–time point x has the field property $\psi(x)$," as we do not say "the space region there has the property of rotten-eggness." The difference is not a "sentiment," as Earman said; it is the basic structure of our way of thinking. (More on reference in field theories is found in the next section).

Earman did not explain the difference between the dematerialized ether and the luminiferous ether, which physics banished, as he acknowledged. The ether is a variant of the old metaphysical notion of substance, a substratum that remains when all qualities are peeled away. I see no reason that physical fields need the support of such a substratum. The fields, with their complicated structures, are self-sufficient. They are concrete enough to take all structural differentiation upon themselves. The extra notion of dematerialized ether is superfluous and does no work in field theories. My primitive spatio-temporal structure M is a manifold, and I have shown an analysis of field theories in which it is significant without being a substratum. As discussed in § 9, the spatio-temporal parameter x has the same function as the discrete index n for beads. $\psi(x)$ as little entails the existence of space–time points as ψ_n entails the existence of integers.

Earman discussed the theoretical construction of removing regions from the manifold.[196] A theoretical calculation is not a record of the process of creation. Many theoretical steps admit no literal interpretation; a notable example is the perturbation expansion, the process of hole construction is another. I cannot make sense out of Earman's assertion that the resultant hole in the manifold represents "space–time points that do not exist."

For an antisubstantivalist absolutist, a topological hole means the world is not simply connected. (A manifold is simply connected if any closed loop on it can continuously shrink to a point. A doughnut is not simply connected; a loop around it cannot shrink to a point because of the hole in the middle). We assume that the physical world is connected, for even if there is a disconnected piece, we would have no empirical access to it. However, we need not assume that the world is simply connected. Its specific topology can only be determined empirically. If the world turns out to be like a torus, we say its topology is not simply connected and not that it has an empty domain. Substantivalists may say there is a topological hole in the material world that is filled by empty space. However, unless they are willing to argue *a priori* that the substantival space is necessarily simply connected, they also have to face the topological question, for it could turn out that empty space is itself not simply connected.

The main issue for substantivalism is the independent existence of space–time. Criteria for existence are mainly empirical. Earman and Field interpret the bare manifold M as the empty space–time for a world devoid of all matter and energy, including the gravitation field energy. Others may like to include the Lorentz metric, so that the substratum is $(M, \eta_{\mu\nu})$. I protest because the empty world is totally without empirical significance. This does not imply a crude verificationism, which I reject. The fields, which I regard as fundamental ontology, are unobservable and unreal according to positivist standards. I am ready to accept the most indirect empirical ramification as discussed in § 12. However, there is a limit. *What distinguishes physical theories from pure mathematics?* Suppose we strip a physical theory such as general relativity layer by layer by setting term after term to zero, not as approximations but literally meaning what they represent vanish. At what point does the theory cease to have physical significance? Can we claim that a concept such as M or $(M, \eta_{\mu\nu})$ is physically significant all by itself simply because it was once employed in a more comprehensive theory that has empirical ramifications? I do not think so. At some point, the Kantian principle of significance must be applied to deny objectivity to a stripped-down theory that loses all empirical ramifications, direct and indirect. Perhaps I cannot delineate the point sharply, just as I cannot define precisely when a person expires, but this does not obliterate the conceptual distinction between life and death. I gladly admit that M is real and objective because it is an irreducible structure of the interacting field system. However, I do not admit M by itself stands for empty space–time when all fields are removed.

Contemporary physics has a vacuum, whose meaning is *the state of lowest energy*. The vacuum is bubbling with quantum energy fluctuation and does not answer to the notion of empty space or dematerialized ether. In many theories the vacuum is defined by a field responsible for symmetry breaking, usually called the Higgs's mechanism (§ 9). I. J. R. Aitchison concluded his review of the concept of vacuum: "Quantum theory teaches us that the 'classical vacuum'

state, empty of all matter and free of all fluctuations, is not physically realizable."[197] Our ancestors had found the notion of nothingness incomprehensible. Quantum theories tell us there is no coherent formulation of nothingness.
In our exploration of nature, we seem to arrive at where we started. Hopefully
we know the place better now.

§ 21. The Referential Structure of Field Theory

Physical theories are about things. They enable us to refer to particular things,
or to pick out a specific event and discuss its properties. By themselves, field
equations and other equations of motion do not refer, for they are universal
statements totally abstracted from particularities. However, it is crucial that
field theories contain the means to refer, for their solutions with specific
boundary conditions purport to describe concrete situations and refer to particular objects. Experimentally testable solutions alone ensure the physical
significance of the theories.

Reference is where ontology, epistemology, and semantics meet. The problem of reference can be regarded as the semantic counterpart to the epistemological problem of perception. They deal with the presence of things in our
experiences, thoughts, or speeches. This section examines the referential structure of field theories. I argue that references to events are *direct* although they
are frame dependent, as perceptions of objects are *immediate* although experiences are subjective. Both involve a complicated conceptual structure that
extends beyond the single sentence or perception.

So far in this chapter we have been considering the coordinate-free form of
field equations. As discussed in § 15, an empirical theory is not complete in the
coordinate-free form, for coordinate systems must be invoked to yield numbers
that can be compared to experiments. Coordinate systems are neither dogmatically predefined nor arbitrarily added on; they constitute an integral part of
the structure of a physical theory. Quantum field theory and general relativity
each has *two* symmetry groups, a local symmetry group and a space–time
group; (§ 7). In § 19, I interpret the local symmetry group as the *kind* of the
field. Now I unfold the structure of the space–time group and interpret it as the
referential structure of field theories; (Fig. 6.4). The coordinate $\{x^\mu\}_\alpha$ for the
event $\psi(x)$ is its relative position, as distinct from its absolute position x. The
absolute position is not informative; the relative position is.

In field theories, the identity or absolute position x is the independent variable that singles out the local field $\psi(x)$ for investigation. However, x is like a
floating pronoun "it" without any context. By itself x is not specific enough for
the general purpose of referring. Spatio-temporal coordinate systems provide
the reference frames in which the floating "it" becomes an anchored "this."
The coordinate value $\{x^\mu\}_\alpha$ for x in the coordinate system $f_\alpha(U_\alpha)$ is a definite
designation by which we refer to the event $\psi(x)$. In natural language, two major
classes of referring expressions are proper names and demonstratives such as
"this" and "that." Proper names correspond to constants in an interpreted

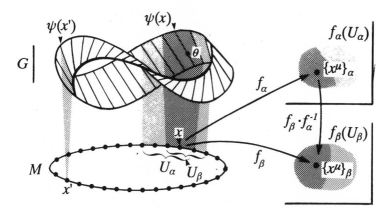

Figure 6.4 A schematic of field theory with its two symmetry groups. The structure of the *local symmetry group G* is mirrored in the internal state space of each event $\psi(x)$ with identity or absolute position x. The totality of identities constitutes the spatio-temporal structure M. The structure of the *spatio-temporal group* is shown explicitly. The coordinate function f_α and f_β assign to x the coordinates $\{x^\mu\}_\alpha$ and $\{x^\mu\}_\beta$ in the coordinate systems $f_\alpha(U_\alpha)$ and $f_\beta(U_\beta)$, respectively. The coordinates can be regarded as names by which we refer to the event $\psi(x)$. The coreferentiality of $\{x^\mu\}_\alpha$ and $\{x^\mu\}_\beta$ is established by the transformation function $f_\beta \cdot f_\alpha^{-1}$.

formal language. *Semantically, the values of coordinates work as proper names.* Specific coordinates $\{x^\mu\}_\alpha$ and $\{x^\mu\}_\beta$ of x can be interpreted as different names of the particular event $\psi(x)$ in different reference frames $f_\alpha(U_\alpha)$ and $f_\beta(U_\beta)$. All points x can be represented in coordinates; thus all events are nameable, although the names are not always invoked.

The referential structure of field theories is elaborate. First of all, it is *intensional*. Since names, that is, values of coordinates, occur in particular reference frames, co-referential names cannot be interchanged. We cannot simply interchange the values of $\{x^\mu\}_\alpha$ in $f_\alpha(U_\alpha)$ and $\{x^\mu\}_\beta$ in $f_\beta(U_\beta)$, though both $\{x^\mu\}_\alpha$ and $\{x^\mu\}_\beta$ refer to $\psi(x)$. Analogously, the name "Cicero" in "he believes that Cicero was a Roman orator" cannot be replaced by a co-referential name "Tully" without altering the truth value of the sentence, for he may have never heard of Tully. *Reference is frame dependent.* The degree of freedom provided by the variety of reference frames means a name can have multiple denotations. The same value, say, $\{1, 2, 3, 4\}$, can designate different events in different frames.

The coordinate transformation $f_\beta \cdot f_\alpha^{-1}$ is not a mere relation between names. It is a composite map that points to the *referent* $\psi(x)$ via the individual maps $f_\alpha^{-1}(\{x^\mu\}_\alpha) = x$ and $f_\beta(x) = \{x^\mu\}_\beta$, which read "$\psi(x)$ is the denotation of $\{x^\mu\}_\alpha$" and "$\{x^\mu\}_\beta$ is the name of $\psi(x)$." Thus the composite coordinate transformation $f_\beta \cdot f_\alpha^{-1}(\{x^\mu\}_\alpha) = \{x^\mu\}_\beta$, reading "$\{x^\mu\}_\beta$ is the name of what $\{x^\mu\}_\alpha$ denotes," has the natural interpretation of a *co-reference statement* analogous to "'Morning Star' and 'Evening Star' are names of the same planet." It may be confusing to call $f_\beta \cdot f_\alpha^{-1}(\{x^\mu\}_\alpha) = \{x^\mu\}_\beta$ an identity statement. For in philosophy, identity of terms usually means substitutivity *salva veritate* (preserving the true), which is not satisfied by co-referring coordinates.

The co-referentiality of the names is established not by substitution but by a proper coordinate transformation. This contradicts the extensional theory of reference, which argues that co-referential names must be freely interchangeable. Proponents of the extensional theory may argue that substitution is a transformation in which all coordinates except the one at issue is unchanged. However, to insist on substitution is to stipulate *a priori* a specific kind of coordinate transformation. It is much more restrictive than our scheme, in which the *a priori* notion is the general concept of transformations, leaving specific transformations to be determined empirically for each particular case. Physical theories provide no evidence that the specific transformation of substitutions is widely applicable. Even worse, the introduction of interchangeability *salva veritate* may have dire consequences for the purpose of reference (§ 23).

The referential structure, containing many reference frames, covers the entire world. However, the reference frames themselves are *local*. A coordinate system need not cover the world, it need only cover a finite patch. It is natural to interpret the group of events identified in the coordinate patch U_α containing the identity x of the event $\psi(x)$ as the *environment* of $\psi(x)$, $x \in U_\alpha$. The finite patches imply that an event never stands alone but always occurs within an environment. A coordinate patch U_α can admit different coordinate functions f_α and f_β, resulting in different reference frames $f_\alpha(U_\alpha)$ and $f_\beta(U_\alpha)$, which can be regarded as various *contexts*.

When coordinate patches overlap, coordinate transformations identifies the points in the overlapping region. The identification "sews" the patches together. Thus when two environments overlap, their contexts are meshed by the establishment of co-referential names through the transformation $f_\beta \cdot f_\alpha^{-1}$. This is analogous to two persons speaking about the same things; the speakers' understandings are united by the objects.

Direct Reference and the Intelligibility of Names

The referential structure of field theories presented is controversial, if only because of its intensional construction and the modal concepts it involves. This is no place for anything resembling a defense. I can only remark briefly on how it answers the standard linguistic puzzles that must be resolved in any satisfactory theory of reference. Reference is a major topic in analytic philosophy. Some philosophers opposed to intensional constructions and argued that they are not suitable for scientific discourses.[198] However, I cannot find any attempt by these philosophers to analyze how references are made in actual scientific theories.

There are two contending classes of reference theories, descriptive and direct. Descriptive theories issue from the works of Frege and Russell.[199] They asserted that we refer by providing a definite description of monadic or relational characteristics. Strictly speaking, we are not talking about a specific thing but about whatever or whoever alone fits the description. Direct theories were advanced by John Stuart Mill and Kripke.[200] They argued that we refer to things directly via proper names and other singular terms, so that identifying descriptions are not required. Direct theories pinpoint a thing by its identity without invoking its qualities, descriptive theories cast a net for some qualities and draw in whatever fish it catches.

Our investigation shows that *reference is direct in field theories.* A set of coordinates designates the identity of an event without involving its qualities.[201] This is consistent with the result of § 19; the descriptive theory is associated with the bundle view of things, which was found insufficient for field theories.

We see in § 19 that the kind to which an event belongs includes all its *possible qualities*, and the concept of kinds is included in the concept of the event. Since possibilities are intrinsic to the concept of events, the construction of possible worlds easily follows. To be a world, the qualities of the events must be compatible with each other. It is possible that Judas betrays Jesus, and it is also possible that Judas dies in infancy, but the two events cannot occur in the same world. "θ is possible and θ' is possible" does not generally imply "(θ & θ') is possible." When the implication obtains, we say θ and θ' are compossible. A specific assignment of compossible qualities to all events constitutes a *possible world* or a *possible state of affairs*. To capture the notion of compossibility, we can find a rule that assigns a specific quality to each event. In the construction, a coordinate designates the same event in all possible worlds, and the event can assume different qualities in different worlds. Transworld identity is no problem. Thus coordinates as names agree with Kripke's theory that a proper name is a *rigid designator*, which rigidly designates the same entity in all counterfactual situations or possible worlds.[202]

Russell had objected to the interpretation of coordinates as names. He argued that coordinates are not arbitrary; a set of coordinates describes a point by its relation to the origin.[203] A name designates a specific object, but there is no reason why the designation must be generated by lottery. "Second Avenue" suggests that its denotation is between the First and Third Avenues, yet it is no more a tag than "Paradise Street," which happens to designate a street in a hellish neighborhood. Hence we should follow Mill and distinguish *accidental information* in a name that does not contribute to its role of designating from the *meaning of a name* that allegedly determines its referent or describes it.[204] The functions of *designation* and *relation* are clearly separated in differential geometry. As discussed in § 5, coordinatization involves only the concepts under D1, the representation of points by sets of numbers. Without further assumption, the representation does not convey any explicit relation such as contiguity, orientation, and distance, which are required for definite descriptions.

Russell is right that coordinates can have descriptive meaning and that their accidental information has important roles. However, even when additional concepts are invoked so that coordinates become relational, the role of the extra information lies only in helping to fix the referent. Kripke distinguished between giving the definition of a term and fixing its referent. Suppose my companion says "the girl in blue is pretty," the "in blue" helps me to fix the referent but does not define the girl, and I understand even if her dress is actually turquoise.

Russell pointed out that coordinate systems are ultimately based on demonstratives. I agree, and argue that this does not undermine but strengthen the thesis of direct reference. Two issues are involved. First, direct reference theories assert that at least *some* names must be bestowed by ostension. Demonstratives express the primary way in which we are acquainted with the world. A speaker must have some empirical contact with the world if he is to understand what it means to refer. Then there is the problem that demonstratives are subjective. The empirical connection is subjective and partial, a fact accounted for by the intensional construction of the referential structure. The important point is that the referent is objective although its name is conventional. For the conventionality is neutralized by coordinate transformations, which constitute an integral part of the referential structure.

A standard linguistic puzzle is the significance of identity statements such as "Cicero is Tully." The statement, as "*A* is *A*," should be vacuous, but it is not, and we gain knowledge from the identification. How come? In answer, Frege distinguished between the meaning and the referent of a term. Kripke distinguished between the ontological concept of necessity and the epistemological concept of apriority. Thus a logical distinction is required, although the exact nature of distinction is disputed. The conceptual complexity is important, for questions such as "is Tully Cicero?" signify a partial mode of knowledge, which is not feasible in some categorical frameworks. In frameworks without sufficient complexity, one must know everything if a referent is "given" and nothing if it is not, thus making a puzzle out of the question "is Tully Cicero?"

The conceptual complexity required to accommodate defective mode of knowledge is satisfied our referential structure, which includes various contexts, reference frames, and coordinate transformations. For instance, one defective mode occurs when we know the names $\{x^\mu\}_\alpha = f_\alpha(x)$ and $\{x^\mu\}_\beta = f_\beta(x)$ but do not know the specifics of the transformation $f_\beta \cdot f_\alpha^{-1}(\{x^\mu\}_\alpha)$. Both names are intelligible, but since the two reference frames are isolated, knowledge is incomplete. Knowing the possibility of a transformation rule enables us to seek it empirically or theoretically. The substantive transformation rule meshes two previously isolated frames by establishing the co-referentiality of $\{x^\mu\}_\alpha$ and $\{x^\mu\}_\beta$. The result is a more comprehensive referential structure and an increase in knowledge.

The greatest difficulty of direct reference theories is to explain the intelligibility of names. If names are not descriptive and are originally bestowed by ostension, how can they make sense to someone who is never in the presence of the referent? This problem is usually addressed by the causal theory of reference, which traces the chain of communication that links the original namer to various hearers of the name. The theory is weak in explaining the intelligibility of names in each transaction. Furthermore, the chain of communication is external to semantics and totally irrelevant to the referent.

Field theory suggests that the relations that secure the intelligibility of absentee referents obtain not between speakers but between the objects named. Thus they are not external but internal to the semantics of reference. The systematic definition of events and their designations implies that neither referents nor names are isolated. Events stand in their environments and names occur in their contexts. The environments and contexts ensure that understanding can be achieved if *some* objects, which need not include the particular referent in question, are designated by ostension.

Suppose I hear the name $\{x^\mu\}$, which falls within the context of my understanding, $f(U_\alpha)$, but whose denotation $\psi(x)$ I have not seen. I understand the name because I assume its referent is related to other objects in the environment U_α, say, the one designated by $\{y^\mu\}$, with which I am acquainted by ostension. I may be unable to single out $\psi(x)$ definitely without finding out some explicit relation between $\psi(x)$ and $\psi(y)$. However, the logical structure of my understanding let me know that it is possible to find out. More important, what I find out is the relation between the objects $\psi(x)$ and $\psi(y)$. The relation is objective, relevant to the referent, and internal to the semantic structure. This is in stark contrast to the usual causal theories in which the relations invoked are among speakers who happened to mention $\{x^\mu\}$.

Suppose $\psi(x)$ falls outside my familiar environment U_α. I can still manage to understand it by finding some item $\psi(z)$ that falls within both $\psi(x)$'s environment and mine. This is represented by finding a coordinate patch U_β that contains x and z such that z falls in my familiar U_α. The coordinate transformation at z sews the two patches together. This is how fragments of conversations in cocktail parties, "this friend of Henry's blah blah blah" make sense.

Field theory presents a simple world. However, its conceptual structure shows that the notion of referents cannot be taken for granted in a theory of naming and reference. The concept of referents is explicitly incorporated in the concept of names, so that the category of relations among objects makes unfamiliar names generally intelligible. Co-referentiality is not prescribed extralinguistically from an omniscient position but is an integral part of the general theoretical structure. To echo a principle of Kant, the general conditions for the possibility of intelligible names are likewise the general conditions for the possibility for the denotations of names.

§ 22. The States of the Field as a Whole: Field Quanta

A field is a whole comprising individual events. So far we have concentrated on the individuation and identification of events within the field. Each event has its own state space, in which its characteristics can be definitely described. The events are the locus of point dynamical interaction, which determines their characteristics. Interaction is discussed in the next chapter. Here we turn to examine the characteristics of the quantum field as a whole.

The topic of a quantum theory is a single unit, the isolated physical system whose characteristics are summarily represented by a state vector, which evolves according to some equation of motion. In quantum field theory, the unit is generally an interacting field system, which may be the entire fundamental microscopic world. That is how physicists study quantum cosmology.

Here we neglect interaction and consider the idealized case where the unit of investigation is a free field, say the electron field. As discussed in § 11, to talk about definite properties we need a specific representation or a set of basis states in which a general state of the field can be expressed. The occupation number representation is a choice that can be intuitively interpreted in terms of the modes of field excitations. In this representation, a state $|n(\mathbf{k}_1), n(\mathbf{k}_2), \ldots, n(\mathbf{k}_i), \ldots \rangle$ of a free field is one for which $n(\mathbf{k}_i)$ quanta in the mode $\mathbf{k}_i$ are found in a measurement of the occupation number observable $N(\mathbf{k}_i)$ (§ 9).

A field quantum is a discrete increment in a mode of field excitation and is more often called a particle. In ordinary usage, a particle is paradigmatic of an entity to which we refer as the subject of a proposition. It is *a spatial unit* and it is *relatively noninteracting* with other entities. These two features give it a distinctive individuality that enables us to pick it out even when it enters into mild relations. They are often jointly satisfied in ordinary things and are rolled together in the usual conception of a particle, a pebblelike thing. Of the two features, spatial unity dominates, for it is the basic means for us to refer to the entity uniquely. Ordinarily, few people would call the harmonics of a violin string "particles," though the harmonics do not couple without anharmonic oscillations. Also, some people are reluctant to call quarks particles because they are forever confined within the hadrons and cannot be spatially isolated, though they are relatively "free" or noninteracting within their confinement.

Field quanta alias particles are not spatially localized, but they are relatively noninteracting. They are categorically distinct from events or local fields $\psi(x)$, which are localized. The technical terms "particle" and "field" mean something diametrically opposite to the ordinary images of particles and fields. In several textbooks on quantum fields, the relation between fields and field quanta is explained by an analogy with solid-state physics. Many solids are lattices of atoms. The atoms in the lattice vibrate. The collective atomic motion sets up an elastic wave in the lattice. The energy of the elastic wave consists integral multiples of certain definite energies as if it is the total energy of many independent harmonic oscillators. Physicists call a unit of the energy a "phonon." Phonons do not stand alone apart from the atomic lattice, yet they are neither parts nor constituents of the lattice. As modes of excitation of elastic waves in a lattice, they are characteristics of the lattice as a whole. A field quantum created by the creation operator $a^\dagger(\mathbf{k})$ is analogous to a phonon, while a local field $\psi(x)$ is analogous to the motion of an individual atom. C. Itzykson and J. Zuber said: "Here we see clearly the relation between particles (phonons) and fields (ψ_n is the displacement of the nth atom)."[205] Let me repeat the differences. The atomic index n is discrete; the field parameter x is continuous. Atoms can be detached from the lattice; the local fields cannot be detached from the integral field. Atomic displacements ψ_n are spatial, the variations of the local fields or events $\psi(x)$ are not spatial; spatial character-istics are all contained in the parameter x.

A local field engages in many modes of excitation, and a field quantum involves many events. The state of the field is result of the configuration of the events, and the dynamics of the field is the collective motion of the events. Generally, field quanta pertain to the excitatory states of the whole field. However, often the energy of a field quantum is sufficiently concentrated that we can say it describes a small patch of the field or the collective behavior of a group of events. In any case, the description is not compositional; neither the field or its events are "made up" of field quanta. To say the field is in a state $|n(\mathbf{k}_1), n(\mathbf{k}_2), \ldots \rangle$ is not to say that it is composed of $n(\mathbf{k}_1)$ quanta in mode $\mathbf{k}_1$ and so on but rather $n(\mathbf{k}_1)$ quanta show up in an appropriate measurement. However, the state itself exhibits all the phase entanglement peculiar to quantum systems that is not accounted for by the occupation numbers. The *vacuum state* $|0\rangle$, in which no field quantum can be found, is not nothingness but a definite state of the field, namely the state with the lowest energy.

Field quanta facilitate descriptions of free fields. However, when fields inter-act, the number operator $N(\mathbf{k})$ is no longer a constant of motion and field quanta can be created and annihilated with great ease. Thus the occupation number representation becomes unclear in an interacting field system. Generally, quantum fields need not be characterized in terms of field quanta. Their essential quantum nature is embodied in the commutation relation of the local field operators $\psi(x)$. As Weinberg said, in quantum field theory, "the essential reality is a set of fields," and particles are "mere epiphenomena." Rudolf Haag said: "The primary physical interpretation of the theory is given in terms of local operations, not in terms of particles."[206]

Are field quanta entities? If an entity is a this-something, then field quanta are not entities; they lack numerical identities. As discussed in § 19, there is a rival theory that regards an entity as a cluster of qualities without the notion of identity. Are field quanta entities that support the cluster theory? I argue not. The theory demands that two clusters as distinctive entities must have some different properties. This condition is not generally satisfied by field quanta. Many quanta of a boson field can be found in the same state; therefore, there is no way to distinguish one quantum from another. Field quanta are not well defined in the presence of interaction, which always occurs in realistic situa-tions. Moreover, the number of quanta $n(\mathbf{k})$ and the quantum number $\mathbf{k}$ that differentiates various modes of field excitation are both eigenvalues and cannot be ascribed as the properties of the quantum field (§ 13).

Are field quanta real? Within their region of validity, they contribute to objective descriptions of the physical world; hence they are objective. Certain numbers of field quanta with certain quantum numbers are most often the quantities measured in scattering experiments. For instance, in an electron–positron scattering experiment, an initial state $|i\rangle$ is prepared; $|i\rangle$ comprises free electron and positron fields with definite numbers of quanta with certain momenta and spins. The fields interact (scatter) and are decoupled again. Various final states $|f\rangle$ are observed to have various numbers of electron and positron quanta with various momenta and spins. The relative numbers of quanta in the final states give the transition probability from $|i\rangle$ to $|f\rangle$. Note the empirical accent in the account. Like all quantum representations, the eigenvalues $n(\mathbf{k})$ of the number operator $N(\mathbf{k})$ are not the properties of a free field but only their partial manifestations in experiments. We cannot say a quantum field *has* a certain number of field quanta just as we do not say a nonrelativistic quantum system *has* certain numerical momentum. Like the numerical momentum, the quantum number $\mathbf{k}$ and occupation number $n(\mathbf{k})$ are empirically determinable indices of quantum coordinate systems in which quantum properties, phases and amplitudes, are expressed. Thus the number of field quanta is as objective as the numerical position or momentum, no more and no less.

Are virtual particles real? "Virtual particles" arise in perturbative calculations; they cannot be observed because the point couplings in which they participate do not conserve energy. I think it is misguided to argue whether virtual particles are real just as it is misguided to argue whether an electron actually goes through one of the holes in the two-hole experiment. Both try to cramp quantum physics into the straitjacket of observing classical concepts. Creation and annihilation of virtual particles may offer a good mental picture, but quantum properties are not susceptible to mental imageries. Thus instead of virtual particles it is more appropriate to talk about various *processes* in which the fields couple. These processes are characterized directly in terms of the local fields.

§ 23. Identity and Diversity: Quantum Multiple-Particle Systems

Quantum fields are distinctive for being units that are expressly composite. The notion of a multitude of events within the unity of the field is explicitly addressed *within* quantum field theory. This is possible because quantum field theory treats the concept of space–time rigorously.

In *nonrelativistic quantum mechanics*, where the meaning of space is different and the notion of events absent, individuation obtains only outside the theory. The system may be an ensemble, but it is treated as a single unit summarily described by a density operator. We do not count the entities within the ensemble until the quantum problem is through and we switch to the statistics of experimental outcomes. We cannot refer individually to the entities even in the statistics.

A *quantum multiparticle system* is one in which a number of previously isolated systems—particles—are brought together to form a unitary system in quantum mechanics. Here particles are not field quanta because no field is considered.[207] The most familiar examples of multi-particle systems are the atoms, which are made up of nucleons and electrons. We can also regard the two electrons in a helium atom as a two-electron system in the potential of the nucleus. Despite the name "multiparticle system," the particles are not distinctive individuals within the system. The "electron cloud" in the helium atom is an undifferentiated whole; the two electrons are all entangled and cannot be individually referred to without explicitly invoking observed results. This is different from the composite nature of quantum fields, where the individuality of the events are defined independently of observational ideas. The difference stems mainly from the different concepts of space in quantum mechanics and quantum field theory. Here we will focus on the case of quantum mechanics.

Conceptually, quantum and classical mechanics are similar in their dependence on preindividuated entities. However, there is no conceptual difficulty in recognizing individuals within a classical multiparticle system. The spatial position, which uniquely identifies a particle, is regarded as one of its fundamental qualities. By virtue of their positional properties, particles within a classical multiparticle system are recognized as individuals. Quantum mechanics does not have this advantage, for quantum qualities are not spatial (§ 12). The property of a quantum system in the position representation is the wavefunction and not a specific location as in the classical case. Wavefunctions can overlap and get entangled; hence they cannot individuate as classical positions do. Thus when several isolated entities are brought together to form a quantum system, the problem of keeping them distinct within the system arises.

The meaning of individual particles within a quantum multiparticle system is related to philosophical problems concerning individuality and diversity, identity and discernibility. More important, it raises questions about the logical principle of compositionality. Its relation to the identity of indiscernibles have attracted much philosophical attention.[208] This section examines this cluster of questions.

In these topics, some terminologies used by physicists and philosophers are quite different. I will adopt the philosophers' usage. *a* and *b* are *identical* if they are numerically one and the same. *a* and *b* are *indistinguishable* or *indiscernible* if they are alike in all respects, so that whatever description is true of *a* is true of *b* and vice versa. Both identical and indistinguishable are stronger than the meaning used by physicists, who call all particles of the same kind, for instance, all electrons, identical or indistinguishable. I will simply call them particles *of the same kind*. Particles of the same kind share the same constant properties such as mass, charge, spin, and other quantum numbers characteristic of the kind. They also share the same symptomic structures such as the same energy spectrum, so that they are expected to yield similar data under similar experiments. The peculiarity of quantum particles is that their bound-state spectra are discrete and the spectral values are often sharp. Thus the observed similarity is almost exact.

Permutation Invariance and the Cleansing of Names[209]

Consider the isolated physical system of a *particle*, by which I mean an entity, a this-something that belongs to a kind and has its identity to which we can uniquely refer. In quantum mechanics, the particle can be characterized in terms of a complete set of basis vectors $\{|\alpha_i\rangle\}$ of the Hilbert space $\mathscr{H}$ (§ 11).

Now consider *an aggregate of N particles of the same kind,* say N electrons. The state space for the aggregate is the tensor product of the individual particles, $\mathscr{H}_N = \mathscr{H} \otimes \mathscr{H} \otimes \ldots \otimes \mathscr{H}$, N times. The basis for $\mathscr{H}_N$ is the tensor product of the single-particle basis. An aggregate basis vector is

$$|\alpha_1\rangle_1|\alpha_2\rangle_2 \ldots |\alpha_N\rangle_N \equiv |\alpha_1\alpha_2 \ldots \alpha_N\rangle \qquad (6.5)$$

There is no mention of the position of the particles. The index i in $|\ \rangle_i$ is the name of the ith particle and is the sole device that distinguishes it from others in the aggregate. For it is possible that two particles i and j are in the same state. The name is tacitly represented by the position of α_i in the right hand side of Eq. (6.5).

If particles of different kinds are involved, then dynamical observables are sensitive to particle indices; a proton is different from an electron. For particles of the same kind, experiments are generally insensitive to which particular particle they observe. Therefore, observables, including the Hamiltonian, are generally symmetric under the permutation of particle indices. If $|\alpha_1\alpha_2 \ldots \alpha_N\rangle$ is a solution for a Hamiltonian, then $P|\alpha_1\alpha_2 \ldots \alpha_N\rangle$ is also a solution, where P is an operator that permutes the positions of the α_is or equivalently shuffles the particle indices. This is the *permutation symmetry* of the aggregate of particles. Conceptually, it is just a kind of coordinate transformation, where the coordinates are the particle indices.

Permutation invariance is general and not confined to quantum mechanics. It says that specific particle labels have no physical significance. The philosophical significance of permutation is not different from that of symmetries in general. As discussed in §§ 15 and 21, to say something specific, we must use some demonstratives, labels, or names, such as this or that, left or right, i or j. These are admittedly conventional, yet we cannot simply discard them because without them we cannot express experiences in which we encounter particular things. Yet we know labels are too specific and say too much. We have to "unsay" a lot of what has been said in the labels to give an objective account that is nevertheless empirically supportable. Unsaying something is much harder than saying, and it is achieved by symmetry transformations, of which permutation is an example.

Permutation invariance purges conventionality and wipes out the artificial distinctiveness hidden in the particles' names, but it does not extinguish the concept of individual particles within the aggregate. To draw the parallel with the discussion in § 21, symmetry transformations erase the specifics of the coordinates $\{x^\mu\}_\alpha$ but leave the coordinate-free identity x of the event $\psi(x)$ intact. Also, permutation invariance does not restrict the forms of the solutions $|\alpha_1\alpha_2 \ldots \alpha_N\rangle$.

State Symmetrization and the Declaration of Identity

For an aggregate of N particles all of the same kind but in different states, permutations yield $N!$ states. Based on empirical observations, we know that only two of the $N!$ states are ever realized. One of the chosen two is totally *symmetric* with respect to the exchange of particle names; the other is totally *antisymmetric*. Moreover, only one of the two applies to each kind of particle. Particles whose states are symmetric are called *bosons* and satisfy Bose–Einstein statistics. Particles whose states are antisymmetric are called *fermions* and satisfy Fermi–Dirac statistics. In quantum mechanics, the two kinds of statistics are usually taken as empirical facts. This contrasts with quantum field theory, where the statistics are consequences of the commutation relations of local field operators, which in turn follows from relativistic invariance and microcausality.

The cornerstone of symmetrization is the simplest permutation operation, that of *transposition*, P_{ij}, which interchanges the names of the ith and jth particles in the composite state, $P_{ij}|\alpha_1 \ldots \alpha_i \ldots \alpha_j \ldots \alpha_N\rangle = |\alpha_1 \ldots \alpha_j \ldots \alpha_i \ldots \alpha_N\rangle$. Two transpositions of the same labels return the state to itself, $P_{ij}^2 = 1$. Thus the transposition operator can have only two eigenvalues, 1 or -1. Symmetrized states are eigenstates of systematic transpositions. If $|\alpha_1 \ldots \alpha_i \ldots \alpha_j \ldots \alpha_N\rangle$ is a symmetrized state, then

$$P_{ij}|\alpha_1 \ldots \alpha_i \ldots \alpha_j \ldots \alpha_N\rangle = \pm|\alpha_1 \ldots \alpha_i \ldots \alpha_j \ldots \alpha_N\rangle \qquad (6.6)$$

where the $\pm$ sign denotes two eigenvalues of P_{ij}. *Totally symmetrized states* satisfy the condition that the transpositions of *all* pairs of labels have the same eigenvalue. The rationale is that transposition is a general criterion; thus the qualitative description it teases out should be a typical characteristic and not peculiar to the two specific particles involved. Since P_{ij} has only two eigenvalues, there are only two totally symmetrized stated. If the eigenvalue is 1, completely symmetric states for bosons obtain; if it is -1, completely antisymmetric states for fermions obtain.

For the positive eigenvalue, Eq. (6.6) equates the states $|\alpha_1 \ldots \alpha_j \ldots \alpha_i \ldots \alpha_N\rangle$ and $|\alpha_1 \ldots \alpha_i \ldots \alpha_j \ldots \alpha_N\rangle$. The equation is an independent criterion and not part of the original symmetry of the physical systems. It is a stronger condition than permutation invariance. Generally $P|\alpha_1 \ldots \alpha_i \ldots \alpha_j \ldots \alpha_N\rangle \neq |\alpha_1 \ldots \alpha_i \ldots \alpha_j \ldots \alpha_N\rangle$, although both states are solutions to the Hamiltonian. Both the permutation symmetry and Eq. (6.6) are assertions of identity, but with different targets and consequences. Permutation symmetry asserts that the multiparticle system is the same regardless whether a particle is called i or j; it erases the conventionality of names but leaves the abstract identities of the particles. The transposition eigenstate equation asserts that the names i and j are the same. Since the names alone distinguish the particles, their identification erases the abstract identities of the particles. Consequently the possibility of individually referring to the particles is closed and everything is fused into a single system. Physicists often say the particles "lose their identities" in symmetrization. A symmetrized state represents a single system without definitely differentiable parts. With symmetrization, an aggregate of N particles is turned into a unitary N-particle system, in which the concept of individual quantum particles vanishes.

Transpositions interchange terms and their eigenvalue statements embody the *salva veritate* principle of identity. The *salva veritate* principle says two terms are *identical* if one can be substituted for the other without changing the truth of any proposition, and vice versa. It applies to terms or concepts, not to things. It goes back at least to Leibniz, and is closely associated with two other Leibnizian principles, that of the identity of indiscernibles and that of the indiscernibility of identicals. The exact relations among the three principles are still under dispute. Some philosophers generalize the *salva veritate* criterion for application to things. Two things a and b are identical if and only if every attribute of a is an attribute of b, and vice versa. In the generalized version, the criterion becomes a biconditional. Going one way, it reads in second-order logic $(F)(Fa = Fb) \Rightarrow a = b$, which is the *identity of indiscernibles*. Going the other way, it reads $a = b \Rightarrow (F)(Fa = Fb)$, which is the *indiscernibility of identicals*.[210]

The symmetrized states employ the *salva veritate* principle in the original narrow sense to eliminate a discriminatory concept. For brevity consider an aggregate state of two particles. The eigenstate equation of transposition for the eigenvalue $+1$ is $P_{12}|\alpha\alpha'\rangle = |\alpha\alpha'\rangle$, or

$$P_{12}|\alpha\alpha'\rangle \equiv P_{12}|\alpha\rangle_1|\alpha'\rangle_2 = |\alpha\rangle_2\,\alpha'\rangle_1 \equiv |\alpha'\alpha\rangle = |\alpha\alpha'\rangle \equiv |\alpha\rangle_1|\alpha'\rangle_2 \qquad (6.7)$$

The equation literally identifies the names 1 and 2 by announcing their interchangeability *salva veritate*. The names are merely the same label applied twice and become redundant. The eigenstate equation for the eigenvalue -1 is

$$P_{12}|\alpha\alpha'\rangle = |\alpha'\alpha\rangle = -|\alpha\alpha'\rangle \qquad (6.8)$$

It negates identity, but does so in a special way. It separates out the difference and expresses it in the minus sign, so that the names are again redundant. It is not surprising that consequently, α and α' are forbidden to be the same in the antisymmetric system state $|\alpha\alpha'\rangle_-$. This is the ground of the Pauli exclusion principle for fermions. *State symmetrization is the incorporation of the concept of identity in the definition of some typical quantum characteristics.* It is a declaration of identity, not a consequence of some predefined identity.

The *salva veritate* principle in the symmetrized states does not identify the two particles. It only eliminates whatever notion we have for individual reference to the particles. Thus it does not involve the identity of indiscernibles.

To recap, consider three particles in states $|\alpha\rangle$, $|\alpha'\rangle$, and $|\alpha''\rangle$ respectively. Permutation yields $3! = 3 \times 2 \times 1 = 6$ aggregate states that satisfy the Hamiltonian: $|\alpha\alpha'\alpha''\rangle$, $|\alpha\alpha''\alpha'\rangle$, $|\alpha'\alpha''\alpha\rangle$, $|\alpha'\alpha\alpha''\rangle$, $|\alpha''\alpha\alpha'\rangle$, and $|\alpha''\alpha'\alpha\rangle$. In these states, α and α' still pertain to distinct particles. Symmetrization yields two system states,

$$|\alpha\alpha'\alpha''\rangle_\pm = (|\alpha\alpha'\alpha''\rangle \pm |\alpha\alpha''\alpha'\rangle + |\alpha'\alpha''\alpha\rangle \pm |\alpha'\alpha\alpha''\rangle + |\alpha''\alpha\alpha'\rangle$$
$$\pm |\alpha''\alpha'\alpha\rangle)/\sqrt{6} \qquad (6.9)$$

where the $+$ sign denotes the totally symmetric state of a boson system and the $-$ sign the totally antisymmetric state of a fermion system. There are four other symmetrized states, but they are never observed. The α in $|\alpha\alpha'\alpha''\rangle_\pm$ no longer means the state of an individual particle. As a linear superposition, the states

$|\alpha\alpha'\alpha''\rangle_{\pm}$ establish unbreakable phase relations among the single-particle states so that they are all entangled. As D. Dieks showed, it is generally impossible to associate $|\alpha\alpha'\alpha''\rangle_{\pm}$ with pure states in single particle state spaces.[211] Individual particle states are no longer viable. The multiparticle system takes over.

Individuation and Observation

Although symmetrization forces us to adopt a holistic view of multiparticle systems, the qualification "multiparticle" is meaningful. In principle we can prepare an N-particle system by bringing together N separate particles, which we have independently studied. The state space of an N-particle system is a product space of N single-particle states. The product construction conveys some idea of composition. Due to the dimensionality of their respective state spaces, a two-particle system is different in kind from a three-particle system. The variable N provides a systematic way to account for the difference.

Quantum mechanics is not complete without the Born postulate, which incorporates empirical connections. Experiences are always of particulars. Particular position or momentum eigenvalues, more generally particular sets of quantum numbers, are observed in a measurement on the system. The distinct quantum numbers distinguish individual particles. I do not mean that measurements uncover some pre-differentiated quantum particles or properties; I mean that we individuate by the observed results. Consider a two-electron system with position and spin eigenvalues. Any measurement must return two different sets of values; either the spins or the positions are different. The observed results, not the description of quantum characteristics, individuate the two electrons: "*the* electron observed at x_1 and *the* electron observed at x_2." The observed electron is a genuine individual and a particular.

The individuation by observed results figures heavily in the Einstein–Podolsky–Rosen argument. The two electrons in a two-electron system are not individuated without invoking their observed eigenvalues or some approximation based on the eigenvalues. Standard textbook treatment is clear to the point that a two-electron state remains a single unit although it extends over miles. It becomes two units only when we discard the coupling terms, arguing that they must be negligible when the difference between the position eigenvalues becomes macroscopic.[212] The approximation may be good in most cases, but not for delicate situations such as the Bell experiments.

We are intrinsically spatial and temporal beings. We can only observe particular entities, handle spatial things, and our thinking depends on the concept of individuals. How to reconcile nonspatial quantum characteristics with spatiotemporal individuation is perhaps the next great problem physics has to overcome.

7

Explicit Relations and the Causal Order: Interacting Fields

> The differentiation between a neutron and a proton is then a purely arbitrary process. As usually conceived, however, this arbitrariness is subjected to the following limitations: once one chooses what to call a proton, what a neutron, at one space–time point, one is then not free to make any choices at other space–time points. It seems that this is not consistent with the localized field concept that underlies the usual physical theories.
>
> C. N. YANG AND R. L. MILLS
> "Conservation of Isotopic Spin and Isotopic Gauge Invariance"

§ 24. Relation and Constant Predication

In Chapter 5 we considered a single object. An important concept for an object is its state space, which encompasses all its possible states. The state space can be described by any one of a class of systems of predication, usually called coordinate systems in physics. Various systems are related by transformations, under which the objective state is invariant. The conceptual structure of representation-transformation-invariant embodies the logical distinction between the objective state and its representations or definite descriptions. In Chapter 6 we derived a host of objects called events, which, although situated in a spatio-temporally structured world, are totally disjointed. There is neither spatio-temporal nor qualitative relation among the events. The absence of qualitative relations is rigorously expressed by the fact that the convention of definite description for each event is chosen independently of that of any other.

When we bring together the ideas of definite description and individual entities, we find ourselves confronting a philosophical problem that worried philosophers since Plato: By virtue of what do we use the same predicate to describe different things, which may be widely scattered? In quantum field theory, the question becomes: By virtue of what do we call two events both protonlike? In this chapter we examine how the issue is brought to focus and resolved in theories of interacting fields. Briefly, the theories expose the fact that the familiar predicates that we apply globally hide a conventional relational structure, which they discard and replace with the physically significant concepts of interaction and interaction fields.

Predicates are conventions, but if we keep changing our conventions, calling snow "white" and polars bears "green," or calling "red" what was called "blue," the result would make what happened at the Tower of Babel orderly by comparison. Therefore once we have chosen what to call white and what to call green, the convention is enforced throughout the world. In so doing we have made an additional assumption besides the conventionality of predicates, we have declared the conventions *global*. Constant predication is equivalent to the subscription of a globally enforced convention. Practically, constant predication is indispensable for our everyday thinking. Philosophically, a justification of globalism needs to be provided.

Most physical theories are like our everyday discourse, in which conventions of predication are automatically regarded as global. In physical theories, global predication is represented by global symmetries, in which coordinate systems apply across the world and their transformations are independent of the spatio-temporal parameter. The parallel grid in a Cartesian coordinate system is a global convention that enables us to give definite descriptions of directions anywhere in the world. The replacement of Cartesian systems by another kind of systems in special relativity does not change their global nature.[213]

Globalism is the default mode of representation and is also found in field theories that are *not* expressly concerned with interaction dynamics. In field theories with global symmetries, a common state space applies across the field. Thus once we have chosen to call a certain state protonlike and another state neutronlike at one point in the field, we are not free to make another choice at another point. Note that the concepts of discrete events and spatio-temporal positions as their identities are not well defined in global theories. The reason, as will be made clear in the following, is that the global convention hides many assumptions and rolls together several concepts that should be kept distinct. The conflation is sloppiness rather than simplicity; global theories either leave out certain effects or need the supplement of extra concepts such as external coupling mechanisms.

When physicists zeroed in on the fundamental dynamics of physical interaction, they found something awry. The adoption of a convention involves certain information, and its global enforcement implies the instantaneous promulgation of the information throughout the world. This violates the idea of locality, in which information can only be transmitted from point to point with finite velocity. The rectification of the problem leads to the *localization of symmetries*, by which the state space is localized to each point in the field. The localization crystallizes the field into a system of discrete events, each with its own identity and state space. An event's state space can be coordinatized separately from that of its neighbors; thus the convention of predication is no longer globally imposed.

This is where we now stand in our analysis of field theories. It is good to purge the theories of the unwarranted assumptions, but the consequence of the purge must also be addressed. As mentioned earlier, global conventions have vital functions in our thinking. A global convention establishes a tacit relational structure that maintains the intelligibility of speech by enabling us mentally to compare different entities. 6 and 8 inches are monadic predicates, but they are implicitly related through their common numerical scale. The abolition of global conventions discards not only the dubious assumptions they harbor, but also the tacit relations. The functions of the tacit relations must be fulfilled by adequate replacements. *Explicit relations* must be introduced to cement the universe.

Explicit relations can be constructed upon global predicates. Mount Everest is higher than Mount McKinley because the former is 8,848 meters and the latter 6,194 meters above sea level. Most familiar relations are introduced in this way. However, such relations rest on weak foundations, they cannot satisfy scientific theories aiming to study fundamental relations. Interacting field theories are cases in point, examples are also found in other sciences.[214]

Interacting field theories admit only those relations that are physically significant. Consequently, interaction fields that transmit action from event to event are introduced. Furthermore, the relata must be sharply defined. Totally disjointed entities are conceptually different from relatable entities. Thus to obtain legitimate relations, the notions of individuals and their properties must be modified to include the notion of relatability. Physically, the charges of individual events become the source of the interaction field.

Relatable events and the field relating them are introduced in the principle of local symmetry, which demands the global invariance of the dynamical equations under local transformations. Mathematically, the relational concepts are represented by a *curve* and a *connection*. An event has two aspects, its numerical identity and its qualities characterized in its private state space. Curves are mainly associated with numerical identities; they gather groups of events and underlie *spatio-temporal relations*, including that of endurance or extensiveness. Connections are associated with qualities and underlie *causal relations*; the connections on principal fiber bundles represent potentials of interaction fields that reconcile the qualities of various events. The spatio-temporal and causal relations so introduced are not external to the events, for they entail a more comprehensive conceptual structure in which the concept of events is enriched with relational properties. Briefly, the comprehensive structure does not curtail the autonomy of the individuals by dictating a global convention. Instead, it increases the complexity of the individuals by endowing each with a mechanism to reconcile its choice with those of others in the community. The next section considers a few spatio-temporal relations, mainly those centering around curves. The remainder of this chapter considers casual relations.

Field theories show that the conjunction of the object-representation distinction and the multitude of individual objects results in a heap of totally unrelated entities that hardly constitute a world. To complete the theory of an objective world comprising individuals we need extra concepts.

A similar situation is found in Kant's "Analogies of Experience," where he combined the pure conceptual distinction between objects and experiences and the plurality of objects made possible by the intuition of space and time. ("Analogy" meant proposition in the mathematical parlance of his time.) Kant argued in experiential rather than linguistic terms, but the general idea is not far off. Self-consciousness logically separates the object from its representations in experiences and paves the way for the synthesis of the representations at various times into coherent knowledge of the objective world. The actual synthesis of the multitude of representations requires more concepts, which he argued are the categories of quantity, quality, relations, and modality. The three analogies are the principles of relations: permanence, causality, and reciprocity. Some people see in them the shadows of Newton's three laws of motion.[215]

§ 25. Permanence, Endurance, and the Concepts of Time

The only functions of the primitive spatio-temporal structure M presented in § 20 are the individuation and identification of events in the field. It has no temporal connotation. In general relativity, M will have no more structure than that of a blank differentiable manifold. In quantum field theories, M will acquire the structure of a Minkowski space, which is determined together with the causal relations among events. As we stand now, M is as structureless as it is in general relativity. M is the concept by which the world is conceived as comprising individual events. In the following, we investigate relations among events and not among spatio-temporal points, although we focus on the identities of the events. Constructions upon the identities constitute spatio-temporal relations among events. They will be connected with causal relations later.

Permanence and the Constancy of the Number of Events

The spatio-temporal structure M is four dimensional; sets of four indices are required to coordinatize it. One of the four indices designating an event is usually called its temporal moment, and the dimension constituted by the temporal indices is called time. The four-dimensional conception is not new to the twentieth century; time has long been considered as a dimension besides space.[216]

We should be careful not to assign too much significance to the time so defined. The time index x^0 is no more than a part of the identification tag of an event. Its temporal connotation is minimal and should not be confused with the time parameter t, which carries the load of temporal significance and which is yet to be introduced. The fourth dimension in M provides an additional degree of freedom in our thinking. Whatever it is called, it is not different from the other three dimensions. The structure M is too primitive to confer special meaning on the time dimension. The lack of specification in M is made clearer by the hierarchy of geometric concepts presented in § 5. M is the concept on the lowest level, D1, while the separation between space and time needs constructions on the higher levels.

It is often said that space is changeless and timeless. These attributes have more than one sense, which we must clarify before we decide which if any is applicable to space–time. Changeless or being constant can mean either having zero change or having nothing to do with the notion of change. As discussed in § 11, a property type includes a range of values, and zero is a definite value. "The thermodynamical system has zero temperature" makes perfect sense; we have a sharp definition for the temperature of absolute zero. However, "the atom has no temperature" is senseless if by "no temperature" we mean zero temperature, for the concept of temperature simply does no apply to single atoms. Concepts generally have finite ranges of applicability; propositions invoking a predicate beyond its range are senseless. Change is no exception. "The spatio-temporal structure is changeless" is sensible only if "changeless" means the inapplicability of the concept of change. To say M has zero change is senseless. Change involves time; colloquially, we say changes occur in time. However, M is not in time; it is all times. More correctly, it is the condition for the possibility of introducing the time parameter and the notion of being "in" time.

Several senses of "timeless" are found in the literature: tenseless, instanta-neous, eternal, permanent, and atemporal. Tenselessness means the exclusion of the notions of past and future; it ignores many logical complexities in the temporal structure of our everyday thinking. Instantaneity and eternity can be understood in a tenseless temporal scheme. Instantaneous means having zero duration or occurring in no time. Something is eternal if it exists in all time, at least it is without end if not without beginning. Atemporality means the inap-plicability of temporal concepts. Permanence means the inapplicability of the concept of change and hence that of time. The permanent is atemporal, as numbers are permanent and atemporal.

The primitive spatio-temporal structure is permanent; it is independent of tem-poral concepts. It contains the time dimension as one aspect and makes possible the introduction of the time parameter, but is itself beyond time and change. M is a continuum of points, which are the numerical identities of events in the world. The number of points in M and hence the number of events in the world are as permanent as M. The constancy is most prominent in quantum field theory, where creation and annihilation operators create and annihilate field quanta. The operators are descriptive of the qualities of individual events, and they do not touch the identities of the events. The permanence of numerical identities should not be mistaken as a conservation law of some kind; it is more fundamental. Conservation usually applies to processes, as indicating some-thing that remains constant amid changes. The permanence of the number of events paves the way for the introduction of conservation laws.

Endurance, Change, and the Temporal Parameter

Time is a most important concept in our thinking. The temporal vocabulary in ordinary language can be roughly divided into two groups. The first consists of past, present, future, and the tense words. The second consists of before, simultaneous, after, and what are representable by a temporal parameter t. The logical structure of the concepts in the first group is more complicated than that in the second group. We do not consider them, not because they are less important, but because they are not involved in field theories. Here we do not even take account of the full temporal connotation of words in the second group. The temporal significance involved in describing the simple systems in fundamental physics is rather thin. Important features such as irreversibility appear only when we consider more complex systems in statistical mechanics and beyond. However, the concepts we are seeking should be general enough to allow the introduction of higher-level temporal notions.

Before and after signify a *temporal sequence*. The idea of sequences is not contained in the primitive spatio-temporal structure but can be readily introduced. Mathematically, a sequence is represented by a parametric curve or simply a *curve* $\gamma(t)$, which is generated by mapping a segment of the real line or part of the real number system into the manifold *M*. The numbers constituting the real line bring out the idea of sequences (Fig. 7.1).

There is no provision in the notion of curves that the sequences they represent are temporal. However, with the help of causal concepts to be introduced, we can separate curves into timelike or spacelike. The numbers parametrizing a curve constitute the domain of the *time parameter* in the former and the *space parameter* in the latter. When we say things change we usually mean variation with respect to the time parameter. However, this need not be so; when we say the landscape along the road changes, we mean variations with respect to a spatial parameter. There is a strong similarity between spatial and temporal notions; that is why people often say time can be visualized spatially.[217] Here we concentrate on temporal notions.

I am too hasty to talk about changes; at least two more concepts are required for such talk. First, we must identify that which changes; second, we must be

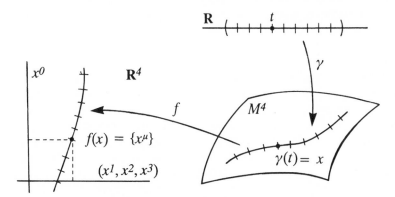

Figure 7.1 The four-dimensional spatio-temporal structure M^4 and its coordinate system, including the time axis x^0, are *permanent* and do not include the concept of time that underlies changes. The *proper time t*, which is closer to the common-sense notion of time, is defined only with the specification of a curve. A curve $\gamma(t)$ is obtained by mapping a segment of the real line $\mathbb{R}$ into M^4. A timelike curve $\gamma(t)$ captures the *endurance* of things; its parameter *t* "ticks" off events and is associated with the notion of change.

able to recognize similarity and difference. The second requirement is not at all trivial, and it is explicated in the next section. For now we pretend that we understand the meaning of being different and concentrate on the entity that changes.

Suppose there are hosts of distinct events with different qualities. We say one event is different from the next or one event occurs instead of another, but we do not say a change occurs. The events themselves do not change; nothing is yet capable of changing. Change involves the idea of something that remains selfsame despite the differences, something that endures. Mathematically, the concept of *endurance* is captured by a timelike curve in M that picks up a set of identities and conceptually binds the set of events they identify into a unit. The unit will be acknowledged as an enduring thing, and the curve will represent the world line of the thing, when it is properly substantiated by qualitative relations. Analogously, a spacelike curve captures the idea of *extensiveness*.

The variable t that parametrizes a timelike curve $\gamma(t)$ is technically called the *proper time* to distinguish it from the time component x^0 of the coordinates that identify events. The proper time partially answers to the common idea of a time that "ticks" and "changes." When we say time changes, we mean that the value of the time parameter t changes. Various values pick out various events whose qualities may be different. Since the events constitute an enduring thing, we say the thing is different at different times; the thing changes.

Endurance and the time parameter underlie the notion of sequences, succession, process, change, duration, and time lapses. The parameter t successively picks out various events encompassed in the unity of the curve. It introduces the idea of *temporal succession*. Analogously, the parameter in a spacelike curve introduces the idea of *contiguity*. Note that succession and contiguity of events are defined only with the specification of a curve. Succession and contiguity in general are not defined; these notions are absent in the bare structure of M.

If the successive events that constitute an enduring thing have different qualities, we say the enduring thing represented by the curve *changes* or *alters*. Thus the notion of change depends on that of the time parameter. It can happen that the events picked out by a timelike curve are all similar in qualities. In such cases we say no change occurs. This is what people mean when they say that time does not necessarily imply change. Note that "changeless" here means having zero change, not independent of the concept of change. The introduction of proper time has enlarged our conceptual scheme and radically alters the meaning of words.

The Category of Quantity

So far in our analysis, we have discussed the concept of qualities or properties of single objects, the modal concept of possibilities, which is represented by the state spaces of objects, and we are beginning to consider the category of relations. There is another important category, that of quantity. The concept of quantity has its share of philosophical difficulties. As an answer to the question of "how many" or "how much," it demands the notion of objects that are explicitly composite; only so can we talk about the number of its constituents (§ A1). The entities represented by parametric curves are composite objects that support the concept of quantities.

The length of the curve, obtained by integration or summation over its points, is a quantity or magnitude. A spacelike curve embodies the *extensiveness* of a momentary one-dimension unit and its length *extension*. A timelike curve embodies the *endurance* of a spatially extensionless unit and its length *duration*. Technically, duration is also called the proper time interval.

The Philosophical Significance of Permanence and Endurance

We have separated two concepts with distinct mathematical representations: permanence as independent of temporal concepts and endurance as descriptive of entities in which change is manifested. Endurance, represented by parametric curves, is necessary for the quantity of duration, the relation of succession, and the notion of qualitative change. These attributes are manifest only in physical things and processes. The time parameter underlies these concepts but is not identical to them. As Aristotle argued, time is neither independent of changes nor identical to them: "Time is not a sheer process but is a numerable aspect of it."[218]

Permanence is the more general and fundamental concept and is presupposed by endurance. As early as Plato if not Parmenides, philosophers had the budding idea of another time that has nothing to do with change. The distinction between a permanent time and a changing time was drawn by the Neoplatonists.[219] However, the precise meaning of the distinction and the definition of permanence remained elusive. Kant, who strived to uncover the most general and primitive concepts in our thinking, pushed hard for them. In the first edition of the *Critique of Pure Reason*, what is called the permanent in the "First Analogy" is mainly the enduring; time is viewed as a time series. Six years later in the second edition, Kant replaced the statement with something that comes close to permanence. He said: "The time in which all change of appearances has to be thought, remains and does not change." However, he did not succeed in clearly separating the two concepts.[220]

The permanence of the spatio-temporal structure resolves several traditional philosophical questions. One problem regards radical existence-change. Philosophers had found the idea of things popping in and out of existence puzzling, for radical changes leave temporal voids before and afterward. We are relieved from the notion of radical existence-change; all changes are conceived as alterations in the abiding four-dimensional framework. The excitation and deexcitation of fields, also known as the creation and annihilation of particles, are understood this way. The common-sense notion of things forming and disappearing can be explained with the help of the concept of kinds included in the concept of things (§ 19). The world contains many kinds of events, $\psi(x)$, $\chi(x')$, etc. A curve picking out events of the same kind represents an enduring thing, say a ψ-thing. A curve can also pick out a set of ψ-events followed by a set of χ-events. When a change in kind occurs in a curve, we say that a ψ-thing perishes and a χ-thing comes into being.

Another problem concerns the possibility of creation. Here we can follow Leibniz's answer.[221] Since M is no more than a structure of the world, there is no event before the Big Bang. When we talk about what happens before the Big Bang, we are thinking not of an empty time but a possible world that contains the Big Bang as an event. We do not conceive the world in a big container space, but the world in a bigger possible world with a correspondingly bigger spatio-temporal structure. That is also how we understand space outside the universe and the worlds depicted in science fictions.

The concepts of permanence and endurance provide the basis for the introduction of other temporal notions. They enable us to conceive of ourselves as three-dimensional beings in a four-dimensional framework. Therefore, they underlie remembrance and anticipation in the moment when we are in the presence of things. However, such complicated notions are way beyond the scope of this work.

Endurance and the proper time associated with it are manifest only in physical things and processes. Philosophers have sensed that the time so manifested can vary from thing to thing. The Neoplatonists said that the permanent time is one and changing times are many; Kant distinguished between substance and substances in his explication of permanence and endurance. The distinction is made clear in the mathematical representation. The spatio-temporal structure M is an integral unit. It can support countless curves, each parametrized in its own way. Proper time is specific to the curve. Each thing sets its own clock, so to speak. And there is yet no means for synchronization; there is no universally applicable time parameter or standard of duration. The notion of simultaneity, which is crucial for our comprehension of the world, is lacking.

Since Einstein's exposition in the special theory of relativity, simultaneity has generated a mountain of philosophical literature. Plenty of reviews are available and I do not repeat them.[222] I only want to mention a couple of points. First, simultaneity presupposes the concepts of permanence and endurance discussed. Thus it is not a relation among atemporal events. Second, the specific spatio-temporal structures, whether they conform to the Galilean or Poincaré groups, are more important for physics than philosophy. Philosophically, the important revision of simultaneity lies not so much in the concept of *time* as in the concept of *sameness*. Relativity does not dismiss the notion of simultaneity; it rejects only the global convention that imposes a standard for the sameness of time. In relativity, simultaneity is established in each case based on physical mechanisms, including the propagation of light. This idea is expounded in various causal theories of time. These theories overlook the basic philosophical problem because they take the notion of causality for granted. An adequate philosophical inquiry cannot ignore the criticism of Hume.

§ 26. How are Regular Successions of Events Possible?

The concept of endurance represented by curves is abstract. The concept of enduring things requires much more than mere curves. Suppose I make a movie by sweeping the lens of my movie camera randomly across some landscape. The movie can be represented by a curve, each point of which stands for a frame of exposure. The movie is shown as a unit, but no one would admit that the conglomerate is a thing. We must also be able somehow to relate the actual qualities of the events in a rulelike way before we can identify the events as belonging to the same persisting thing. Thus we turn to examine relations among qualities.

Hume classified relations into two kinds: those that depend entirely on the things that we compare, and those that may be changed without any change in the thing. The first kind of relations, including resemblance, contrariety, degrees in quality, and proportions in quantity, can be objects of knowledge and certainty. The second kind, including causation, identity, and the relations of space and time, is problematic. As a result of his criticism of causality, many philosophers subscribe to the Humean thesis that causality is no more than the regular succession of events.[223]

Philosophers criticize the notion of causality and bank on the relations of similarity and difference. Several positivists take similarity as the basic primitive concept for the logical structure of the world. Science seems to go the opposite way. In the development of the sciences that are concerned with fundamental relations, similarity relations long taken for granted give way to causal relations. Simultaneity as merely the sameness of time is criticized and replaced by causal relations in the special theory of relativity. "Same value" or "same utility" is discarded by modern economics. "Same state" is supplanted by causal mechanisms in field theories with local symmetries.

The Problem with Regular Succession

Let us find out the presuppositions of the austere notion of causality as a regular succession of events. To avoid the ambiguity of words, we can construct a mathematical model using the concepts we have developed so far. An event has a unique identity and realizes a specific quality among the possibilities circumscribed by its kind (§ 19). The succession of events is represented by a curve (§ 25). To be more specific, let us consider a kind of entities similar to those in general relativity. Consider an entity whose identity is x and whose actual quality is a specific vector v_x in the tangent space T_x, which contains its possible qualities. The actual qualities v_x and $v_{x'}$ are what we call "events" in ordinary speech. A curve γ joining the identities x and x' represents the succession of the events as illustrated in Fig. 7.2.

We cannot say whether the qualities at x and x' exhibit any regularity. To recognize regularities, we must be able to compare the qualities, but we cannot. The mathematics is unambiguous. Comparisons of vectors over different points are not generally possible. Let me refer to § 5. Our model uses only

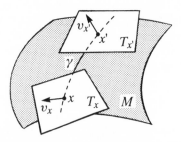

Figure 7.2 As I move along the curve γ that represents my world line, I see qualitative features v_x and $v_{x'}$ and so on. The features are separate and disconnected. They cannot be compared.

the concepts in D1, D2, and D3. To compare vectors we also need the affine structure in D4. Similar results obtain for other types of events characterized by other local symmetry groups. *The sharply formulated concept of disjoint individuals does not support the notion of similarity. The qualities of the individuals cannot be compared without additional concepts.*

Without comparison, we cannot tell if the events are different. Without the relations of similarity and constancy, we cannot have any notion of changes. Without change, the notion of succession loses much of its meaning. No sense of a world obtains from such a heap of disjoint events.

To bring our inquiry closer to the traditional philosophical debate, we can use the same mathematical model to represent the empiricist notion of experiences. In this case the curve γ represents my world line and the events represent a successive sequence of sense perceptions. I successively have the sense impressions v_x and $v_{x'}$, etc. The impressions are loose and separate, conjoined but not connected, as Hume described.[224] Since no relation obtains among the impressions, the result is the solipsism of the moment. Obviously it fails to describe our ordinary experiences. To account for intelligibility, empiricists further assume the possibility of observing regularities across the perceptions. The assumption rests on the relations of similarity and difference, but empiricists seldom explain how they can justify these relations in view of the transitory nature of sense perceptions.

It is not enough to argue that the things do have similar properties. They may have, but we cannot tell. We are concerned with the meaning of "being the same" and the general conceptual framework of our experiences through which we first recognize similarities. Since theories often employ auxiliary elements to facilitate thinking, we want to make sure that the auxiliary elements are not unwittingly ascribed to objects.

We routinely use some kind of conventional scheme as the global standard for the comparison of qualities. The most common global schemes are monadic predicates that are applicable to many individuals. Often we explain why two things are similar by saying they are both white. Such explanations are circular when we realize that the predicate "white" is arbitrary and the legitimization of its multiple application is the similarity of the things. Because superficial relations based on global predicates are so indispensable in our everyday thinking, they often make us forget their hidden conventionality. The above mathematical model shows that when the global convention of predication is purged, the notion of similarity collapses.

Skeptical Doubts about the Constancy of Predication

Suppose a skeptic argues that our ascription of the same predicate to different entities is a mere habit that is unjustifiable. He does not take issue on whether entities have definite properties. He agrees that we have worked out some satisfactory scheme of predication at some moment, say, for now the predicate "white" applies to snow and "green" to emeralds. Thus we can legitimately say

$$\text{at time } t, a \text{ is } F \text{ and } b \text{ is } F. \tag{7.1}$$

However, he doubts this is the same scheme we used yesterday, and he challenges the extrapolation of the scheme to tomorrow. He argues that we can find no objective justification for familiar statements such as

$$a \text{ is } F \text{ at time } t_1 \text{ and } b \text{ is } F \text{ at time } t_2. \tag{7.2}$$

Even if we do use the same predicate F, we cannot justifiably assert that their meaning is the same at different times. We cannot regard the universals F as Platonic forms, for even if they are, we have no empirical access to them. Aristotelian realists holding the predicate F to be actually in the particulars a and b fail to explain what makes F a universal. Conceptualists cannot justify their application of the same concept separately to different entities. Thus no satisfactory justification for consistent predication is available. All we are doing is following some habits. The challenge is an instance of Wittgenstein's skepticism on the criterion of following rules, here the rules of predication. A criterion is just another rule. Wittgenstein argued that ultimately we can find no justification for following a rule than to say we are trained to do so; it is our practice and way of life.[225] The argument is more general than is required of our topic. I will consider the case within a restricted Humean skepticism. In field theories, the skeptic's challenge amounts to a demand for the grounds for ascribing, say, the proton state to different events.

To avoid unnecessary complications, we use a loose meaning for similarity, which requires neither exact resemblance nor ordering of the degrees of likeness. Also, we assume the relata have only one qualitative aspect. Thus all we need to answer the skeptic is to establish objectively whether two entities are qualitatively similar or different. That is easy; we compare them.

Let us distinguish between *comparisons of co-present objects* and *comparisons of noncontemporaneous objects*. Co-present objects are those that can be grasped in a single perception. Since this is an epistemological investigation, I count as noncontemporaneous simultaneously existing objects that we separately perceive. Hume said semblances are "discoverable at first sight, and fall more properly under the province of intuition than demonstration."[226] I agree if the entities compared are co-present. Comparisons of co-present objects are grounded in immediate experiences and constitute the ultimate epistemological ground for relations. We accept as primitive the co-present relation

$$a \text{ is similar to } b \text{ ; } a \text{ and } b \text{ are co-present.} \tag{7.3}$$

This relation is included in our mathematical model. Since our concept of space–time allows the coincidence of enduring things in an event, the vectors in an event may represent features of coincidental things (§ 20). And we can compare qualities of the same event, which are represented by different vectors in the same tangent space.

The comparison of noncontemporaneous objects is something else. It takes a leap of faith to go from (7.3) to the relation:

$$a \text{ at time } t_1 \text{ is similar to } b \text{ at time } t_2. \tag{7.4}$$

The noncontemporaneous relation is not grasped in any perception; a may have long perished when b is observed. What are the grounds for accepting statements such as (7.4)?

The easiest answer is to dismiss the problem summarily by postulating non-contemporaneous comparisons as primitive. This is the move of many resemblance nominalists. It amounts to postulating a Laplacean–demonic view in which we have a whole stream of experiences present to us for comparison. However, we are not Laplacean demons, and we have no access to God's database to get a listing of pairs of qualitatively similar instances. The metaphysical speculation of resemblance nominalists is unsatisfactory.

Another quick answer is to say that a and b stimulate our senses in the same way. The answer is factually false; similar things need not stimulate similarly, and dissimilar things can produce similar images, all depending on the circumstances of perception. Furthermore, we still have to compare the sense stimulations that occur at different times, and that comparison is as doubtful as the comparison of things.

Co-presence need not imply a temporal singularity. We can insert intermediaries with infinitesimal time extents, $c_1, c_2, \ldots$, so that we can establish a series of co-present comparisons linking a and b. Thus:

$$a \text{ is similar to } c_1 \text{ during } t_1 + \delta t, \text{ and } c_1 \text{ is similar to } c_2 \text{ during}$$
$$t_1 + 2\delta t, \text{and}, \ldots, \text{and } c_n \text{ is similar to } b \text{ during } t_1 + n\delta t \equiv t_2. \tag{7.5}$$

Since we lack a universal scheme to quantify features, the comparisons are qualitative and the procedure cannot account for the possibility that the accumulation of minute differences can be glaring. Suppose we want to compare the orientations of two sticks normal to the surface of the earth, one at the North Pole and another at the Equator. We compare the first stick to the one next to it and conclude the two are parallel. We continue the process, arriving at the erroneous conclusion that the stick at the Pole is parallel to the stick at the Equator.

In daily activities, we often bring along a specimen m when we want to compare distant things:

$$a \text{ is similar to } m \text{ at } t_1 \text{ and } b \text{ is similar to } m \text{ at } t_2 \qquad (7.6)$$

This move works only if standard m has not changed during the time lapse, which we have no way to ascertain. The choice of m is conventional, we may obtain contrary conclusions if we had chosen a borderline paradigm that is close enough to a but not to b. More important, *the standard m must be an enduring thing, and the very concept of enduring things is lacking.*

The concept of enduring things is not trivial. It is not grasped by newborn infants, who do not search for objects that are covered up or displaced. Only when the infant is between 12 and 18 months old does it display the understanding that objects endure when they are absent from immediate perceptions.[227] Philosophically, Hume found that the concept of enduring things depends on coherence and constancy of their features. The relation of identity, which enables us to ascribe various perceptions as that of a single thing, is as suspicious as that or causality. Based on sense impressions, enduring things are no more than the regularity and invariableness of perceptions.[228]

The concept used by many empiricists to justify regularities is recollection. Memory is invoked to "bring the things together." Hence we have the relation:

$$b \text{ at } t_2 \text{ is similar to the recollection of } (a \text{ at } t_1) \text{ at } t_2. \qquad (7.7)$$

Physiologically and psychologically, memory is indispensable for our being conscious and intelligent. Philosophically, memory cannot be taken as primitive, for the task is to analyze memories and find out what concepts are presupposed by them. How does a memory know that it is a memory? Recollection is a kind of imagination occurring in the present; it involves at least two distinctions: first, the distinction between *perception* and *imagination*, between "I see a" and "I am having the image of a in my mind"; second, the distinction between my *past* state and my *present* representation of that state, between "I saw a" and "I remember seeing a."

As far as sense impressions are concerned, there is no difference between perception and imagination; both are modifications of mental states, and to neither does the concept of corrigibility apply. Clarity alone is not sufficient for distinction; imaginations can be vivid, and perceptions blurred. Only by applying the general concept of the objects can we differentiate certain representations by their lack of objects and call them imaginations. The minimal conceptual complexity required for the distinction of perception and imagination is discussed in § 15.

Even greater conceptual complexity is required for *enduring* objects. The distinction between perceptions and imaginations does not entail the distinction between past happenings and present recollections. To know that my imagination is a recollection, the imagery's "pastness" must be intrinsic; that is, my experiences must be time-stamped. We have no purely mental clock;

even our heartbeats refer to something physical. *The required time stamp is provided by the general concept of enduring, rule-governed objects*—time ticks only in concrete things, as discussed in the preceding section—enduring objects are constructed upon the basic spatio-temporal framework, so that any representation falling under the concept of enduring object is automatically linked via it to others occurring in the framework. The daffodil greeting me by the river tells me it is April. The oak supporting my back tells me it was there yesterday and will be there tomorrow, barring something extraordinary. In the concept of the enduring object I not only see the past and future in the present; I date my perceptions whenever they occur.

The frequent failure of memories alone would convince us that the record of memory cannot be on the same epistemological footing as the present testimony of the senses. When we want to remember something, we write it down. When I remember putting my keys on the table but they are not there, I doubt my recollection. Without the concept of enduring objects, memory is worse than a mysterious action-at-a-distance, for it conjures up only ghosts. If the recalled image is a mere present occurring, it is not a recollection and has nothing to do with *a* at t_1. Even when we find some criterion to mark it as a recollection as distinct from perceptions, we are still only comparing a present thing to the recollection of something. We have no way to compare the recollection to the thing in the past. The concept of enduring things changes the situation. An objective happening, though past, has fallouts, some of which persists to the present. The present world provides independent checks to my memory so that I am able to recognize the possibility of errors and to distinguish "I remember being thus" from "I was thus." Things are not only more reliable than memory; *the concept of enduring things is necessary for the concept of memory.* Since enduring things mean no more than regularity for empiricists, their argument based on memory is viciously circular.

The result of these considerations agrees with our mathematical model of empiricist theories. When the absolute primacy of sense perceptions is taken seriously, we cannot breach the solipsism of the moment. Similarity and regular succession of events are as shaky as identity and causality. The typical skeptical solution to such dilemmas is to assert that an unjustifiable belief is grounded on nothing more than a brute human fact. For instance, Hume argued our belief in causality is untenable, and explained it in terms of the brute fact of observed regularities. Now that we find that all determinate relations, including regularity in succession, are uncertain, what fact is left? Hume suggested an answer: He said principles such as the customary relation of causes and effects are "the foundation of all our thought and actions, so that upon their removal human nature must immediately perish and go to ruin."[229] Thus we arrive at the ultimate skeptical answer: It is a brute fact that human nature has not perished and we do have intelligible nonsolipsistic experiences, or else we could not be arguing. We do not and cannot justify this fact. We do not ask why; we just ask how it is possible. This is the Kantian turn.

The Uniformity Principle and the Local Symmetry Principle

We are looking for the ground of consistent predication and the general concept of relations, but the tactical moves (7.4)–(7.6) all aim to extrapolate or interpolate specific facts. This approach is common to many reflections on causality, as is apparent from most statements of the uniformity principle in philosophy. The uniformity principle usually applies to instances, or specific features of entities, which are given beforehand independently of the concept of uniformity. For example, Hume said: "That instances, of which we have had no experience, must resemble those, of which we have had experience, and that the course of nature continues always uniformly the same."[230] Some instances, notably those of which we have experiences, are taken to be the standard that others are supposed to resemble. Causal analyses usually focus on the assumptions behind why resemblance obtains between certain instances. Hypotheses such as "rules of working" gear towards the question of establishing specific relations between facts that are separately defined.

The austere logical structures of many epistemological theories do not allow analysis of the concept of empirical objects. The lack of conceptual complexity for objects underlies the notion that they are given and self-evident. Unfortunately, the self-evident are also self-secluding; they come with a stop sign that bars further investigations. The self-seclusion of the given bars justifiable relations. *Our failure to justify consistent predication stems from our having assumed a concept of objects that precludes it.* If objects are predefined to be Teflon-coated marbles, then cementing is destined to fail; if they are ordained to be disjoint, then relations must collapse. Empiricists regard sense impressions as given and transient, and it is not surprising that they find the concepts of enduring objects and causal relations untenable. Kant once summarized his reflections on these topics: "When Hume took the objects of experience as things-in-themselves (as is almost always done), he was entirely correct in declaring the concept of cause to be deceptive and an illusion."[231]

Just as surface preparation is the prerequisite for successful gluing, we must ensure that objects are relatable if we hope relations hold. To incorporate noncontemporaneous comparisons, we have to rework and expand the general structure of our understanding of objects instead of taking part of it as unanalyzable facts. The reworked concept of objects makes explicit the comparability and relatability of the objects. Since the conditions for relations are now included in the concept of objects, they are no longer arbitrary but acquire objective significance. The order and regularity that we entitle nature are not given in sense perceptions. What institutes regularity and brings events together is neither memory nor habit but the general concept of enduring objects including the notion of causality, by which we present what we perceive to ourselves as features of the objective world. I think this is not far from the gist of Kant's reply to Hume.

In physics, the recommendation of a satisfactory conceptual framework that includes the relatability of events is embodied in the principle of local symmetry, also known as the gauge principle. The principle of general covariance is a local symmetry principle (§§ 7, 10). *The local symmetry principle* says that:

1. The parameters of symmetry transformations should be allowed to vary freely from point to point in the world.
2. The entire field system should be invariant under such local transformations.
3. The global invariance should be achieved not by extraneous stipulations but by physically significant concepts.

The first point demands the articulation of the concept of *individuals*, which we have examined in discussing the significance of the local symmetry group in § 19. The second point demands the articulation of the concept of *the world*, or the whole of which individuals are parts. The third point demands the articulation of the concept of *causality*, so that integrity of the world is achieved not in conventional but in physical terms. The three points together repeal universal conventions found in global symmetries and replace them by physically significant concepts.

The logic of local transformations and global invariance is most clearly exhibited in gauge field theory. Instead of repeating the material in § 10, let me briefly discuss the case of general relativity. The local transformations of general relativity define a kind of event whose possibility structure is a Minkowski space. The disjointedness of the events means the tiny Minkowski spaces are free to rotate independently of their neighbors (Fig. 3.3c). Without the local freedom, the tiny Minkowski spaces would be forced to assume a uniform orientation and become essentially the same. They would then be represented by a global Minkowski space, and the whole situation would relapse into that of special relativity. Therefore, the rotational freedom of each tiny Minkowski space is not an arbitrary feature; it stems from the explicit formulation of the idea of individual events.

The concept of disjoint events, when expressed crisply, does not support comparisons and relations. To restore comparability without extraneous rules is the job of global invariance, which relaxes some restrictions imposed by the idea of strict disjointedness in a peculiar way. The rotational freedom of the individual Minkowski spaces lead to the appearance of extra terms, such as the affine connection $\Gamma^\lambda_{\mu\nu}$ in the field equation. Because of these terms, the entire system changes whenever an individual Minkowski space rotates. The result is a heap of individuals that does not amount to a whole with its own laws of motion.

The idea of a whole can be achieved in two ways. The first is akin to the implementation of some kind of uniformity principle, which dictates the conformity of the individuals by external rules. The Lorentz metric $\eta_{\mu\nu}$ is an established fact in special relativity. We extrapolate the fact and decree that the metric tensor $g_{\mu\nu}$ must be restricted to $\eta_{\mu\nu}$. The restriction forces the extra term $\Gamma^\lambda_{\mu\nu}$ to vanish. It effectively imposes a "baseline" across the tiny Minkowski spaces, snuffs out their individuality, restricts the theory to an inert global Minkowski space, and brings us back to special relativity.

I argued in § 7 that the principle of general covariance is a local symmetry principle. It recommends an alternate method that opposes the first. It rules against imposing arbitrary restrictions on specific terms but enjoins the expansion of the theoretical structure so that the extra terms can assume full physical significance and global invariance is restored. The extra terms turn out to represent the interaction field in the expanded theory. The physical and philosophical significance lie in the idea of interaction made possible by the principle of general covariance, which overthrows the restriction of $g_{\mu\nu}$ to $\eta_{\mu\nu}$. Weinberg said: "The difference is that we do not require that these quantities [the metric tensor $g_{\mu\nu}$ and the affine connection $\Gamma^{\lambda}_{\mu\nu}$] drop out at the end, so that we do not obtain any restrictions on the equations we start with; instead, we exploit the presence of $g_{\mu\nu}$ and $\Gamma^{\lambda}_{\mu\nu}$ to represent gravitation fields."[232]

Logically, the connection is a rule that individually acts on each event and modifies its behavior. In quantum field theories, which describe fully interactive systems, the characteristics of the events are so enriched that they themselves become the source and locus of interaction. The concept of individuals is modified to that of individuals-in-the-world.

In linguistic terms, each individual event has its own convention in predication. The predicates are monadic and the individuals are not related. To make the system of individuals intelligible, the uniformity principle imposes a universal convention for the monadic predicates, upon which explicit relations can be constructed. In contrast, the local symmetry principle expands the monadic predicates to account for *relational properties*, more specifically the idea of point interaction, which maintain the harmony of the whole.

The uniformity principle tries to generalize substantive or empirical concepts. The principle of local symmetry tries to expand the general conceptual structure. The principle of local symmetry neither presupposes a set of entities and empirical concepts nor attempts at factual generalization. Its concern is formal. It makes explicit the vague common-sense notion that things are individuals and yet belonging to a world, in which they are interactive and comparable. This general concept has many facets. Narrow theories tend to suppress one of its aspects so that it does not look self-contradictory. For instance, things are presupposed to be discrete; consequently their coupling must be conventional. The principle of local symmetry renounces such arbitrary suppressions by prescribing a broad general framework in which the concept of objects intrinsically has proper generalities.

§ 27. Relational Properties: Phase and Potential

In field theories, each event is free to choose its own convention in expressing definite qualities. The conventions of various events are mediated and their randomness compensated by the interaction potential. In the conceptual framework that incorporates these ideas, the structure of the event itself is modified. The properties of an event, which is designated by a single identity or absolute position, are explicitly analyzed into matter and interaction fields and their coupling. Since dynamical coupling becomes a characteristic of an event, the concept of causal relation is built into the concept of individual events.

Consider various expressions of the relation "*a* is warmer than *b*." The relation is founded upon individual properties such as "*a* is cool and *b* is cold." With more refined calibration, we say:

$$a \text{ is at } 20°C \text{ and } b \text{ is at } 10°C. \tag{7.8}$$

The warmth adjectives and the thermometer ascribe, however vaguely, thermal properties to the entities by instituting a general scheme of temperatures. However, upon close examination we find that "*a* is at 20°C" is not a clean-cut monadic property but is implicitly relational. The internal relation is hidden in the thermometer scale or the relative adjectives, which are conventional. I do not only mean that the Fahrenheit scale is as good as the Celsius. I mean something more fundamental, namely, that we have assumed that the same numerical scale applies to all entities. The idea of a constant scale incurs the difficulty of constant predication discussed in the last section, and we must supply objective justification for this assumption.

In cases such as temperature, which is a high-level concept, we can appeal to more basic physics such as molecular motions for justification. The more basic science provides the absolute "baseline" for the temperature scheme. Since the thermometer scale is *externally grounded* on molecular physics, it is conventional as far as thermal physics is concerned. Most familiar predicates are absolute but conventional in this sense. This is no answer to Hume, who questioned the epistemological justification of science itself. However, let us suppose we are satisfied when more basic sciences are available to supply the backing. We still have the problem in fundamental physics. I think this is also the root of the philosophical problem regarding universals, for metaphysics considers basic situations such as knowledge in general where the disciplinary sciences furnish no appeal court. Fundamental physics and metaphysics have to bite the bullet and find the grounding for the baseline *within* themselves. The internalized baseline acquires physical significance and becomes the mechanism that ties the events together.

Suppose temperature is a fundamental quality and we know nothing about molecules. Can we avoid the conventionality of the thermometer scale? Any attempt must involve an overhaul and expansion of the structure of thermal concepts. Let us introduce the notion of heat transfer. Thus "*a* is warmer than *b*" is expressed as:

a is at temperature T, b is at temperature T', and heat transfers from objects with temperature T to objects with temperature T' when they are in thermal contact. (7.9)

Without the last clause, the temperatures of the objects mean nothing, for T and T' are indeterminate and not ordered in any way. The order is specified only by the direction of heat flow. Similarly, "the same temperature" is expressed by the absence of heat transfer. Thus temperature becomes an incomplete quality awaiting the stipulation of the direction of heat flow.

Practically, this definition is a mess compared to numerical temperatures, but conceptually it is superior. For unlike the thermometer scale that remains outside thermal physics, heat transfer is an integral part of thermal physics. It is important that the heat flow is introduced as a physical concept, not as a mere formalism for ease of characterization and later made to vanish. The temperature becomes a kind of gear operating between two physical quantities, the object and the heat, binding the two into a unitary system. Since the heat flow is defined in terms of the indeterminate temperature, it implies that when two objects with different temperatures are put in thermal contact, heat transfer must obtain. The heat flow can be regarded as a kind of causal mechanism. The general concept of objects has also been enriched because the form of the objective property temperature has been changed; temperature is now intrinsically coupled to the heat flow.

From Velocity and Force to Phase and Potential

When we examine predicates in classical and modern physics, we find they are generally different. Classical qualities such as position and velocity are akin to the temperature in (7.8). From "*a* has a speed of 60 cm/sec" and "*b* has a speed of 90 cm/sec" we can infer that *a* is slower than *b*. The generic quantum quality is the *phase*, and it is akin to the temperature in (7.9). From "the event $\psi(x)$ has phase θ" and "the event $\psi(x')$ has phase θ'" we can infer nothing, for the two events have different conventions in designating their phases. Being events of the same type, they all have the same possibility structure or the same internal rule for ordering various phases. What they lack is a universally recognized "zero" phase as the norm. It is well known that absolute phases are conventional, only relative phases have physical significance (§ 12). To find relative phases, we need to reconcile the phase conventions of various events. The mediation is provided by something akin to the heat in (7.9), and it is called the potential.

Again we note the difference in classical and modern physics. The concept of *force*, which is prominent in classical mechanics, almost disappears in modern dynamics. In its place we find the potential. *Potential* is the integration of force, and indefinite integrals come with arbitrary additional constants. Thus the potential is intrinsically relative and can be determined only up to a constant. Every electric plug with its two prongs testifies to the need of potential difference. Absolute potential has no physical significance; a bird resting on an exposed high-voltage wire and experiencing only the absolute potential is unharmed. It would be cooked instantly if it was somehow connected to the ground or to another wire with a different potential.

Components of the field strength, which is proportional to the force, are the fundamental quantities in classical electromagnetism and the quantities measured in experiments. However, potential is the quantity that appears in the crucial coupling terms in quantum field theories. For example, in quantum electrodynamics, we measure the electromagnetic field component, but the fundamental coupling term $e\bar{\psi}(x)\gamma^{\mu}A(x)_{\mu}\psi(x)$ involves the potential A_{μ} (§ 10).

Many physicists were unhappy about the coupling term because the potential is intrinsically conventional. What is a convention doing in our most fundamental physical terms? Many attempts had been made to express the coupling in terms of the field strength. No way, physical theories are adamant; interactions must be expressed in terms of potentials, not forces or field strengths. For it is the potential, not the force, that couples to the phase factor, the other conventional element in physical theories. The local coupling of phases and potentials can be seen in the Aharanov–Bohm effect (Fig. 4.5d). In the experiment, the solenoid is carefully shielded to ensure that no magnetic field extends into the regions with any significant electron field. Thus the change in the interference pattern can only be produced by the variation in the vector potential A_μ.

Phases and potentials are both what I call *relational properties* or *gear qualities*, in the sense that an isolated gear is incomplete; gears are designed to be coupled into a system. The two relational properties complement each other to form an interacting system. The specific values of the phase and the potential are as inconsequential as the specific teeth that are engaged in a system of gears. Physical significance lies in the changes of the engaging teeth. When the two qualities are coupled, what appear to be conventional in isolation cancel each other and express the dynamics of interacting fields. *Both relational properties and their coupling belong to a single event designated by the same x. Thus the concept of events is enriched.*

As derivatives of potentials, forces account for potential changes. When a change is viewed in isolation as the turning of a single wheel, it appears to be absolute. In an interacting system in which the changes of one quantity are coupled to the changes of another, absolute changes of one partner have little significance. When we do measurements, we effectively substitute one partner of the interacting system with our experimental equipment and then discount the equipment. We think that we are measuring something absolute. It is like the case of a clock. Seen from the outside, the positions of certain wheels have absolute meaning, because we have "coupled the effect out" by putting hands on their axes. The positions of the clock's hands may be all important to us, but they are completely irrelevant to the internal clock mechanism. As far as the mechanism is concerned, it is just movement of the gears, one wheel turning another; the absolute position of none has any meaning.

If the physics of locomotion is like studying the motion of the clock's hands, then the physics of dynamical interaction is like probing into the clock's innards. The clock metaphor does not imply determinism because we have made no assumption about the clock's working principles; all we are doing is to try to look inside and find out the principles, whatever they are. In classical mechanics, both the force and the velocity refer to some absolute external baselines. The two baselines are coupled by another external rule, the law of inertia, which correlates constant velocities with vanishing forces. The law of inertia is much criticized for being basically arbitrary. It is a dangling loose thread because there are no more fundamental grounds. With the concept of relational properties, quantum field theory internalizes the law of inertia in the sense that the coupling of the two fundamental variables is given full dynamical meaning.

§ 28. Causal Relations: Connection and Parallel Transport

The potentials of interaction fields can be represented by connections on principal fiber bundles (Appendix B). The connection and a curve enable us to compare qualities of distant events. To correlate the qualities of different events we must go beyond a singular point. The concept that extends to a tiny region around a point is the *tangent vector*, which captures the notion of *infinitesimal displacements* about a point. Tangent vectors are essential in classical mechanics, where they represent instantaneous velocities of mass points. They play more roles in field theories, where they represent all instantaneous variations, including going from one event to another as manifested in an infinitesimal change of identity or position, and the infinitesimal differences in the qualities of various events. We may grant that the notion of infinitesimal variation is epistemologically primitive because they are local and momentary, something we can grasp in an act of observation.

To make noncontemporaneous comparisons, we need ways to identify qualities of different events. In classical mechanics, the identification is facilitated by the concept of parallelism embodying the notion of *same* direction. The trouble is not with parallelism but with the conventional way it is established; there is no physical significance to the parallel grids except perhaps as the structure of a substantival space. In modern physics, parallelism is generalized, cleansed of its conventionality, and acquires a physical significance that is not geometric. The affine structure, comprising a connection and a specific curve, facilitates the parallel transport of a quality from one event to another, hence to identify states of different events.

Imagine a set of events $\psi(x)$, $\psi(x')$, etc. joined by a curve γ as schematically illustrated in Fig. 7.3a. The state spaces of the events are all the same, and are represented by similar lines. The event $\psi(x)$ has actual phase θ, $\psi(x')$ has phase ζ, etc. We assume that the variation of the phases from one event to another is continuous. Thus the actual phases can be joined by a curve $\hat{\gamma}$. The curve $\hat{\gamma}$ describes the "trajectory" of our experience better than γ; we see variations of qualities, not identities. Our problem is that we cannot decide whether θ is the same as ζ, because each event has its own idiosyncrasy in labeling phases. To reconcile the idiosyncrasies is the job of the connection.

The *connection* is a general rule that is individually applied to each event. Crudely speaking, for each event, the connection separates the direction or the tangent vector along $\hat{\gamma}$ into a "horizontal" part and a "vertical part." This can be interpreted as the classification of all instantaneous changes issuing from θ into infinitesimal changes in identities and changes in qualities. As we go from $\psi(x)$ to its successor, the total change is represented by the vector $\hat{\gamma}$. The connection separates it into two components, the change in identity v_x and change in quality v_θ. The change in identity is the direction indicator that picks

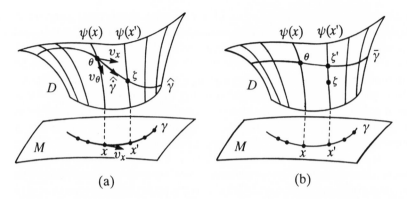

Figure 7.3 (a) The curve γ joins a set of events $\psi(x)$, $\psi(x')$, etc. The section $\hat{\gamma}$ joins actual qualities, θ for the event $\psi(x)$, ζ for $\psi(x')$, etc. The vector $\dot{\hat{\gamma}}$ represents the infinitesimal change pointing away from θ. The *connection* is a general rule that analyzes $\dot{\hat{\gamma}}$ into the change in identities, v_x, and the change in qualities, v_θ. (b) By accounting for identity changes only, we can identify ζ' in $\psi(x')$ with θ in $\psi(x)$ by *parallel transporting* θ along the horizontal lift $\bar{\gamma}$.

out $\psi(x')$ as the succeeding event instead of, say, the event $\psi(x'')$, which is not shown in the figure. It is the same as the direction indicator of the spatio-temporal curve γ, and it is often called positional change. The positional change vector v_x indicates a phase ζ' in the event $\psi(x')$. Since v_x ensures that nothing is changed except the identity of the event, we can identify ζ' in $\psi(x')$ with θ in $\psi(x)$. Technically, θ is said to be *parallel transported* to ζ'. At ζ' we can use the rule of connection again, thus constructing a "horizontal curve" $\bar{\gamma}$ as in Fig. 7.3b. It can be proved that given θ, γ, and a connection, the horizontal curve $\bar{\gamma}$ is unique.

Having found a way of identifying phases of various events, we can compare the qualities of different events. For example,

> $\psi(x)$ has phase θ, which, given a connection, is identified with ζ' in the phase space of $\psi(x')$ along the curve γ; $\psi(x')$ has phase ζ, so the phase difference between $\psi(x)$ and $\psi(x')$ via the curve γ is $\zeta - \zeta'$. $\hspace{2em}$ (7.10)

Since ζ and ζ' are within the phase space of the same event, (7.10) involves only co-present comparisons. The connection, which is the general rule under-lying the identification of θ and ζ', physically represents the *potential of the interaction field*. Intuitively, the potential mediates various phase conventions. It is significant that the potential is not imposed externally; the connection is a general rule that analyzes the tangent spaces of *each individual* event. To cor-relate qualities of different events, the connection must be supplemented by a curve or more generally a vector field. In stark contrast to classical mechanics, field theories admit no path-independent means of comparing and correlating. This is epistemologically satisfying, for universal and path-independent stan-dards of comparisons are conventional.

Suppose γ is a timelike curve. The idea of a unit comprising temporally successive events whose qualities vary continuously in a rulelike manner and have causal power is familiar. It can be realized as the *enduring things* that we handle every day. In the everyday context, (7.10) can be interpreted as the noncontemporary comparison of $\psi(x)$ and $\psi(x')$ with the help of an enduring thing. In quantum field theory, where enduring things as we commonly envisage are not found, the concepts of curves and qualitative relations are still paramount. There are innumerable curves joining two events, and none is privileged. To evaluate the phase difference and the coupling between two events, we have to account for the *contribution from all paths*. Colloquially, we can say everything between two events contributes to their coupling. The pervasiveness of the agents of interaction further affirms the demise of empty space.

The potential of the interaction field is part of the structure of the event, not something added on to ready-made events. Even if we start with an idealized pure matter field without potentials, the enforcement of local symmetry leads to the appearance of coupling to a potential; (§ 10). Physicists do talk about free electrons, but they emphasize it is only an approximation. Similarly, we talk about isolated and decoupled things, but it is a mistake to regard these things as absolute and not approximations we make in our common understanding. The approximation may be good because the coupling between things are weak compared to the self-binding of things, but it is an approximation all the same. If things were really so absolute, we would be totally out of touch; we have to interact with them to see them. The myth that nature consists of a set of ready-made discrete objects, each with its definite stimulus meaning, is no more than treating some customary approximations as sacred. *When we acknowledge our approximations and coherently articulate the logic of what we comprehend, we find that the general concept of interactions is inherent in the general concept of objects.* This is the result of field theories with local symmetries. The explicit spatio-temporal and causal relations among events are not external but internal and well founded. Events have not lost their identities with the introduction of explicit relations. Rather, they have individually acquired richer intrinsic structures accounting for their being members of the community that is the interactive field system. And it is within the system that individual events become fully intelligible.

§ 29. Thoroughgoing Interaction: Renormalization

The world is an interacting whole. In our efforts to comprehend the world, we have conceptualized it as consisting of entities that are related to each other. The analysis is usually based on the assumption that the effects of the relation on the behavior of the entities are relatively minor. Therefore the relation can be treated as some kind of perturbation or approximately ignored in some cases. The analysis is deeply rooted in common sense. However, it spawns some interesting philosophical issues. In conceiving entities, we have separated two groups of features: The first we ascribe to the entities as their monadic properties; the second we regard as relations among the entities. How clean can the separation between properties and relations be made? Can we define entities that are absolutely and not merely approximately free from relations? These questions lead to the philosophical debate on internal and external relations. The problem of infinities in quantum field theory and its solution in renormalization shed light on the debate and on the limitation of conceptualization.

Exact definitions of internal and external relations are still lacking, although philosophers have debated on them for a long time. In one version, *external relations* are independent of the properties of the relata taken in isolation, so that two pairs of entities (a, b) and (a', b') can be alike in all respects except that an external relation holds between a and b and not between a' and b'. *Internal relations* are dependent on the properties of the relata. The problem with the formulation lies in the ambiguity as to what count as properties of the relata, which depends on how the world is conceptualized. For instance, spatial relations are cited by some philosophers as examples of internal relations and by others as examples of external relations. Intuitively, Boston would not be the same city if it were located not 200 miles north of New York but 500 miles west; however, a cup would be the same whether it was on the table or not.

An older version of the two kinds of relations focuses not on entities but on the concepts of the entities. External relations are independent of and not contained in the concepts of the relata, which are unchanged by the relations. Internal relations are at least partially dependent on and contained in the concept of the relata. In Leibniz's words, internal relations are well founded; external relations are not. A concept "contains" other concepts if it is a structure analyzable in terms of them, so that the relata of internal relations must have certain conceptual complexity. In this sense I argue that spatial relations between entities are internal, for the concept of entities is analyzable into the concepts of kinds and absolute spatio-temporal positions, as discussed in § 20.

A conceptualization of the world with more internal relations is more coherent and integrated. Some philosophers of the holist bent argued that all legitimate relations are internal. Leibniz was an example, while others were idealists in the beginning of this century. Since all external relations are explicit and all implicit relations are internal, Leibniz made some effort to resolve explicit relations into the implicit form. Opposing the holists are positivists who argue that internal relations are incoherent and all relations are external. In the positivist doctrine, an absolute distinction can be drawn between the properties and relations of entities and entities can be considered in isolation without any relations.[233]

Probably both internal and external relations apply to things we talk about in ordinary language, for although ordinary things are intuitive, their concepts are often not clearly and sharply defined. Entities in fundamental physics are not intuitive, but their concepts are crisp. They call into question the meaning of truly external relations and isolated entities. "The electron and the proton electrically attract each other" is an internal relation. Any electron and proton separated by a certain distance cannot fail to attract each other with a certain force, for the attraction emanates from the electric charges, which are defining characteristics of the particles. Opponents can argue that the charge is a relational property. However, if we look more closely at elementary particle physics, we find that what appear to be truly monadic properties, the masses of the particles, are inseparably bound to interactions, even when gravity is not considered.

The Perturbation Method and the Problem of Infinities

The Lagrangian of quantum electrodynamics is a typical example of interacting systems (see note 47). It contains three terms: The first describes an ideal electron field in isolation, the second an ideal electromagnetic field in isolation, and the third the coupling between the two. The electron field is characterized in terms of a mass parameter m, and the coupling a charge parameter e. These parameters pertain to a literally bare electron stripped of all electromagnetic field, which is described separately. The formulation appears to support the view that properties and relations are neatly distinguished. The appearance is deceptive.

The standard and most successful method to solve problems of interacting systems is by *perturbation*. The characteristics of idealized free entities are determined, then the interaction among them is treated approximately as a small perturbation, which can be expanded in a series of terms. The effect of the interaction is calculated more and more accurately by including higher and higher-order terms in the perturbative expansion. The method works only when the coupling energy is small compared to the energy of the free entities. It fails in the strong nuclear interaction and physicists must find more holistic methods of solution, which is far more difficult. Fortunately, the condition of weak coupling is satisfied in many interactions, including the electromagnetic interaction.

When quantum field theory was first formulated, the first-order results were quickly obtained for many quantities, and they agree fairly well with experiments. However, when physicists pushed to the second-order terms, the results blew up in their faces; they were all infinities. The problem of infinities surfaced in 1928, and twenty years passed before it was brought under control by the renormalization procedure, which became a central pillar of quantum field theory.

Renormalization and "Dressed-up" Particles

In the fully interactive theory of quantum electrodynamics, two electrons do not interact instantaneously across a distance. They interact via the electromagnetic field, which transmits the disturbance of one electron to the other with finite velocity. One electron excites the electromagnetic field, and the excitation is absorbed by the other. The quanta of electromagnetic field excitation are called photons. Thus the two electrons are said to exchange virtual photons in their interaction. The photons are called virtual because their creation and annihilation in the interaction do no conserve energy and momentum. One electron creates some virtual photons, which are annihilated by the other. Thus the electron and electromagnetic fields form an integral system. Given enough energy, an electron can easily create photons. Conversely, an energetic photon can excite the electron field and create electron–positron pairs.

A problem arises because an electron spontaneously creates photons. There is nothing to prevent the electron from absorbing a photon that it itself creates. In fact the electron cannot escape from interacting with the electromagnetic field it itself generates. The self-energy of interaction gives rise to infinities.

The solution to the problem of infinities is based on the insight that the bare electron, with which the theory starts, is fictitious. For the electromagnetic field is not something external but is generated by the electron. Thus the electron is always accompanied by some electromagnetic field excitation or surrounded by a cloud of virtual photons. The electron dressed up in electromagnetic field excitations, not the bare electron, is the physically significant entity. Thus the mass parameter m and charge parameter e, which pertain to bare electrons and which appear initially in the equation make no physical sense. They must be replaced somewhere down the line by the physically significant parameters pertaining to dressed-up electrons.

In *renormalization*, all infinities are absorbed into the fictitious bare parameters, which are replaced by the real electron mass and charge as measured in experiments. The rationale is that what experiments measure are not bare electrons but electrons in constant and spontaneous interaction with the electromagnetic field. The real and empirical parameters are usually called the "free-electron" mass and charge. However, the renormalization program shows that they are *not* absolutely free. Their concepts have already absorbed some meaning of dynamical interaction, or causal power in traditional philosophical terms. Since the causal power is encoded in the monadic properties of mass and charge, the causal relations into which the electrons enter are strictly speaking not external.

We have seen in § 27 that individuals must have relational properties if they are to relate to one another. The renormalization program demonstrates again that reliability depends on definite presuppositions. Individuals in isolation are different from individuals in a community; the former must be enriched with relatability to become the latter. The free electron is free only within the electron-electromagnetic field system, in which it is obliged by its own characteristics to relate causally with other electrons and the electromagnetic field. As for the absolutely isolated and unrelatable electron, it turns out to be a ill-defined fiction.

The Limitation of Conceptualization

Renormalization is crucial for the triumph of quantum field theory and the standard model in elementary particle physics. It also shows the limitation of our conceptual ability to analyze the world. The theory is unable to specify the real parameters for the physically significant entities; finally it has to appeal to experiments. Physicists regard this as a blemish of the theory. A truly fundamental theory should be able to specify all dimensionless parameters, leaving only a few dimensional parameters to be empirically determined. Quantum field theory falls short. Glashow said: "Today, the standard theory requires the specification of nineteen fundamental dimensionless parameters. Surely, these are too many arbitrary constants to describe a truly fundamental theory. Most attempts to compute some of these parameters have led to the introduction of even more."[234]

This is where we stand as the twentieth century is slowly drawing to a close. We have extended our knowledge a long way into the microscopic world and gained much insight into the origin and evolution of the universe. However, there are glaring gaps in our knowledge: The measurement problem that highlights the difference between quantum and classical characteristics; the unresolved tension between the single quantum system and the statistical nature of experimental data; the action-at-a-distance in ground state fluctuations and distant correlations in Bell-type experiments. These anomalies are revealed against a generally causal and intelligible background of physical events. We try to fill the gaps, regarding them as tasks set by the postulation of the objective world. Perhaps we will not succeed; nevertheless we attribute them to our deficiency and refuse to mystify nature.

8

Epilogue

§ 30. The Intelligibility of the Objective World

The relation between knowledge and object, concept and experience, we and the world, is a central problem in philosophy and fascinates many people besides philosophers. The mind–body relation shows up in quantum interpretations, mainly through the writings of physicists. In many interpretations, the mind is conceived of as an observer that is peculiar for being beyond the laws of physics. They echo the philosophies in which the mind is an exotic object that is somehow affected by plain objects. These interpretations have been undermined by the research in the quantum measurement problem. We can find causal relations among objects, but these objects do not answer to the definitive qualities of observers; such qualities are outlawed in quantum theories.

The interpretation of quantum field theory presented in the preceding chapters implicitly contains a notion of the mind. The mind is understood not as an intelligent object but as the intelligibility of objects. The eternal mystery of the world is its comprehensibility, and the most profound astonishment lies in the fact that anything is experienced at all. Experiences are intelligible, but the world is not given to us intelligibly through the senses; it is made comprehensible by spontaneous mental efforts. The objects of cognition, the sun and the moon, trees and flowers, are not raw data but processed information because they are already conceptualized. The general concept of the world, which has already been used in immediate experiences so that their contents are comprehended as objective, reflects the partial structure of the mind for theoretical reasoning. The mind is found in the general conceptual structures of experience, common sense, and scientific theories, as the meaning of life is found in living and art in art works.

Some philosophies assert that knowledge must be founded upon primitive elements that have no conceptual complexity, for these elements are given, certain, and free from theories. The role of theories is merely to organize the given elements for more efficient reckoning. We have found in our investigation that every notion they deem given is conceptually analyzable. One common assumption they make is that the world consists of preindividuated entities that constitute the domain of discourse, over which the variables of theories range. This assumption is shown to be dubious in quantum field theory, in which individuation of events is a major theoretical task that is achieved only with the concepts of space–time and private state spaces. The result of individuation, not the givenness of the entities, enables us to comprehend the world as comprising discrete events. Consequently the concept of individual entities is complex and incorporates intrinsic spatio-temporal notions. The concepts of properties and relations are equally analyzable. Even the basic relations of similarity and difference have important presuppositions that are tied to the concepts of enduring things and causality.

Whenever we think, talk, or observe, we have used some specific substantive representations that are conventional. However, contrary to relativism, the general concept of objective reality sweeps across various representations and appropriately unites them. It condones conventions for definite description, accommodates the idiosyncrasies of experiences to secure empirical evidence, stipulates the systematic abstraction from conventions and idiosyncrasies, and acknowledges what they describe as objects. It is embodied in the representation–transformation–invariant structure of most modern physical theories and becomes even more prominent in quantum theories, which expressly state the form of observations.

Concepts are the products of our intellect. If all knowledge and experience, even the most immediate ones, have a certain minimal conceptual complexity, then the reality that is meaningful to us is structured by the mind. This does not imply that reality is constructed and relative to specific theories, for the general concept of objects, which is our own posit and reflects the structure of our mind, is formal and says nothing about the substantive features of world. It merely carries the notion that the substantive contents of our empirical knowledge are objective and not fabricated.

Once upon a time we listened to God speaking through burning bushes. We came to realize that even if a voice is heard, it is still up to us to decide whether it conveys the words of God. Things we handle every day are so obvious that we think objects announce themselves as if Nature has spoken, perhaps through some stimulus meaning. The naive notion of the given has been criticized by many philosophers; even if an image is seen, it is up to us to pose the rules for judging its objectivity. Macbeth's encounter with the dagger illustrates the common-sense notion of reality. Objectivity is not the state of givenness but a basic problem for theoretical reason. The unfamiliar topics of modern physics prompt us to take up the criticism anew. After a lifelong reflection upon objective reality, Einstein said in what may be his last apology: "The real is not given to us, but is set us as a task (by way of a riddle)."[235]

Appendix A

Measure and Probability:
Quantity, Quality, Modality

§ A1. Measure: The General Concept of Extensive Magnitudes

Quantum theories are probabilistic, but the interpretation of probability is itself controversial. Notions such as "objective probability" or "ignorance interpretation" have caused much confusion in quantum interpretation. We need the concept of probability in our analysis of quantum mechanics to avoid being stuck with a preselected conception, whose specifics may turn out to be irrelevant and obtrusive. Distinctions have to be clearly marked between statistics and probability, and between attributes of objects and our ways of making judgments. Fortunately, the measure-theoretic concepts underlying the probability calculus have natural general interpretations in terms of *quantity* and *quality*, which are less problematic. In this Appendix, I try to push these concepts as far as possible to see exactly how and where the more disputed notion of *modality* comes in.

The Kolmogorov axiomatization of the probability calculus based on measure-theoretic concepts enjoys almost universal acceptance.[236] The meaning of the calculus is another story. There are at least a dozen authorities advocating more than as many interpretations. The *classical probabilists*, who developed mathematical probability, made heavy use of the principle of insufficient reason, which invokes symmetry considerations and assigns equal probability to alternatives when no reason justifies favoring one alternative over the other. The *logicists*, who try to make probability into the foundation of inductive logic, argue that probability is a logical relation between two propositions; it measures the degree of support or confirmation a hypothesis receives from available evidence. Contemporary philosophers of physical science mostly agree that probability describes facts of nature. There the agreement stops. The *limiting frequentists* identify the concept of probability with the limit of relative frequency of the occurrence of a type of events in an infinite sequence of independently repeated experiments. The *propensitists* maintain that probability is a dispositional property of experimental arrangements, whose tendencies to produce certain results are measured by their propensities. Statisticians distinguish themselves for being end users of probability and for formulating broader frameworks of inference or decision in which probability is employed and interpreted. They too are divided. The *statistical frequentists*, who admit only statistical data in inference, restrict

$$ i\hbar\, \frac{\partial \psi}{\partial t} = H\,\psi $$

Even when God plays dice, He gambles according to mathematical rules.

the values of probability to statistical frequencies of events. The *Bayesians* argue that evidence beside statistics should be admitted. They assert probability is the index representing a particular idealized rational person's opinion about an event, or a measure of the degree of belief the person has in the truth of a proposition about an event. The statistical frequency view is perhaps the most widely accepted among physical and biological scientists. The Bayesian view is gaining acceptance among statisticians and human scientists.[237]

History offers a reason for the diversity of interpretations. The development of the mathematics of probability was motivated by various practical problems, notably the games of chance, the risk-bearing finance of the rapidly expanding commerce, and the need to find rational guides for decision making under uncertainty. Thus from its infancy the calculus was imbibed with diverse meanings, from which it was never clearly separated. A similar case is geometry; despite the tremendous expansion and abstraction of mathematics, there is still a tendency to call everything to which geometry applies "space" or "space–time." The tendency is much stronger for probability. Land, if not space, has weighty meaning independent of the metric. However, it is fair to say that the

Table A.1 Interpretation of Probability and Measure-Theoretic Concepts

Probability theory	Measure theory	General interpretation
Probability space	Normed measure space (Ω, S, μ)	Aggregative system
Outcome	Point in space $\omega \in \Omega$	Individual in the system
Event	Measurable set $E \in S$	Extension (attribute)
Probability	Normed measure $\mu(E)$	Relative magnitude of E
Random variable	Measurable function f	Individual attribute of ω
Expectation value	Integral $\int f \, d\mu$	Mean value of attribute f
Distribution	Induced measure μ_f	Attribute of the system Ω

concept of a quantitative probability, as distinct from various vague notions of the probable or the likely, is dependent upon the calculus.

To extract the concept of probability, I highlight the structure of the probability calculus and simultaneously interpret its elements in terms of the traditional philosophical categories of quantity and quality. The first two columns of Table A.1, adopted from M. Loéve's textbook on probability, give the correspondence between some probabilistic and measure-theoretic terms.[238] The third column summarizes my general interpretation of the measure-theoretic concepts.

The Concept of Magnitudes and Its Difficulties

Nowadays many qualities, including all qualities in physics, are represented numerically or quantitatively. To avoid confusion I use *magnitude* for the traditional concept of quantity and use "quantity" loosely. In thermodynamics and statistical mechanics, a distinction is made between extensive variables such as energy and entropy and intensive variables such as temperature and pressure. Both kinds of variables are quantitative. Extensive variables are proportional to the amount or size of the system under consideration. Intensive variables are independent of the amount. Doubling a system doubles the energy but not the temperature. The definition of extensive variables depends on a more fundamental concept, that of amount. *The abstract concept of amount I call magnitude.* The formal concept of magnitudes agrees with Aristotle's primary sense of quantity.[239] Magnitudes answer questions of "how many" or "how much." Therefore magnitude is a *collective* concept, as opposed to quality, which can pertain to *individual* elements. Kant called a magnitude extensive when the representation of the whole is made possible by the representation of its parts.[240] In the continuous case, sizes such as the length of a line segment generally involve integration, and only so are sizes magnitudes, as opposed to the size of a dress, which is an individual quality.

There are speed bumps everywhere in categorical analysis; we cannot accelerate even with such an innocent notion as magnitude. In common language, magnitude signifies plurality and is expressed in plural terms and number words, which have caused considerable difficulty in analytic philosophy. The existential quantifier ranges over singular items in the domain of discourse, and the predicates are individual qualities or relations. "There is x such that Fx" is a proposition about an individual. This form of propositions appears to have little room for a unitary concept of magnitudes. Consider the meaning of "three" in: (1) There are three elephants in the zoo. (2) There are three-ton elephants in the zoo. The "three" in (2) is part of the predicate describing the quality of individual elephants, but the "three" in (1) is not. "Three" in (1) is a quantity or magnitude. The troubling questions are: What does it describe? What is the subject of which "three" is the predicate? The questions are important because the subject pertains to ontology.

The status and role of number words have been extensively studied by analytic philosophers, especially in the philosophy of mathematics. An obvious interpretation of "three elephants" is "the number of elephants is three." But in so saying, we run the risk of reifying numbers. We have to seek alternatives if we do not want to subscribe to a realist interpretation of mathematical entities. Suppose we construe elephants-in-the-zoo as a unitary whole with parts; we find that "three" can be attributed to neither the whole nor the parts. For a unity is one and there are six differentiable parts; two-elephants is a definite part. A most sensible way is to construe three as a universal describing groups of three. However, if the interpretation requires the definition of the set of all sets with three members, we run into difficulties in consistent set theories. Thus some analytic philosophers argue that number words should be discarded. Thus "two elephants" should be expressed solely in terms of individual qualities and relations: "There is x, there is y, x is an elephant and y is an elephant and x is not identical to y." Similarly for more items. In the regimentation, the question "how many" is ruled out of court.[241]

The banishment of the concept of magnitudes is disastrous for scientific theorizing. We want to introduce magnitudes in such a way that quantities and qualities are categorically distinct but assume the same logical status in propositions as attributes of entities. We need a proper definition of composite entities that are at once unitary so that they can stand as the subjects of propositions and manifestly aggregative so that they can take magnitudes as attributes. This is achieved by measure theory.

Magnitude is a basic structural element in our understanding of the world. In abstracting the general concept from the specifics in which it is embodied and making it precise, measure theory has unleashed its power. It enables us to apply the concept of magnitudes not only to concrete things as in ordinary experiences, but to abstract terms. It underlies the statistical thinking in which we talk about the amount of risk, the measure of possibility, the weight of evidence, the magnitude of a Hilbert space, or the size of a generation, not poetically but rigorously. Not surprisingly it finds wide applications in the physical, biological, social, and human sciences.

Measure Spaces: Systems of Magnitudes[242]

Magnitude is the magnitude of something. That something I call *extension*, which is understood abstractly with no spatial or substantial connotation, although spatial extension is its paradigm. A theory of magnitudes involves two aspects: the *delineation of extensions* and the *assignment of numbers* to the extensions to signify their respective sizes. Ordinarily, delineation is taken for granted. Things in our daily experiences are self-evidently individuals. For stuff such as water and air, we have familiar auxiliary means to divide them into parts, for instance, by buckets or geometry. We appreciate the significance of individuation when we confront abstract concepts. What is a chunk of possibility or a bunch of confidence?

Consider an abstract set Ω with elements ω. What are the entities to which the notion of size applies? The seemingly obvious answer is the elements ω. The answer is wrong. For Ω can be a continuum, in which a point ω has zero size. That points are extensionless had been realized by Leibniz, and played no small role in his rejection of Descartes' identification of matter with spatial extension. We need an extra concept to support magnitudes. A candidate is a subset E of elements. A subset can contain only one element, but the singleton subset $\{\omega\}$ and the element ω are completely different concepts. The relation between a set and its subsets is inclusion, $\{\omega\} \subset \Omega$; the relation between a set and its elements is membership, $\omega \in \Omega$. A subset corresponds to a segment of a line instead of a mere point. It is promising, but it needs certain restrictions. For it can be shown that if we arbitrarily include all subsets of Ω, we cannot always define a rigorous concept that captures the ordinary additive notion of magnitudes.[243]

Intuitively, if we combine two magnitudes or strike off part of a magnitude, we should get another magnitude. Thus the subsets supporting magnitudes should be combinable and subtractable. The mathematical structure that captures these intuitive notions is called a σ-algebra. A *σ-algebra S* is a collection of subsets E_i of Ω satisfying certain set-theoretic axioms.[244] The members E_i of S are called *measurable sets*. They are intuitively "parts" to which the concept of magnitude applies. As parts, they necessarily stand in a whole. The set Ω with a σ algebra S of subsets is called a *measurable space*, denoted by (Ω, S).

Magnitudes are additive. If two rods are combined, the length of the resultant rod is the sum of the lengths of the individual rods. This idea is captured by the concept of measure. A *measure μ* on a measurable space is a *set function* that assigns a number $\mu(E_i)$ to each measurable set E_i as its *magnitude*. It differs from ordinary functions because its domain consists of sets instead of elements. The measure has several properties: (1) The entire set Ω has a definite magnitude, and the empty set has zero magnitude. (2) The magnitudes are nonnegative, which does not prohibit some sets from having vanishing magnitude. For instance, the set of all rational numbers embedded in the real line has zero measure. (3) The magnitudes satisfy the intuitive notion of countably additive. If the subsets E_i and E_j are all disjoint, then the magnitude of their union is the sum of their respective magnitudes. A measurable space (Ω, S) equipped with a measure μ becomes a *measure space* (Ω, S, μ). In applications, a measure space represents an *aggregative* or *statistical system*.

A most important measurable space is the Borel line, whose underlying set is the real line or the system of real numbers, $\mathbb{R}$.[245] A class of familiar subsets of the line is the semiclosed intervals $[a, b) = \{x: a \leq x < b\}$, where a and b are two numbers. Let B be the smallest σ-algebra that contains all such intervals. The measurable space $(\mathbb{R}, B)$ is called the *Borel line*, and its measurable sets $\Delta \in B$ are called *Borel sets*. Product spaces of the real line are also measurable spaces, whose measurable sets are also called Borel sets. The Borel line admits many measures. The most intuitive one is the *Lebesgue measure* $\mu([a, b)) = b - a$, which captures the notion of the length of a line segment.

As an example, let (Ω, S, μ) be the aggregate of animals in a zoo and E_1 be the set of elephants. $\mu(E_1) = 3$ reads "the set of elephants in the zoo is 3-numbered," or more colloquially "there are three elephants in the zoo." Several points are worth noticing. First, the set of elephants is defined only as a part of an aggregative system, the zoo. The systematic definition marks it from ad hoc statements and alleviates the reluctance to predicate arbitrary collections. Second, magnitudes are represented by functions in parallel to individual qualities, which are also represented by functions in many physical theories. The only difference is that the domains of magnitudes are sets and not individual elements. Thus statements involving magnitudes and qualities have the same logical form. Third, the number word does not refer to abstract numbers; it is strictly attributive; "3-numbered" means 3 in a scale with dimensionless units, logically not different from 3 tons. In this light "the number of elephants is three" assumes a parallel structure as "the speed of the car is 60 mph." Neither grammatical subject refers, for each is logically a part of the predicate. The logical subject of the sentence is the set of elephants or the car. Fourth, the definition of measures is constrained only by exceeding by general rules; therefore, the assignment of 3 to the aggregate of elephants is not unique. It is only the result of adopting a conventional scale, akin to the scale of tons rather than kilograms. We can equally say "the elephants in the zoo is 6-numbered," provided the measure is consistently assigned to all aggregates of the zoo. Generally, various measures are connected by well-defined transformation rules. We may call the measure that yields 3-numbered "natural." This is similar to physicists calling the unit "natural" in which the speed of light and the Planck constant are set to unity.

Normed Measure and Probability: Relative Magnitudes

Physical quantities are finite. When the total magnitude X of a set Ω is finite, we can divide everything by X and redefine the measure for the entire set to be unity. The result of normalization is a *normed measure space* (Ω, S, μ), whose *normed μ assigns the magnitude 1 to the whole set, so that $\mu(\Omega) = 1$. A normed measure gives the relative magnitudes* of the sets $E_i \in S$; it signifies the *ratio* or *proportion* of the magnitude of E_i with respect to the total magnitude of the system Ω. Normed measures are of special interest because the notion of absolute magnitudes is not significant in many abstract concepts. In quantum mechanics, for example, only normalized vectors of a Hilbert space are important. Since we are only interested in finite cases, from now on μ denotes a normed measure unless qualified.

We can use the magnitude of a certain subset, say E_1, as an effective normalization factor. Suppose the subsets E_1 and E_2 overlap. Then $\mu(E_2|E_1) = \mu(E_1 \cap E_2) / \mu(E_1)$ is the relative magnitude of the overlap with respect to that of E_1. Here we are effectively restricting our attention to certain subsets.

As an example, let Ω represent a collection of balls, each of which is represented by an element ω. The total mass of the balls is normalized to 1. Various groupings of the balls are represented by measurable sets E_i, and the normed measure μ gives the fractional masses of the groups of balls. For instance, if E_1 is the group of all wooden balls, E_2 the group of all red balls, then $\mu(E_2|E_1)$ the fractional mass of wooden balls that are red. The example is not exciting because it is so trivial. However, the groupings can become highly complicated. Suppose each ball ω is marked with a long sequence of letters, H and T, and the measurable sets group the balls according to the patterns of the sequences. Thus E_i can represent the set of all balls whose ith letter is H, and E_j all balls whose first n letters contain r Hs. The normed measure $\mu(E_i)$ and $\mu(E_j)$ then gives the relative masses of the respective groups of balls. Here lies the power of the theory. Because of its abstractness, ω can represent things with the most complicated structure, or it can represent abstract entities that are not things at all.

I now repeat the material presented above, using different terminologies. A *probability space* (Ω, S, μ) is a normed measure space. The elements ω of Ω are called *outcomes*, the members E of the σ algebra S *events*, the underlying set Ω the *sure event*, the normed measure μ *probability*, and $\mu(E_2|E_1)$ the *conditional probability* of event E_2 given event E_1. In the example given, forget the balls and consider only the sequences of H and T. Suppose each sequence represents the possible outcome of repeated flippings of a coin. Then $\mu(E_i)$ represents the probability of the event that the ith flip is a head, and $\mu(E_j)$ the probability that the first n flips contain r heads. The "probability" so defined exactly means a ratio, a proportion, or a relative magnitude.

§ A2. Function and Distribution: Individual and Collective Qualities

The concept of magnitudes is unable to perform alone in any empirical science. Consider a chunk of matter or a patch or red. It is through the solidity that we know the extension of the matter, through the redness that we observe the extension of the patch. Magnitudes are manifested through qualitative features. A task in the human and social sciences, where magnitudes are intangible, is to define and delineate significant qualities so that measurements of magnitudes can be made. This does not imply that qualities are something added on to bare extensions, or that magnitudes are theoretical constructions upon observed qualities. It means the concepts of magnitudes and qualities are both already involved in any cognizance of extensive objects. Magnitudes and qualities are distinct. As Leibniz argued, things without parts and magnitudes can have complicated qualities. Monads are extensionless, yet they have internal complexities. Point particles are extensionless, but they have intrinsic spins.

This section considers qualities, both for individuals and, when coupled with the concept of magnitudes, for statistical systems. When only individual elements are under consideration, qualities are usually represented by *functions* in physical theories. In statistical systems, qualities are represented by *measurable functions*. The measurability of the functions ensures that the concept of qualities is compatible with that of magnitudes. It also serves as a bridge between the representations of qualities by functions and by sets, which is the standard representation in formal semantics. The collective qualities of statistical systems are represented by *distributions*, which are the center of attention of probability theories.

Individual Qualities: Measurable Functions[246]

Let us forget for a while about extensions and magnitudes. Consider simply a set Ω with elements ω, which can be anything, concrete or abstract. An *individual* ω is an entity that can be the subject of a proposition. Individuals have qualities, which are their *properties*. In analytic philosophy, properties are usually represented by *sets*; to say an entity has a certain property is to say it is the member of a certain set. As discussed in § 11, the concept of properties contains a distinction between types and values. This distinction is not intrinsic in the set-theoretic representation. The probability calculus is equipped with measurable sets E_i that are interpreted as qualities such as "being four-of-a-kind." However, it does not stop but proceeds to introduce functions, which are also interpreted as qualities. Given a function, we can define suitable classes of measurable sets by introducing "indicator functions."[247] Unlike sets, functions have the advantage of capturing the type–value distinction of properties.

A *function f* represents a quality type and stipulates a rule that assigns a unique number x to each individual ω in Ω corresponding to its specific qualitative value. For instance, let Ω represent the state space of a classical particle with mass m and $\omega = (q, p)$, the particle's position q and momentum p. The particle's kinetic energy, $K(\omega) = p^2/2m$, is a function. The functional representation of qualities covers all cases in classical physics and probability theory, but not quantum physics.

Let f be a function from a measurable space to the Borel line. f is *measurable* if the inverse image of every measurable set Δ_i in the Borel line is a measurable set $E_i \equiv f^{-1}(\Delta_i)$ in the measurable space. Measurable functions ensure the compatibility of the concepts of magnitudes and qualities.

In the probability calculus, measurable functions are called *random variables* or *stochastic variables*. In fact they are intrinsically random. They are functions whose variables represent outcomes of experiments that can happen to be random.

Induced Measure and Distribution: The System Takes Over[248]

Measurable functions are important because we can use their inverse to project the measurable nature of the Borel line into an arbitrary set Ω and turn it into a measurable space. As the line interval Δ is varied, the inverse function $f^{-1}(\Delta) = E$ effectively carves out various subsets E of Ω, which can be shown to be measurable and capable of supporting the definition of a measure.[249] Given a measure μ and a measurable function f, we can define a measure μ_f for the Borel line such that the measure of a Borel set Δ is equal to the measure of its inverse image E,

$$\mu_f(\Delta) = \mu[f^{-1}(\Delta)] = \mu(E). \tag{A.1}$$

μ_f is called the *measure induced by f*, it depends on f, and it turns the Borel line into a measure space $(\mathbb{R}, B, \mu_f)$. $\mu_f(\Delta)$ gives the relative amount of individuals whose values of f fall within the range Δ. Note that the individuals ω are not explicitly invoked in the induced measure, although they are implicitly assumed via the function f. The induced measure $\mu_f(\Delta)$ is a characteristic of the statistical system as a whole. In most physical cases, the measure μ is unknown. We do not measure μ directly; we measure μ_f by observing the values of f. From empirical data and (A.1), we determine the measure μ of the object system.

In probability theory, the induced measure μ_f is called the *distribution with respect to f*. Distributions $\mu_f(\Delta)$ are set functions, which are awkward to handle. We can define corresponding point functions, called *distribution functions* $F_f(x)$, which are susceptible to powerful analytic tools.[250] Distributions and distribution functions are the locus of the probability calculus, and they are prominent in many physical and statistical theories. For example, the velocity distribution function of a collection of particles is Maxwellian.

Summary and Reflection

In talking about the distributions and statistics of a composite objective system, we have used the inverse function f^{-1} representing a quality type to project the Borel structure of the real line into the abstract set representing the system, thereby partitioning it and turning it into a measurable space. This rationale is most clearly demonstrated in the Lebesgue integral, where the independent variable is not the set elements but their values in the real line. For a specific range of values, the inverse function seeks out the corresponding set of elements and accounts for its magnitude.[251] The procedure makes explicit and articulates precisely the conceptual framework presupposed in the quantitative measurements of composite systems. An arbitrary set is not measurable. When we acknowledge certain numbers as empirical data of the magnitudes of the parts of a system, we have already presupposed that the system has certain general features:

> The object system is formally representable by a measure space and
> its qualities are representable by measureable functions. (A.2)

This is a weaker postulate on the system than numerical representability, because the measure space need not be analytic.

The abstract concepts contained in the measure space are intuitive but non-trivial. Philosophically, the postulate (A.2) means the application of the categories of quantity (magnitude) and quality to the objective world before actual measurements. The Borel structure of the real line is our intellectual construc-tionn. In projecting it, the mind gives form and lights up the intelligible nature. This the sense of "*a priori*" in Kant's writing, where only formal concepts can be *a priori*. The forms posited are abstract and devoid of substantive content, which are filled by actual experiences. We assume the system has the general form of a measure space (Ω, S, μ) but do *not* specify the σ-algebra or the measure; the specific S and μ are left to be determined empirically. Instrumentalists are not immune to these presuppositions. In fact, they often unreflectingly impose stronger forms on nature. For instance, when they use rigid rods to define the meaning of geometry, they have ideologically imposed a specific measure, which is a substantive concept and should be left to the discretion of empirical theories.

§ A3. Individual, Statistical, and Probabilistic Statements

So far there is no notion of probability apart from that of a ratio. Where is probability? To answer this question, let us examine more closely the nature of individuals and aggregates to see what we can and cannot say of them.

Individual Statements

Individual statements describe individuals without invoking the notion of the system. They involve the concepts of particulars and qualities; they need not invoke magnitudes. We have a set Ω of individuals ω, characterized by some quality type f. The subjects of predicate statements are individuals ω or groups of individuals $E = \{x_i\}$. Individual statements have the general form:

$f(\omega) = x$; ω is x; he has an annual income of $50,000;
$f(E) \in \Delta$; the members of the set E have values of f within Δ; they belong to the income group $50,000–75,000.

Statistical Statements

Statistical statements describe the properties or patterns of an aggregate as a unit. They explicitly invoke the concepts of qualities and magnitudes but not the individuals ω, although they implicitly depend on the individuals through the individual quality type f. The properties of aggregates are represented by distributions $\mu_f(\Delta) = \mu[f^{-1}(\Delta)] = \mu(E)$. There are two interpretations of distributions. We can regard the Borel sets Δ as cells of individual qualitative values, and $\mu[f^{-1}(\Delta)]$ as the relative occupation number or the fraction of the system's members that have their qualities falling within the range Δ. Alternatively, we can regard Δ as an autonomous variable that delineates parts of the system, and $\mu_f(\Delta)$ as the relative magnitudes of the parts. The two interpretations, respectively, view the aggregate as a collection of members or a whole with parts. Statistical statements have the general form:

$\mu[f^{-1}(\Delta)] = r$; the relative magnitude of the set $\{\omega : f(\Omega) \in \Delta\}$ with respect to Ω is r; the fraction of Americans in the income group $50,000 – $75,000 is 0.1.

$\mu_f(\Delta) = r$; the relative magnitude of Δ is r; 10% of the American population is in the income group $50,000–75,000.

Logically, individual and statistical statements are similar; both are in the subject–predicate form. They differ only in subject matter, one describing an individual and the other a system, which is treated as a unit whose composite nature is held in view. Both kinds of subjects can range from the concrete to the abstract. Whether a particular statement is factual depends on its specific subject.

The word "statistic" appeared in Italy in the mid-sixteenth century, where it meant information, usually numerical, about the political state. Now its scope extends far beyond the state to cover numerical information of all kinds. Thus the basic meanings of "statistic" and "probability" have little correlation. One reason they are often rolled together is this. The concept of relative magnitudes is applicable to whatever that satisfies the general conditions discussed in § A1. Measure theory provides a powerful way of quantifying abstract concepts. It has been used to formulate theories of relative magnitudes for the intuitive notions of possibility, ignorance, evidence, or degree of belief. Such theories are the roots of several conceptions of probability. However, in these cases the sense of probability is derivative; it lies solely in the subject matter. It would be less confusing if we call them the statistics of possibility, evidence, or whatever. The statistics of possibilities or actualities are both statistics. We still lack the key to why the subject matters convey the sense of probability. The concept of probability, being a modality, should show a logical difference. Its significance is not captured by the substantive topics of statistics.

Statistical statements remain statistical and not probabilistic with the introduction of randomness and idealization for the aggregates at issue. Casino records, actual or hypothetical, are statistics, not probabilities. Quantitative descriptions of exactly formulated random systems, such as "collectives" or "long runs," are statistical. We can idealize an aggregate to be infinite, and evaluate its statistics such as the mean value by taking limits. Limits are familiar in physics. Velocity is the limit of the time rate of change of positions. Physicists routinely evaluate quantities "in a box" and take the limit as the "box" becomes infinitely large. The limit of a sequence, if it exists, is a definite value; it does not imply anything probabilistic. The notions of ignorance, determinism, uncertainty, or the like have no place in statistical statements except perhaps as specific topics of accounting. We are simply interested in the qualities and patterns of the entire aggregate, period.

Statistical statements can be complicated. "Consider a large number of sequences of n throws of a fair die, and let n_1 be the number of aces in each sequence. The variance of the value of n_1/n from $\frac{1}{6}$ approaches zero as n is increased indefinitely" asserts a fact about the outcome distribution of long sequences of die throws. Concepts such as the mean, average, variance, deviation, skewness, in fact, all characteristics of distributions, are factual descriptions of aggregates. Magnitudes can overlap and we can change the normalization by restricting our attention to certain subsets of the system, leading to all kinds of correlative statistical analysis.

The behaviors of aggregates are governed by their own laws. We make predictions about aggregates just as we make predictions about individuals. Both kinds of predictions are empirically testable. "In the next hour, $\frac{1}{38}$ of all roulette outcomes in Las Vegas will be the number 5" is a prediction about the gambling phenomenon. Boltzmann's transport equation governs the evolution of the distributions of many-particle systems just as Newton's equation governs the motion of single particles. Statistical predictions may be lousy because we do not have good rules for the variations of complicated aggregates. However, logically they are not different from poor predictions of projectiles without considering such complications as air drag and the spin of the ball. The philosophical problem of induction afflicts both cases.

Statistical statements, like individual statements, can refer to hypothetical cases. As we can describe individuals in possible worlds, so can we describe the collective patterns of possible worlds by statistical statements. There can be general statistical laws, and as laws, they support counterfactual assertions. The modality does not lie in the statistics.

Probability Statements

As long as "probability" is a normed measure, it can be replaced by "ratio," "proportion," or "relative magnitude" without loss of meaning, resulting in statistical statements. We can substitute "probability" in many philosophical papers by "proportion of E," where E is the appropriate subject matter, and see that valid logical arguments remain intact.

"Probability" has not been replaced by "ratio" or "relative magnitude," not only because fuzzy big words are charming; more important, the commonsense notion of probability is not a mere ratio. To seek the concept of probability that is distinct from statistics, we have to go beyond the normed measure. It is often said in textbooks that probability pertains not to outcomes of trials but to events, not to a throw of a die but to the event that red shows up. I agree, but point out that it is common for textbooks to shift between two senses of "event." In the technical sense, an event, represented by a measurable set E_i, is formally a property such as redness, it is a universal and is abstract. This is not the common usage of "event." A common-sensical event is always a *particular* occurrence, an *instantiation* of the universal, as the event that red shows up in *a* throw; it is more appropriately represented by $\omega \in E_i$.

Consider "of all possible poker hands, a fraction $\frac{6}{4154}$ is full house. Therefore, the probability is 0.0014 that an arbitrary poker hand is a full house." From cases like this, some people identify the concepts of probability and statistics.

They have confused the *concept* of probability with the *values* of probability. The "therefore" approximates or at best equates the values. Yes, poker hands are random, and symmetry arguments and empirical data both give firm support for the equation of values. However, they give no support for identifying the concepts of proportion and probability, which are formally different. The formal difference lies in "a hand," which is absent in the statistical premise. *The idea of particularity, absent in statistical statements, ushers in the concepts of possibility and probability.* The capability to think of an event as an instance of something general brings in the notion of possibilities. But no matter how much we generalize, the cry "why this one in particular?" can never be extinguished. That is why temporal concepts occur so often in philosophical analyses of probability; we can only experience events one at a time. We are easily deceived by the temporal connotation of "relative frequency" to miss the fact that it is just a ratio.

Probability statements require the independent specification of both an individual ω or a group of individuals and a value x or a range of values Δ of a quality type f. They have the general form:

$\pi[\omega \in f^{-1}(x)] = r$; the probability that ω falls within the inverse image of x is r; the probability that he has income \$50,000 is r.

Conceptually, this probability π is not a relative magnitude. Usually, either the domain or the range of a function is varied independently, so that one has either $f(\omega) = x$ or $\mu[f^{-1}(x)] = r$. Probability is unusual for demanding the straddle in which both the domain and the range of a function are independently varied.

The problem of probability assessment is the reverse of the problem of statistical inference. We should distinguish statistics from statistical inference. *The concept of probability is involved not in the statistics but in the inference*, which requires the extra concept of *samples*. A sample is a particular individual or a particular group of individuals of an aggregate. It is important that the samples are random; otherwise the evaluation of probability or its reverse process cannot succeed. In statistical inference, we take a sample, determine its quality, and infer something about the pattern of the population. In probability, we figure out the statistical pattern of some perhaps hypothetical population, and use it to assess the likelihood that a sample of the population has certain qualities. Consider an example in quality control. Suppose a batch of N items rolls off the assembly line, of which a fraction μ is defective. In a probability problem, we take a sample of $n < N$ items and calculate the probability that k of the n are defective, given N and μ. However, in reality we do not know μ and it is often not feasible to examine all N items. We want to make some estimate of μ by examining only a sample, and to gauge how good the estimate is. This is a problem of statistical inference. Both problems involve individuals and statistical patterns, hence the concept of probability.

Unlike individual and statistical statements, probability statements are not verifiable. The assertion "the probability is $\frac{1}{6}$ that this throw of the die is a four" cannot be empirically tested. The outcome of the throw is either four or not, with no hint of probability. Experience can dissolve a probability question but cannot answer it. When people talk about testing a probability proposition, they have quietly converted it to a statistical proposition, which is verifiable.

Since probability statements are not verifiable, they are *not descriptive but judgmental*. Properly stated, they have the form:

We judge that $\pi[\omega \in f^{-1}(x)] = r$; I judge that the probability of my drawing a full house in this poker game is 0.00144.

We may all agree to the specific value cited, and confer on the judgment the honor of intersubjective agreement.

As James Bernoulli announced in the title of his treatise, probability is the art of conjecture. Possibility and probability are modal concepts. They are part of the conceptual scheme with which we think about things, but they are not attributes of things. In many theories, for instance, statistical mechanics, we construct abstract state spaces to reckon with the possible states of large systems, but the reckoning is always acknowledged to be theoretical. This view is akin to the Bayesian school, although I abstain from its heavy psychological overtone. The Bayesian is the most comprehensive of all conceptions of probability. It is usually spurned in the circle of physical science. Therefore, I hasten to clarify several points.

The judgmental nature of probability statements is completely conveyed in their formal structure. As Henry Kyburg remarked, probability statements are intensional.[252] "The probability is r that ω has quality x" occurs as a that-clause, "I judge that $\pi[\omega \in f^{-1}(x)] = r$." The that-clause contains the *content* of the judgment, and *the content is strictly about objective states of affairs*; the judgment is passed on objects and says nothing about the subject. Even when the judgment is about the degrees of belief, as in many decision problems, the beliefs are treated as externalized objects in psychology, not to be confused with the subjective disposition of the person who judges. That is the meaning of the stipulation that the beliefs are ideally rational. The distinction between the internal and external representations of such concepts as beliefs are clearly expounded by Thomas Nagel.[253] What makes probability judgments "personal" is the permission of various verdicts based on the same evidence and the nonsubstitutivity of those verdicts. These conditions are formal. The formal structure of the probability statement at once captures and exhausts any sense of uncertainty. The judge and his dispositions never appear in the content of the judgment. The case is similar to objective experiences. Observations are made by individuals and are personal, but the content of observations are objects.

Probability judgments are rational, and the criteria of rationality have been spelled out rigorously. The criteria are broad enough to allow different evaluations of probability for the same system. However, the values will rapidly converge if sufficient data are available, as B. de Finetti proved in the convergence of exchangeable sequences. There can be *intersubjective agreement in the*

Table A.2 Three forms of statement. In all cases, the individual ω may be replaced by a group of individuals, the number x by a Borel set Δ. The quotation marks on the probability statement indicates that it is the content of a that-clause.

Statement	Form	Fix	Ask the value of	Example
Individual	$f(\omega) = x$	ω, f	quality f	He is aged 5.
Statistical	$\mu[f^{-1}(x)] = r$	x, f	proportion μ	The portion of five year olds in the population is r.
Probability	"$\pi[\omega \in f^{-1}(x)] = r$"	ω, x, f	probability π	"The probability that he is a five year old is r."

probabilistic judgment of objective events. As far as I know, this is the most objective sense of probability that is distinct from statistics.

The forms of individual, statistical, and probability statements are summarized in Table A.2. All three forms of statements require the specification of an individual quality type f. Their subject matters vary not only in specific but in kind. Individual statements describe individuals. Statistical statements describe aggregates. Probability statements assess the relations between individual and aggregate qualities; probability is not an attribute of things, simple or composite.

The Evaluation of Probability

Relative magnitudes can be evaluated for any statistical systems. However, it is not always reasonable to judge that the value of the relative magnitude is the value of the probability pertaining to an arbitrary individual. *Randomness* and *large number*s provide rational grounds for the judgment and conditions for the reasonable evaluation of probability. They are not required for statistics, which can be compiled for the most regular or most irregular qualities. They are necessary only if the statistical results are to serve as the ground for judging single events. The law of large numbers, which depends on randomness, provides such ground.

The evaluation of probabilities is the major practical concern. That is why probability is so intimately associated with randomness. There are two major concepts that capture the notion of randomness, independence and exchangeability.[254] *Exchangeable* events are those that occur randomly in a sequence. For such events, there are theorems proving that we can establish intersubjective agreement on definite values of probability. Two events E and F are *independent* if $\mu(E \mid F) = \mu(E)$. This condition is readily generalized. A family of events $\{E_i\}$ are mutually independent if

$$\mu(E_1 \cap E_2 \cap \ldots) = \mu(E_1)\mu(E_2) \ldots \qquad (A.3)$$

Independence has no natural interpretation in terms of magnitudes; the best we can say is that it stipulates certain ad hoc ways of the overlapping of extensions. Equation (A.3) captures the intuitive idea that the events are unrelated, but the relation between independence and randomness is not exactly clear. Two measurable functions f_1 and f_2 defined on a probability space are *independent random variables* if $f_1^{-1}(\Delta_1)$ and $f_2^{-1}(\Delta_2)$ are independent events for any Borel sets Δ_1 and Δ_2. Kolmogorov said a most important problem in the philosophy of natural science is "to make precise the premise which would make it possible to regard any given real events as independent."[255] The problem has not been satisfactorily solved despite much effort.

Conceptions of Probability

Probability statements, in the general form $\pi[\omega \in f^{-1}(x)] = r$, depend only on the inverse function $f^{-1}(x)$ and *not* on the function f itself. Since a function is generally a many-to-one mapping so that many individuals can have the same value, the inverse f^{-1} is coarser grained than f. If we examine the meaning of $f^{-1}(x)$ by itself, we find that it represents some kind of classification scheme for a statistical system; it groups entities under the variable x or Δ. In the classical case discussed, the classification is made according to the physical properties of the entities. However, since the notion of probability arises precisely because individual property assignment is lacking, we can relax the dependence on properties without harming the meaning of probability. For instance, I can classify things according to how they please me, and the grouping does not explicitly invoke their properties.

The concept of probability involves the judgment about an individual in a system with a classification scheme. The concept is formal. It leaves open the kinds of individuals and classification schemes, the admissible grounds and methods of judgment, and it does not guarantee that reasonable values of probabilities can be obtained. Various conceptions of probability flesh out the formal concept by specifying various individuals, classifications, grounds of judgment, and methods of evaluation.

Let us restrict ourselves to objective states of affairs. What evidence is acceptable for probabilistic judgment? Bayesians admit nonstatistical reasons in inference. Statistical frequentists admit only statistical data and exclude other considerations such as symmetry and experience, for even if there are symmetry laws, their empirical determination is nonstatistical and indirect. They equate or approximate the values of probability by limiting frequencies and use the laws of large number to link the values of probability to observed statistical distributions. The statistical frequency conception is distinct from the more idealistic limiting frequency conception, which goes further and identifies the concepts of probability and limiting frequency.[256] The limiting frequency conception uses the law of large numbers to reduce the concept of probability to the characteristics of a narrowly circumscribed class of sequences that are allegedly observable. It denies any meaning of probability to individual cases. Two common criticisms of it are that it completely distorts the meaning of probability, and it lacks empirical significance, for it is based on infinite sequences and experiments are always finite.[257]

The probability calculus can be used to compute the statistics of rational beliefs or of physical objects, and it can be used to justify conjectures on their particular occurrences. The results may be called epistemic or ontic probability. This sense is innocuous.

There are fallacious senses of ontic and epistemic probabilities. Ontic probability is sometimes said to be some objective property of individuals, such as their tendency or propensity to be something. Such properties are either illusory or irrelevant to probability. If an individual has a certain property, it would be expressed in individual statements. The potential energy of particles is expressed by a function f. Less mathematically, dispositional properties such as solubility are expressed as predicates of individuals. These properties are not probability, and they are not statistics either. If no such properties are described by a theory, then we have no right to posit them, even if the theory employs the concept of probability. Probability signifies a gap in substantive concepts. A physical theory that invokes probability does provoke the feeling of something missing, but the alternative may involve a drastic overhaul of the theory. Before that is available, probability should be left alone and not covered up by occult forces.

Some people say that we make probability judgments without complete knowledge; therefore, the resultant probability is a measure of our ignorance. Such probability is called epistemic.[258] The argument for it confuses the nature of a proposition with the content of the proposition and wrongly pulls the condition or disposition of the judge into the content of the judgment. Suppose we use a simplified model to calculate the moon's trajectory; we get an approximate solution, not a measure of approximations. We can proceed to estimate the errors, but that is a different problem. Physical theories involve many idealizations, but they are not theories of ideals. Similarly, a proposition made in partial ignorance is not a proposition about ignorance.

Determinism and its negation are irrelevant to the concepts of probability or statistics. There may or may not be deterministic laws of nature; we make judgments all the same. Chance is a vague notion. It seems to be more weighty than possibility. Possibility is formal and signifies only the idea of one in many. Chance pertains more to the substantive description of the world; it attempts to tell how the one is chosen. The theory of probability does not answer this question. In the enumeration of possibilities, as in the state space calculation of many physical theories, the possibilities are acknowledged to be theoretical. The novelty of statistics and the probability calculus that emerged in the late eighteenth century lies in the formulation of abstract and complicated aggregates that opens our minds to the vast range of possibilities and enables us to grasp the aggregative world as a unity. The intellectual grasp engenders the feeling of power and control, but it is misleading to say that chance or its overcoming is inherent in the concepts of probability or statistics.

Appendix B

Fiber Bundle
and Interaction Dynamics

The terms "fiber" and "fiber space" first appeared in 1932. However, the idea of associating some mathematical object or variable space to every point x of a space M had been around for some time in differential geometry. The collection of tangent vectors at each point is such a construction, which constitutes what we now call the tangent bundle. Another important source is the theory of moving frames, which expresses mathematical quantities with respect to a frame that is free to rotate from point to point rather than some fixed coordinates. In the 1920s, É. Cartan applied the idea of moving orthonormal basis to Riemannian geometry and general relativity. His theory generally does not include a global isometry group acting on the manifold. Instead, the orthogonal group acts on the orthonormal basis $\{e^i(x)\}$ defined on the tangent space over the point x. The Levi-Civita connection provides a path-dependent linear mapping among the $\{e^i(x)\}$ at different points. Today we call such theories the bundle of orthonormal frames.

In the 1940s, the fruitfulness of fiber spaces in many areas of algebraic topology and differential geometry became evident. In the 1960s, some physicists including E. Lubkin and A. Trautman recognized that interaction potentials can be represented by connections on principal fiber bundles. In 1975, T. T. Wu and C. N. Yang used the fiber bundle method to solve a long-standing problem on magnetic monopoles. Their paper first introduced the power of the method to the community of physicists at large.

Both general relativity and quantum field theory can be formulated in terms of fiber bundles. The fiber bundle is also used intensively in modern formulations of classical mechanics. However, there the bundles do not have a local symmetry group, which is a central feature of field theories.[259] Thus the fiber bundle provides a unified conceptual framework for all our physical theories. The idea of fiber bundles is general and applicable to many areas. I was tempted to construct a model of Monadology using it.

The fiber bundle that finds application in field theories is a mathematical structure (D, M, G, π) with four major elements. The *structure group G* becomes the local symmetry group in field theories. The *total space D* comprises discrete *fibers* $\psi(x)$, each of which is either a copy or a representation of the structure group G. The *projection map* π sends each fiber $\psi(x)$ into a unique point x in the *base space M*. D, M, G, and $\psi(x)$ are all differentiable manifolds. The coordinate covering of M supports the spatio-temporal group in field theories. The potential of the interaction field is represented by the connection on D.

Two views of the fiber bundle will be presented in the following. The first starts from M and G and constructs D as a patchwork of local products (coordinate bundles). The second starts from D and G and derives M as a quotient space (principal fiber bundle). The two definitions of fiber bundles are equivalent. The difference in emphasis exhibits the physical and philosophical significance.

Coordinate Bundles[260]

I will only illustrate the main ideas of the first approach with an example. Let us construct a fiber bundle, the Möbius strip illustrated in Fig. B.1, from the following elements:

1. A circle M with a family of coordinate coverings $\{U_\alpha\}$. Each family contains at least two coordinate patches, which is the minimum required to cover a circle.

2. A line segment, which serves as a typical fiber ψ. A line segment is an instance of a vector space. Generally, fiber bundles whose typical fibers are vectors spaces are called *vector bundles*. The matter field in field theories is represented by a vector bundle.

3. A structure group G acting on the fibers. G consists two elements (e, r), where e is the identity element and r flips a fiber around the midpoint. (If G consists only the identity element, we would get a cylindrical strip instead.)

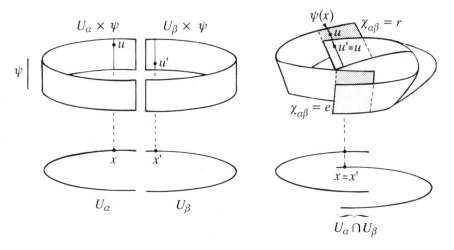

Figure B.1 Left: Two coordinate patches U_α and U_β cover a circle. The two strips are the product spaces of the patches and the typical fiber ψ. A point θ in the strip is the order pair (x, u). Right: The two patches overlap at two disjoint arcs. In one arc the transition function $\chi_{\alpha\beta}$ is the identity element e of the structure group. In the other it is the element that flips a fiber. Points in the two strips, such as (x, u) and (x', u'), are related by $\chi_{\alpha\beta}$ and are identified. They are "sewn together" to yield the Möbius strip.

The construction of the fiber bundle hinges on two ideas: that of a *local product space*, and that of *identifying points according to group-theoretic equivalence relations*. It proceeds as follows:

CB1. We form the Cartesian product of $U_\alpha \times \psi$ of a coordinate patch and the typical fiber. A point θ in $U_\alpha \times \psi$ is the ordered pair (x, u), $x \in U_\alpha$ and $u \in \psi$. This step highlights a defining characteristic of fiber bundles: They are locally product spaces.

CB2. We take the disjoint union of all product spaces so obtained.

CB3. The resultant union is too large for the Möbius strip; it contains redundant points because the coordinate patches overlap. To identify the points we introduce an equivalence relation. Two points θ and θ' are equivalent if and only if $x = x' \in U_\alpha \cap U_\beta$ and $u' = \chi_{\alpha\beta} u$, where the transition function $\chi_{\alpha\beta}$ corresponds to an element of G. The identification essentially "sews" the overlapping patches together. In one of the overlapping regions, $\chi_{\alpha\beta} = e$, and the fibers are sewn to each other as they are. In the other overlapping region, $\chi_{\alpha\beta} = r$, and the fiber $\psi(x)$ is identified with the fiber $\psi(x')$ flipped upside down. The result is a Möbius strip.

CB4. The total space D is defined as the quotient of the disjoint union in CB2 by the equivalence relation defined in CB3.

CB5. The map that takes the equivalent class of points (x, u) in D into a point x in M is the projection map π.

The identification of points in CB3 highlights the function of the two symmetry groups in field theories. The identification of the points x and x' in the base space is achieved by the usual coordinate transformations of differentiable manifolds (§ 5). It corresponds to the action of the spatio-temporal group in field theories. The identification of the points u and u' in the vector space is the action of the local symmetry group. The transition function is the coordinate transformation function of the total space.

In field theories, a fiber $\psi(x)$ is interpreted as an event or an entity in the field and a point θ in the fiber as a possible characteristic of the event. The concept of product spaces reveals that the characteristic of the event $\psi(x)$ has two "dimensions": a qualitative dimension encompassing all possible phases and represented by the variable u, and a spatio-temporal dimension represented by the fixed parameter x. Thus when we talk about the features of a distinctive event, we have already used the spatio-temporal concept.

Principal Fiber Bundles[261]

In principal fiber bundles, the typical fiber is the structure group itself. In field theories, the potential of the interaction field is represented by the connection of a principal fiber bundle. The principal fiber bundle (D, M, G, π) has the following properties:

PB1. An element g of G acts on a point θ in D by mapping θ to another point θg in D. *The action of G on D is free*; that is, the only group element that maps any point in D into itself is the identity element e, so that the group leaves no element in D fixed.

PB2. The action of G on D induces an *equivalence relation* $\sim$ among those points in D that can be mapped into each other by a group element. For $\theta \in D$ and $g \in G$, $\theta \sim \theta'$ if $\theta g = \theta'$. $\sim$ is an equivalence relation because it is reflexive, $\theta = \theta e$; symmetric, $\theta = \theta' g^{-1}$; and transitive, if $\theta' = \theta g$ and $\theta'' = \theta' g'$, then $\theta'' = \theta g g'$. The identity e, the inverse g^{-1}, and the product gg' are all in G according to the group axioms.

PB3. The equivalence relation $\sim$ *partitions D into G-orbits*. For θ in D, the action of all elements $\theta \in G$ picks out an equivalence class of elements $G(\theta) = \{\theta g : g \in G\}$, called the G-orbit through θ. Another point ζ in D determines another G-orbit $G(\zeta)$. The process can be repeated until all elements in D are grouped into G-orbits. The structures of the G-orbits are identical; each is diffeomorphic to G. Any two G-orbits are either identical or disjoint; they are identical if all their elements are the same, disjoint if they share no common elements. The set of G-orbits constitutes a partition of the total space D.

PB4. The partition of D leads to a *quotient space D/G*. Since the G-orbits are all disjoint, we can define a projection map π that sends each G-orbit in D into a point x in the space D/G. With the assignment of x, the G-orbit becomes a *fiber $\psi(x)$*. The inverse projection map $\pi^{-1}(x)$ of $x \in D/G$ is the fiber $\psi(x)$ above x, and x is the index of the fiber $\psi(x)$. The space D/G is the quotient space of D by the equivalence relation $\sim$ induced by G. It is specifically named the *base space* $M = D/G$.

PB5. D is locally a *product*. Around each point $x \in M$ there is a local coordinate patch U_α and a diffeomorphism χ_α that sends the segment of total space above U_α into the product space $U_\alpha \times G$. χ_α is called the *trivialization map*. It unwraps the twists of the total space locally and shows that locally, D is a product. The transformation between the product spaces is the *transition functions $\chi_{\alpha\beta} = \chi_\beta \cdot \chi_\alpha^{-1}$*, (Fig. B.2).

The trivialization map makes possible the coordinatization of the segment $\pi^{-1}(U_\alpha)$ of the total space above the coordinate patch U_α. However, it does *not* introduce a coordinate system in $\pi^{-1}(U_\alpha)$. It establishes a one-to-one relation between the elements of a fiber and the structure group G. However, it does not establish a "horizontal" baseline for us to decide if points in two *separate* fibers correspond to the same element of G. This is a main difference between a fiber space and an ordinary Cartesian product space. The Cartesian space is like a broad staircase with vertical handrails; the fiber space is like a set of identical escalators moving up and down independently. The individuality of the fibers underlies the concept of point coupling in field theories. For the independence

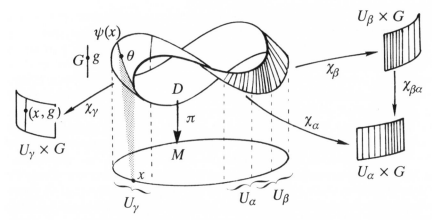

Figure B.2 A principal fiber bundle (D, M, G, π). The total space D is represented by the Möbius strip and the fibers by lines isomorphic to the structure group G. The projection map π sends the fiber $\psi(x)$ into $x \in M$. The base space M is covered by coordinate patches. The local trivialization χ_γ untwists the fibers and reveals the total space as a local product $U_\gamma \times G$. The transition function $\chi_{\alpha\beta}(x)$ is the coordinate transformation between local trivializations, and it depends only on x.

of the fibers means there is no universal class of coordinate systems applicable to all fibers. Whatever coordinate system we pick for one fiber is a convention for that fiber alone, and it is changed by some *dynamical* means as we go from fiber to fiber.

Section, Connection, and Parallel Transport[262]

A *section* σ_α: $U_\alpha \to \pi^{-1}(U_\alpha)$ maps a segment of the base space into the total space. It introduces a local coordinate system for the portion of the total space above the coordinate patch U_α by designating a definite element in each fiber as the identity element. If a single section can be introduced for the whole fiber bundle, the bundle is called trivial. A cylindrical strip is a trivial bundle but a Möbius strip is not.

While a section maps the base space M into the total space D, a *principal connection* Γ maps the tangent spaces over the base space into the tangent spaces over the total space of the fiber bundle. More precisely, a connection maps the tangent space $T_x(M)$ over a point x in M into the tangent space $T_\theta(D)$ over a point θ in the fiber $\psi(x)$ above x. Since D is much larger than M, the dimension of $T_\theta(D)$ is larger than that of $T_x(M)$. By mapping $T_x(M)$ into $T_\theta(D)$, the connection picks out the part of $T_\theta(D)$ that is isomorphic to $T_x(M)$; this part is called the horizontal subspace. Thus the *connection* Γ is a rule that splits the tangent $T_\theta(D)$ at each point θ into a *horizontal subspace* and a *vertical subspace*. The definition of the connection is independent of coordinates. The horizontal space is invariant to the actions of the structure group; all horizontal spaces in the same fiber project isomorphically into the same $T_x(M)$.

The connection enables us to decompose any vector on the total space into a horizontal part along the base space and a vertical part along the fiber.

As in the case of a differentiable manifold, there is no general correlation among various points. Correlations must be established along a curve. Given a connection, a curve γ in M can be *horizontally lifted* into a curve $\bar{\gamma}$ in D such that all vectors tangent to $\bar{\gamma}$ are horizontal; that is, the tangent vectors of $\bar{\gamma}$ lie in the horizontal spaces at all points it passes. The horizontal lift is unique. Through any point θ in the total space above the curve γ, there passes only one horizontal lift $\bar{\gamma}$. The uniqueness of γ provides the way to define the *parallel displacement* of fibers along any curve in M. Suppose γ passes through x and x' in M and the horizontal lift $\bar{\gamma}$ that passes through the fiber $\psi(x)$ at the point θ passes through $\psi(x')$ at ζ' (Fig. 7.3b). The uniqueness of $\bar{\gamma}$ allows us to identify θ and ζ'. By varying θ, we obtain a mapping of the fiber $\psi(x)$ onto the fiber $\psi(x')$.

Interaction Dynamics in the Fiber Bundle Formulation[263]

A *quantum-dynamical system* comprises a matter field that couples at every point to an interaction field. For instance, an electron field coupled to an electromagnetic field. Mathematically, the system comprises a principal fiber bundle (P, M, G, π) that accounts for the interaction field and its associated vector bundle (D, M, G, π_D) which represents the matter field (Fig. B.3). Here D and P are the total spaces and π and π_D the projection maps of the respective bundles. The "associated" signifies that the vector bundle shares the base space M and the local symmetry group G of the principal bundle. The typical fiber of the principal bundle is G itself, and the typical fiber of the associated vector bundle is a vector space representation of G. G is often called the gauge group of the field theory.

The wavefunction of the matter field is represented by a section θ of the vector bundle. A fiber $\psi(x)$ in the vector bundle represents a point of the matter field, and the points $\theta(x)$ and $\theta'(x)$ in the fiber characterize its possible phases. The actual phase that occurs at the point $\psi(x)$ depends on its coupling with the potential of the interaction field. The interaction potential, for instance, the electromagnetic potential, is represented by the connection over the fiber $\phi(x)$ in the principal fiber bundle. The two fibers $\psi(x)$ and $\phi(x)$ share an identity x, and together they represent an *event* in the dynamical system, which includes the idea of point interaction. The system of identities M constitutes the parameter space of the field theory, and is usually called space–time.

Consider a timelike curve γ in M. The curve picks out a certain spatio-temporal segment of the dynamical system, and we are interested in the dynamical variation as we move in this segment.

All possible internal states of the dynamical system are contained in the total spaces. A change occurs when we move from one point to another in the total spaces. The change can be qualitative or spatio-temporal or both. We want to separate the two and see how they are correlated. We assume that the portions of the total spaces above γ admit a section. A section maps γ into $\hat{\gamma}$ in the total

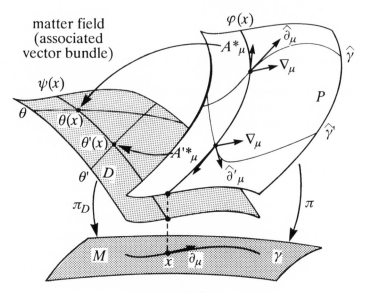

interaction field
(principal fiber bundle)

matter field
(associated
vector bundle)

$\varphi(x)$

A^*_μ

$\hat{\partial}_\mu$

$\hat{\gamma}$

∇_μ

P

$\psi(x)$

θ

$\theta(x)$

$\theta'(x)$

A'^*_μ

∇_μ

$\hat{\gamma}'$

θ'

D

$\hat{\partial}'_\mu$

π_D

π

M

x ∂_μ

γ

space-time
(shared base space of the two bundles)

Figure B.3 A dynamical system of interacting fields in the fiber bundle formulation. We study the variations of the system along a timelike curve γ in M. Only the portions of the total spaces above the curve γ are shown. They are marked P for the principal fiber bundle accounting for the interaction field, and D for the associated vector bundle representing the matter field. A section θ in the vector bundle represents the wavefunction of the matter field with specific phase factors. $\hat{\gamma}$, a section of γ in the principal fiber bundle, introduces a coordination in P. The connection Γ at a point in P decomposes the derivative $\hat{\partial}_\mu$ of $\hat{\gamma}$ into the covariant derivative ∇_μ and the fundamental vector A^*_μ. All three are components in a given local coordinate system. ∇_μ registers changes due to spatio-temporal variation. A^*_μ is directly related to the component of the interaction potential, A_μ. The potential determines the phase of the matter field by picking out a point $\theta(x)$ in the associated vector bundle. Thus the two fields are coupled at the point x. The point coupling is highlighed by the fact that the two fibers $\phi(x)$ and $\psi(x)$ share one identity, x. The system of identities M constitutes a parameter space usually called space–time. The section $\hat{\gamma}$ is coordinate dependent. A change of coordinate to $\hat{\gamma}'$ changes the potential to A'_μ. This in turn picks out a different phase $\theta'(x)$ for the matter field. The whole system is invariant under coordinate transformations.

space of the principal fiber bundle. The partial derivative of the section $\hat{\gamma}$ is given by $\hat{\partial}_\mu$, which measures the total change of the system along γ. Given a connection on the principal bundle, the derivative can be decomposed into two parts,

$$\hat{\partial}_\mu = \nabla_\mu + A^*_\mu \qquad (B.1)$$

The covariant derivative ∇_μ of $\hat{\gamma}$ is the horizontal lift of the partial derivative ∂_μ of γ. It measures the changes due to spatio-temporal variation by tracking the directional variation of γ in M. The difference between the covariant derivative ∇_μ and the partial derivative $\hat{\partial}_\mu$ is called the fundamental vector A^*_μ. It determines the dynamical variation of the system and is directly related to the interaction potential A_μ. The fundamental vector A^*_μ uniquely determines a point $\theta(x)$ in the vector·bundle. $\theta(x)$ specifies the phase of the matter field. Thus the matter and interaction field is coupled at the point x.

The decomposition (B.1) is particular to the section $\hat{\gamma}$. If we pick another section $\hat{\gamma}'$, we get a different fundamental vector $A^{*\prime}_\mu$ and a different phase $\theta'(x)$. However, various sections can be transformed into each other according to the symmetry group G. Local symmetry transforms $\hat{\gamma}$ and θ simultaneously, so that the interaction potential and the phase of the matter field are transformed simultaneously, leaving the dynamical system invariant.

We can move to a nearby point x' in γ and repeat the whole argument. The general rule of the connection allows us to compare the fundamental vector and the potential at x and x'. The presence of the matter field and the variation of its phases induce a variation in the potentials, hence a variation in the specific values of the connections. The changing values of the connection lead to a nonzero curvature, which is the exterior derivative of the connection. Physically, the curvature in the principal fiber bundle represents the intensity of the interaction field.

Tetrads, Orthonormal Frame Bundles, and General Relativity[264]

Gravity, like the other three fundamental interactions, can be represented by the connection on a principal fiber bundle, the bundle of orthonormal frames. In general relativity, there is no phase factor, which is a quantum characteristic. What corresponds to the phase and what the gravitation potential couples to can be interpreted as the orientations and deployments of our measuring equipments.

The tangent space T_x over each point x in the spatio-temporal manifold M is a vector space. The union of all tangent spaces constitutes a fiber bundle called the tangent bundle. The tangent bundle underlies the Lagrangian formulation of classical-dynamical systems. However, in these applications, it does not have a local symmetry group.

In general relativity, the tangent space T_x over each point x in M is equipped with orthonormal reference frames called *tetrads*, $\{e_i(x)\}$. The tetrads are directly introduced by mapping the natural basis of $\mathbb{R}^4$, $(1, 0, 0, 0), \ldots, (0, 0, 0, 1)$ into T_x. Their orthonormality is expressed by the inner product

$(e_i, e_j) = \eta_{ij} = \mathrm{diag}(-1, 1, 1, 1)$. Thus the introduction of a system of tetrads naturally induces a metric for the manifold. Each tetrad transforms according to the Lorentz or Poincaré group, depending on whether one wants to include spin and torsion. These groups are the local symmetry groups for general relativity and are often called tangent space groups.

The tetrads $\{e_i(x)\}$ are often called nonholonomic or noncoordinate bases to distinguish them from the coordinates $\{x^\mu\}$ defined in the coordinate patches around the point x. The orientations of the tetrads correspond to the phases in the quantum fields, and like the phases, they vary with x. Physically, a specific tetrad represents a particular experimental configuration. The time basis, e_0, is aligned with the tangent vector of the experiment's world line. The other three bases, e_1, e_2, e_3, characterize the spatial orientation of the arrangement. Conceptually, our experimental arrangements are infinitesimal.

The collection of all tetrads at all points in M constitutes a principal fiber bundle, the bundle of orthonormal frames. The gravitational potential is represented by the connection on the bundle of orthonormal frames.

We know from special relativity that all experiments connected by the transformations in the Poincaré group yield the same result. The same holds in general relativity, only that the experiments are now confined to those that occurs at a single point x. They are represented by the tetrads on T_x that are invariant under the Poincaré transformations. Since the tetrads at different points rotate independently, this means that the same matter distribution will yield different results depending on the choice of local experimental configurations. This is clearly unacceptable. To ensure that the local Poincaré transformations yield equivalent measurements, the gravitation potential is introduced.

The orientation of a tetrad axis can be expressed by its components with respect to a coordinate $\{x^\mu\}$ at x; $e_i(x) = e_i^\mu(x)(\partial/\partial x^\mu)$. In a gravitational field, the actual rotations of the tetrads are derived by replacing the partial derivative by the covariant derivation,

$$\partial_\mu \to \nabla_\mu = \partial_\mu + \Gamma_\mu^{ij}(x)A_{ij}$$

where A_{ij} is the generator of the Lorentz group. $\Gamma_\mu^{ij}(x)$ is the coefficient of the affine connection. It represents the gravitational potential that reconciles the orientations of the tetrads. Going from x to $x + dx$, we find the tetrads have rotated by $e_i\Gamma_\mu^{ij}dx^\mu$.

Appendix C

The Cosmic and the Microscopic: An Application

A most fascinating application of quantum field theory is in cosmology, where it helps to explain the evolution of the early universe. Cosmological problems are out of the scope of this book. I briefly and loosely describe the meeting of the cosmic and the microscopic as an afterword.[265]

Man leaves the city lights behind and drives into the desert. There he lifts his eyes and looks into the depth of time. He sees in the constellation Andromeda a faint patch that is the spiral galaxy closest to us, our neighbor two million light years away. The light he sees left the Andromeda Galaxy when his ancestor, *Homo*, first walked on earth. Man is not satisfied with what he sees with his naked eye. He builds telescopes, radio antennas, and other detectors to scan the firmament in various frequencies of the electromagnetic spectrum. Radio astronomy lifts the sky's veil of tranquility and reveals it to be a place of violent activities: galactic collisions, supernova explosions, cosmic jets bursting with energies equivalent to converting the mass of a million suns. The heaven reveals itself to be vast and mighty, but blind and indifferent. It has absolutely no regard for Man's hopes and longings, and will in its own time reduce him and all his achievements to void.

Despite the truth that scalds like molten lead, Man's mind rises to embrace the wheel of fire. He mobilizes the physical laws he has formulated and verified in terrestrial laboratories and assumes the laws are the same anywhere in the universe, because the universe is homogeneous and isotropic on the large scale. This assumption he calls the cosmological principle. Thus armed with laws and astrophysical data he collected with instruments designed according to these laws, he constructs theories about the universe and tests them with further observations. The story of the universe begins to unfold.

Historically, Edwin Hubble found in 1929 that the speed with which a galaxy moves away from us is proportional to its distance from us. Since the cosmological principle demands that the universe should look the same to observers in all galaxies, Hubble's observation points to an expanding universe. The expansion implies that the universe was smaller in the past. Using the measured rate of expansion, physicists estimated the time when everything was close together. The birthday of the universe was approximately sixteen billion years ago. The Hubble expansion provides the first leg of the observational triad supporting the big bang model. It is based on direct observation and simple physics—the conventional type of astronomy. The other two legs of the triad are the three-degree cosmic microwave background radiation and the abundance of relic elements in the universe. There the astronomer's instruments, telescopes, antennas, and the like function more like archaeologist's spades, uncovering the relics of ages long gone and probing deep into the universe's past (Fig. C.1).

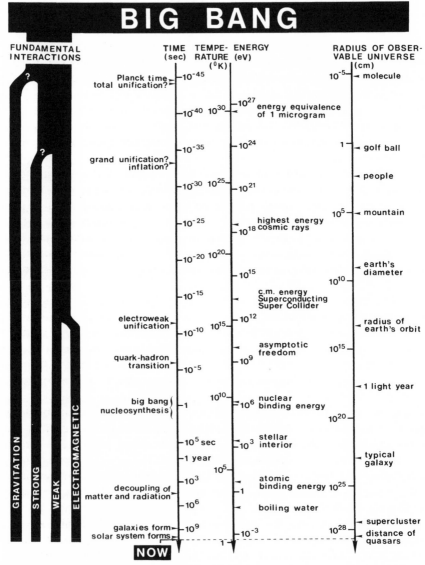

Figure C.1 The early history of the universe. Born in perfect symmetry in a primeval fireball, the universe cooled and expanded, and symmetries were broken as the four fundamental interactions were separated.

Quasars, those brilliant and supermassive objects at the centers of galaxies, sit at the edge of the structured universe. With quasars we look back more than three-quarters of the way into the life span of the universe, and that is almost all that we can directly see. As we go further back in time the universe became smaller, denser, and hotter. Around its 500,000th birthday, the universe was a cloud of atoms bathed in a bright yellow light. Its temperature was around ten thousand degrees Kelvin, slightly higher than the surface temperature of the sun; this temperature corresponds to an energy of about one electron volt. It was the time when stable atoms first formed and light decoupled from matter, henceforth to travel freely and to cool. When light "cools" its wavelength becomes longer. Today, the yellow light has cooled to slightly below three degrees Kelvin and was detected in 1964 by Arno Penzias and Robert Wilson as the cosmic microwave background radiation. The spectrum of the three-degree background radiation agrees with the prediction of the big bang theory. It confirms to better than one part in ten thousand the homogeneity and isotropy of the universe. Recently, faint ripples are detected in the microwave background, suggesting the first formation of the large-scale structures of the universe.

The idea of the "quantum ladder" will help as we travel further back in time. In studying atomic and subatomic properties we find there are widely separated energy levels exhibiting distinct and unique phenomena. Each level can be seen as a rung of the quantum ladder. Three rungs are now firmly established— atomic, nuclear, and subnuclear. Enormous energy gaps separate the rungs. Within each energy range there is a distinct set of stable particles with their characteristic interactions. Each of these rungs also corresponds to a distinct period in the evolution of the early universe.[266]

We live in the atomic rung of the quantum ladder; stable atoms constitute our part of the universe in our epoch. The characteristic energy of the atomic rung is a few electron volts, or equivalently, slightly below a hundred thousand degrees centigrade. Atoms can remain stable only below this energy. As we trace the history of the universe back from its 500,000th birthday, we find its temperature soon rose above the atomic binding energy, and atoms were stripped of their electrons. The ingredients of the universe became electrons, nuclei, and photons. The charged particles coupled strongly to the photons, and a veil dropped over the universe. Light could get nowhere because it was constantly scattered by electrons; the universe became opaque. The opaque universe was in perfect thermal equilibrium, containing a uniform hot soup of particles and radiation constantly colliding with each other. Scars associated with the pain of its violent birth, if there were any, were soothed and erased by the thermal equilibrium.

Traveling back in time, we find nothing much happened for 500,000 years except the temperature of the universe had increased a hundred thousand fold. Finally, when it was about thirty minutes old, the universe reached the nuclear rung of the quantum ladder. At temperatures around a billion degrees the nuclei became unstable; they would soon be torn apart into protons and neutrons. The big bang nucleosynthesis started when the universe was about three minutes old. For thirty minutes or so the universe was like a gigantic hydrogen bomb in which neutrons and protons were fused into stable nuclei. Physicists draw on known nuclear reaction cross sections and predict that the chemical ingredients of the early universe should be roughly three-quarters hydrogen and one-quarter helium, with traces of deuterium and lithium. These predictions are borne out by the observed composition of first-generation stars. The relative abundance of the remnant elements is the deepest probe we now have into the early universe. It is the third major evidence for the validity of the hot big bang model.

What happened before nucleosynthesis? We have not yet found and identified relics of the first seconds of the universe; we have to rely on theories. Astrophysics and nuclear physics guide us up to this point. Beyond lies the regime of elementary particles, and the advancement of particle physics during the past two decades has opened a window on the early universe. Based on experimentally confirmed relativistic quantum field theories, we infer that the universe reached the subnuclear rung of the quantum ladder when it was a millionth of a second old. At temperatures above a billion electron volts the strong interaction becomes weak, and protons and neutrons decompose into free quarks. We can extrapolate back to within one ten-billionth of a second of the big bang, when the entire observable universe could fit into the earth's orbit, and its temperature was a hundred billion electron volts. A little before that instance the symmetry of the electroweak interaction broke; the electromagnetic and weak "froze out" into two distinct interactions.

Before that? Physicists believe nature exhibits higher symmetries and physical laws are simpler as energy increases. There are grand unification theories, or GUTs, conjecturing the unification of the strong interaction with the electroweak at energies ten trillion times above the electroweak unification. The grand unification occurred when the whole universe was the size of a baseball. Other theories extrapolate to energies a million times beyond the GUT level. Such high energies correspond to distances so minute that even the extremely weak gravitational interaction becomes significant. These theories attract much attention; they have predicted the existence of many exotic particles, but none of their results can be checked experimentally yet.

The grandly extrapolating theories are not unbridled flights of fancy. They address real questions. Magnificent as they are, the standard models in particle physics and cosmology contain many problems. Their solutions may also shed light on mysteries in other branches of physics. For instance, an intriguing puzzle in astronomy is the "case of the missing mass." Galaxies and clusters of galaxies are held together by gravitational force, whose strength is proportional to the masses of the interactants. Physicists find that the mass of the galaxies must be larger than is apparent for gravity to be strong enough to hold them together. Over the past few years all evidence points to the proposition that most of the matter in the universe is dark, that is, invisible to existing telescopes and detecting devices. The luminous part of the galaxies accounts for less than ten percent of their actual mass. There is also evidence that the dark matter is not ordinary protons or neutrons.[267] What can this missing mass be? Can it be some of the exotic particles predicted by the GUTs?

A very unsatisfactory feature of the big bang model is its sensitivity to initial conditions. To obtain the solution corresponding to the present universe, the initial state from which the universe evolved must be specified to incredible precision. Can we explain the initial conditions in a more fundamental way? This is one question addressed by the inflationary model of the universe, which is closely tied to the grand unification theories. The inflationary model and its extensions conjecture that all the matter and energy in our universe emerged from almost nothing. Originating as a quantum fluctuation of the vacuum, our universe expanded more than a trillion trillion trillion trillion fold to the size of a golf ball in almost no time. At the same moment the strong interaction was frozen out. I say "our universe" because in the inflationary model, our universe is only a small part of the entire universe. It is only one bubble in the sea of foams that is the quantum fluctuation of the vacuum. The idea is speculative, but many problems in the standard big bang model are "inflated away."[268]

Another puzzle associated with the initial conditions is the imbalance between matter and antimatter in the universe. We never encounter antiparticles outside laboratories, not even in cosmic rays that come from all over the universe. The imbalance is fortunate for us because everything would be annihilated into a sheet of light if there is a little more antimatter to offset the excess matter. But while physicists thank their stars they cannot help asking, why does nature favor matter? The physical laws we know do not prefer matter to antimatter. Did the universe happen to start with the asymmetry? Or maybe it contains segregated regions of matter and antimatter? The GUTs have an explanation for the imbalance in a fundamental way. The universe was born with equal amounts of matter and antimatter, but violent collisions in very early times led to an excess of matter. This imbalance was locked into the universe when the symmetry broke around the time of the grand unification.[269] The GUTs are not idle plays but have great explanatory power. The theories are intellectually attractive because of their beauty and simplicity; their only trouble is the lack of empirical verification.

The kind of high energy required to verify the grand unified theories is impossible to achieve with our present technology. The superconducting supercollider, the largest accelerator in the world for the coming decades, if it were built, would achieve a maximum energy of 40 trillion electron volts. This is still thirty billion times short of the energy scale of the GUTs. Even an accelerator built around the globe would be far from adequate. Some low-energy manifestations of the grand unification are predicted, notably proton decay. These experiments have been eagerly tried. Unfortunately, they, too, run into sensitivity limitations. We may be running out of artificial means to achieve very high energies, but there is still the universe. Physics had begun by studying the heavens, then physicists invented the terrestrial laboratory. Now some physicists are back, realizing they have to rely more on the heat of the very early universe to provide evidence for their theories. The remnant elements and the three-degree microwave background radiation are relics of the big bang in its later stages. Physicists believe that the present universe should also contain relics of very early times when the temperature of the universe was near or above the GUT energies. They hope that the "missing mass" of the universe will turn out to be some of those exotic relics, monopoles, photinos, axions, strings, or others.

Particle physicists aim at something more fundamental than the evolution of the universe. They believe that a set of simple and beautiful principles underlies all physical processes, and they want to know it. The universe is but one instance of the principles. There are still many questions in the standard model of particle physics. The model requires 19 dimensionless parameters to be determined empirically. This is regrettable because pure empirical inputs are accidental at the fundamental level; an ideal fundamental theory should not contain arbitrary dimensionless parameters. The model cannot account for the origin of the particle masses; it fails to explain the relative strengths of the various interactions; it leaves as a puzzle the existence of several generations of particles; it includes only three interactions, leaving gravity in the cold. There are too many contingencies in the theory. Physicists know their theories are not perfect, and they believe the arbitrariness can be eliminated in a completely unified theory. With limitations on our experimental capabilities and various socio-economical constraints, it may turn out that we are unable to find the unifying principle even if there is one. But then it will not be due to the lack of vision or effort.

After describing the evolution of the universe, Weinberg reflected that it is hard to realize that our cozy little world is just a tiny part of an overwhelming hostile universe. "It is even harder to realize that this present universe has evolved from an unspeakably unfamiliar early condition, and faces a future extinction of endless cold or intolerable heat. The more the universe seems comprehensible, the more it also seems pointless. . . . "

"But crushing truths perish from being acknowledged." "And human life begins on the far side of despair." "In spite of Death, the mark and seal of the parental control, Man is yet free, during his brief years, to examine, to criticize, to know, and in imagination to create." And Weinberg continued, "if there is no solace in the fruits of our research, there is at least some consolation in the research itself. . . . The effort to understand the universe is one of the very few things that lifts human life a little above the level of farce, and gives it some of the grace of tragedy."[270]

The more the universe seems pointless, the heavier the weight of meaning rests on us. Opening to the starry heavens above awakes within us the full consciousness of our fatal freedom, at once liberating and binding. Kant wrote in his analytic of the dynamically sublime: "In the immensity of nature we find our own limitation. . . . At the same time in our rational faculty we find a different, nonsensuous standard, which has that immensity or infinity itself under it as a unit. In comparison with this standard everything in nature is small. Thus in our mind we find a superiority to nature even in its immensity."[271]

NOTES

Chapter 1

1. Since its introduction in the theory of electromagnetism, the concept of fields revolutionizes what we deem as the basic ontology of the world. However, it has received little philosophical attention. Stein (1970) surveys the notion of fields in classical physics. There are some philosophical papers on quantum field theory, but the number is negligible compared with the papers on quantum mechanics or relativity. The only anthology I know is Brown and Harré (1988), whose editors said: "such issues (the history and foundation of quantum field theory) have to date enjoyed far less attention than they deserve."

2. Feynman (1982), p. 471. Also, Murray Gell-Mann said: "We suppose that it (quantum mechanics) is exactly correct. Nobody understands it, but we all know how to use it and how to apply it to problems, and so we have learned to live with the fact that nobody can understand it," ("The Search for Unity in Particle Physics," 1977, quoted in B. Cohen, *The Newtonian Revolution*, Cambridge University Press, p. 147).

3. Bohr said: "There is no quantum world. There is only an abstract quantum physical description. It is wrong to think that the task of physics is to find out how nature *is*. Physics concerns what we can say about nature" (quoted by A. Petersen, Petersen, 1963). Heisenberg said: "The conception of the objective reality of the elementary particles has thus evaporated" (1958, p. 100). An antirealist interpretation of quantum mechanics, which avoids many pitfalls of positivism, is recently offered by van Fraassen (1991).

4. Einstein, Letter to J. Rothstein, May 22, 1950, in Fine (1986), p. 87; Penrose (1987), p. 106, his italic; Wigner (1971), p. 4. Many physicists do not consider indeterminism the central issue of quantum interpretations. In the same references Penrose said: "I do not regard *indeterminacy*, in the ordinary sense of that word, as being necessarily objectionable." Wigner said: "The principal difficulty is not the statistical nature of the regularities postulated by quantum mechanics." Einstein thought similarly. His quip that "God does not throw dice" has become famous, especially in the more popular literature. This tends to obscure the real issue. Einstein's major concern is objective reality, not determinism. He complained: "The Talmudic philosopher sniffs at 'reality' as at a frightening creature of the naive mind" (letter to Schrödinger, June 19, 1935). In the height of his argumentative correspondence with Born, Pauli visited him and afterward explained to Born: "It is misleading to bring the concept of determinism into the dispute with Einstein." Pauli reported that Einstein denied energetically he had ever postulated that the sequence of events must be machinelike and deterministic, and he disputed that he used as criterion for the admissibility of a theory the question "is it rigorously deterministic?" Pauli concluded that Einstein's point of departure is "realistic" rather than "deterministic," and said it is "*exactly the same* in Einstein's

printed work." (Letter to Born, March 31, 1954, in Born, 1968, p. 221, his italics.)
5. Augustine, *Confessions*, Bk. 11, Ch. 14. Leibniz said: "When I am able to recognize a thing among others, without being able to say in what its differences or characteristics consist, the knowledge is confused. . . . It is when I am able to explain the peculiarities which a thing has, that the knowledge is called distinct" (*Discourse on Metaphysics*, xxiv).
6. The other highest principle is the principle of noncontradiction. The *Critique of Pure Reason* considers theoretical reason and knowledge in general. Kant applied his philosophy to mechanics in the *Metaphysical Foundations of Natural Sciences* and sections of the *Opus Postumum*.
7. Schrödinger (1961), p. 10, his italics. Einstein, "Physics and Reality," reprinted in 1954, pp. 290–323.

Chapter 2

8. The rival axiomatizations have different primitive concepts. Usually the primitive concepts of a formulation are given preponderant significance. Thus various axiomatizations often channel us toward one interpretative outlook or another.

Feynman's *path-integral approach* is in a class by itself; it has had tremendous impact in the development of quantum field theory. The *algebraic approach* was first conceived by Jordan and von Neumann and developed by I. E. Segal. Its primitive concept is the observables, which are the elements of an involuted algebra. The states are positive linear functionals on the algebra. The algebraic approach is also applicable to statistical mechanics and quantum field theory, where its impact is greater. Haag (1992) gives a good account of the algebraic approach. In the *convex set approach* of B. Mielnik, the primitive concept is the states, which are the elements of a convex set. The closely related *operational approach* was developed by G. Ludwig, E. B. Davies, J. T. Lewis, and C. M. Edwards. It generalizes the concept of observables, which advances our understanding of quantum-mechanical measurements (Davies, 1976). The generalized concept of observables, represented by positive operator measures, is used in practical communication problems; see Yamamoto and Haus (1986). Busch et al. (1991) adopts it in a general theory of measurement. The *lattice-theoretic approach* originated with G. Birkhoff and von Neumann. It was developed by G. W. Mackey, J. M. Jauch, and V. S. Varadarajan. Here neither the state nor the observable is primitive; both are derived from propositions, which are the elements of an orthocomplemented lattice. The lattice-theoretic approach is a favorite among philosophers, and forms the basis of the quantum logic interpretation of quantum mechanics. Srinivas (1975) compares the concepts and empirical significance of the operational and lattice-theoretic approaches.

Some of the developments are better regarded as variants or generalizations of quantum mechanics. For example, the lattice-theoretic approach predicts many states for stationary spin-$\frac{1}{2}$ systems that have neither a counterpart in the Hilbert space formulation nor experimental evidence. The axiomatics posit structures more general and abstract than the Hilbert space. To link highly abstract structures such as a C*algebra or an orthocomplemented lattice to experimental data, concrete representations must be found in the Hilbert space. For the algebraic approach, the representation is adequately given by the GNS construction. For the lattice-theoretic approach, they are given by the Piron representation theorem. However, both the Piron theorem and the Gleason theorem work only for systems with dimension greater than two. Thus the theory overdetermines the spin-$\frac{1}{2}$ system without translational motion, whose state space is two dimensional (Beltrametti and Cassinelli, 1979, §§ 1, 6).

For a critical summary of various axiomatizations of quantum mechanics, see Gudder (1979). Wightman (1976) gives a more detailed account. Both include extended bibliography.

There are also the *hidden-variable theories* of David Bohm and others, which frankly claim to be variants of quantum mechanics. Belinfante (1973) gives a good survey of hidden-variable theories. The debate between the proponents of hidden variables and quantum logic has been a strong driving force in the philosophy of quantum mechanics. It is discussed in great detail in recent philosophical expositions of quantum mechanics. See Redhead (1987), Hughes (1989), and van Fraassen (1991).

9. For example, the random transitions between metastable states of a single Ba$^+$ ion in a radio-frequency trap are observed (Nagourney et al, 1986). Earlier single-photon experiments were done only with highly attenuated beams. Recently, genuine single-photon states have been produced (Hong and Mandel, 1986).

10. See, for example, Dirac (1930), Ch. 1; Sakurai (1985), §§ 1.2, 1.3, 1.5. Dirac distinguished between a *ket* $|\phi\rangle$ and a *bra* $\langle\phi|$. The kets are vectors, and the bras are linear functionals that map kets into complex numbers. By the Riesz representation theorem, to every linear function F there corresponds a unique vector $|\psi\rangle$ such that $F(|\phi\rangle) = \langle\phi|\psi\rangle$. Hence the bras and kets can be combined into a single concept, the vectors of the Hilbert space. In this case the notion $|\ \rangle$ is clumsy. I use it to separate clearly the quantum state $|\phi\rangle$ from the field operator ψ.

11. The *adjoint* $A^\dagger$ of A is defined as $(\langle\phi|A^\dagger)|\psi\rangle = \langle\phi|(A|\psi\rangle)$, where $A^\dagger$ acts on the vector on its left and A acts from the vector on its right. A is *self-adjoint* if $A^\dagger = A$.

12. "Transformation function" is Dirac's term. $\langle\beta_j|\alpha_i\rangle$ is also often called the "transition probability." However, "probability" applies only to $|\langle\beta_j|\alpha_i\rangle|^2$. "Transformation" is rather theoretical, while transition implies a physical process. I will reserve "transition" for amplitudes such as $\langle\beta_j|V|\alpha_i\rangle$, where a physical coupling between the states $|\beta_j\rangle$ and $|\alpha_i\rangle$ is induced by the observable V.

13. Beltrametti and Cassinelli (1981), p. 79.

14. Ballentine (1990).

15. Busch et al. (1991), p. 138ff.

16. Belinfante (1973) gives a good review of various hidden-variable theories. Greenberger et al. (1990) summarizes the Bell-type theories and experiments together with proposal for further experiments. Quantum logic is clearly explained in Beltrametti and Cassinelli (1981); Hughes (1989) gives a more popular account.

17. Another conception of reality that can be rejected with little regret is the criterion in the paper by Einstein, Podolsky, and Rosen: "If, without in any way disturbing a system, we can predict with certainty (i.e., with probability equal to unity) the value of a physical quantity, then there exists an element of physical reality corresponding to that physical quantity." By mixing up disturbance and prediction, the stipulation implicitly assumes some conditions of verification, plausibly that of faithful measurement, which is problematic. It is interesting that Einstein was unhappy about the paper, which was penned by Podolsky. He complained that "the essential thing was, so to speak, smothered by the formalism" (quoted in Fine, 1986, p. 35). His own article, "Physics and Reality," written some months after the EPR paper and collected in Einstein (1954), presents a completely different conception of reality.

18. Galilei (1629), pp. 234, 445; (1638), pp. 166f. Newton, letter to Bentley, February 11, 1693.

Chapter 3

19. Weyl (1921), p. 92. Philosophers have paid much attention to Euclidean and non-Euclidean geometries. By using only part of the Euclidean axioms, two distinct geometries can be extracted: *absolute geometry*, which ignores the parallel axiom, and *affine geometry*, which ignores the congruence axiom. Historically, the parallel axiom was the suspect that led to major developments in mathematics. Some people erroneously think that the concept of parallelism has been discarded; therefore absolute geometry with its congruence axiom is the one with physical applications. Parallelism appears in a more general form in differential geometry, and absolute geometry plays little role in physics.

20. Several developments in the first half of the nineteenth century motivated Riemann's work. The rise of non-Euclidean geometries and their impact are well known. Gauss's geometrical representation of complex numbers pointed to the possibility of geometrically representing all continuously varying objects. Riemann's own research on functions and integrations had led him to the incipient idea of a functional space. All these currents converged on what Riemann called "the general concept of multiply-extended magnitudes," or what we call "manifolds."

It is commonly accepted that space is a continuum, which was a vague notion. Unlike the case with discrete entities, the concept of quantity in a continuum was lacking. This was one of the reasons why space was so mystifying. Riemann noted the difficulty. However, *he did not assert that physical space is therefore metrically amorphous*, as some philosophers imputed to him. He limited himself to the conceptual difficulty, regarded it as a problem, and set himself to solve it by making distinct previously confused notions. In the process he spawned the precise formulations of the manifold and the metric. He showed that manifolds and metric spaces are different mathematical structures. He defined a topological manifold and showed that topological and metrical properties are distinct. A topological manifold M is not a metric space; however, it can support an unambiguous metric g, resulting in a Riemannian manifold (M, g). The topological manifold is metrically amorphous, the Riemannian manifold is not.

Having invented new concepts applicable to any continuum, space included, Riemann delivered them to physicists. Space may or may not have the structure of M, he said, but if it does and the physical world has the structure (M, g), then the specifics of g has to be manifested dynamically. As to what particular structures space and the physical world actually have, he expressly left them for physics (Riemann, 1854, § III.3). Riemann made no *a priori* assumption on the procedure of measurement; his position is clear and impeccable.

21. Einstein (1949a), p. 67.

22. See, for example, Helgason (1978), §§ 1–9.

23. See, for example, Wald (1984), Ch. 1.

24. An intriguing question is the origin of the minus sign in the Lorentz metric $\eta_{\mu\nu} =$ diagonal $(-1, +1, +1, +1)$. If all four signs were positive, then the Lorentz transformations would be rotations in a four-dimensional space. The minus sign leads to phenomena such as time dilation and length contraction. The most common explanation for it is that no signal can travel faster than the speed of light. However, as Hawking and Ellis argued, the speed of light is specific to electromagnetism, while the Lorentz metric is generally valid even when electromagnetic fields are absent (1973, p. 91).

Consider a set of coordinate systems moving relative to one another with constant velocities along the x direction. The y and z components are constant and suppressed for brevity. Let u be the velocity between the coordinate systems $S = (x, t)$ and $S' = (x', t)$. The transformation between S and S' is $S' = f_u(S)$. Berzi and Gorini (1969) have shown that if: (a) the transformations f_u form a group; (b) f_u are linear; (c) reciprocity holds, $f_u^{-1} = f_{-u}$; (d) space is isotropic, then the coordinate transformations have the form

$$x' = (x - ut)/(1 - u^2/w^2)^{1/2} \, , \; t' = (t - ux/w^2)/(1 - u^2/w^2)^{1/2}$$

where w is a constant. The sign of w^2 is *not* determined by the conditions (a)–(d). If w^2 is negative, then the above transformation becomes a simple rotation of the coordinate (x, t) through an angle $\theta = \tan^{-1}(u/\sqrt{-w^2})$. A negative w^2 implies the complete isotropy of space and time. The isotropy of time means the two directions of time are equivalent. We know from experiments that the transformations are not rotations. Thus w^2 is positive. For positive w^2, it can be shown that w has the meaning of a limiting velocity. Thus it seems that some sense of "directionality of time" has been built into the Lorentz transformations via the concept of velocity in (c) and (d).

25. In the Maxwell equations, the speed of light c is related to the dielectric constant and the magnetic permeability, and both can be determined by static means. To test the constancy of c directly is difficult because it needs reference frames with speeds close to that of light. The independence of c on the motion of sources was not directly tested when posited in 1905. It was confirmed up to velocity $c/2$ in an experiment on gamma rays emitted from electron–positron annihilation in 1963.

26. Plato said: "The qualities of measure and symmetry (*symmetron*) invariably . . . constitute beauty and excellence" (*Philebus*, 64e). Aristotle said: "The chief forms of beauty are order and symmetry and definiteness" (*Metaphysics*, 1078a). Polyclitus of Argos, a fifth-century Greek sculptor, wrote in the *Canon* that beauty is a matter of *symmetria*, which is an elaborate system of comparative dimensions or proportion. Later, the Stoics said beauty depends on measure and proportion, and symmetry regulates the agreements of the parts. Plotinus probably had the Stoics in mind when he said beauty was generally regarded as "the symmetry of parts toward each other and towards a whole" (*First Ennead*, VI, 1–2). This formulation was popularized by Heisenberg among physicists. However, Plotinus rejected the formulation of beauty in terms of symmetry because it is not holistic enough; it makes the parts too prominent. Practice predated theorizing. For the manifestation of symmetries in ornamental arts, see Weyl (1952).

27. A *group G* comprises a set of elements $\{g_i\}$ with a single rule of composition such that: (a) G is closed: for any two elements $g_1, g_2 \in G$, $g_1 g_2 \in G$; (b) the composition is associative; $(g_1 g_2)g_3 = g_1(g_2 g_3)$; (c) G contains an identity element e such that for all $g \in G$, $ge = eg = g$; (d) for every $g \in G$, there exists an inverse element $g^{-1} \in G$ so that $gg^{-1} = g^{-1}g = e$. A group G is commutative or Abelian if $g_1 g_2 = g_2 g_1$ for all $g_1, g_2 \in G$. There are numerous books on group theory. Elliott and Dawber (1979) is among those geared toward physicists.

28. Particles participating in the strong interaction are called *hadrons*. Hadrons were thought to be elementary. However by the early 1960s, hundreds of hadrons were produced in particle accelerators, mocking the epithet "elementary." Now all hadrons are accounted for by six quarks. The development of theories for the strong interaction can be seen in the three *SU(3)* symmetries associated with it: the *SU(3)* group of isospin and strangeness, also known as the "eightfold way,"

flavor-$SU(3)$ introducing the quarks, and color-$SU(3)$ introducing the color quantum number of the quarks. Color-$SU(3)$ came last and is fundamental. From it sprang the name quantum chromodynamics.

The properties of the hundreds of hadrons exhibited some regularities. In 1961 Murray Gell-Mann and Yuval Ne'eman independently classified the hadronic spectrum according to their quantum numbers of *isospin* and *strangeness*. Not all positions in the $SU(3)$ symmetry classification are filled. The discovery of predicted particles silenced skeptics of the eightfold way. Despite its predictive power, the eightfold way is a phenomenological classification reminiscent of Mendeleev's classification of chemical elements into the periodic table. It presents an order, which cries out for explanation. Fortunately, the order also enables us to pose questions. A significant feature of the hadronic spectrum is that the fundamental representation of the $SU(3)$ group, from which all higher-dimensional representations can be constructed, remains unfilled. The throne is left vacant for the quarks.

In 1963, Gell-Mann and George Zweig independently proposed the quark model. They explained the entire eightfold way by postulating a more fundamental constituency, which Gell-Mann called *quarks*. The quarks are classified by their *flavor quantum numbers* according to the flavor-$SU(3)$ group. The six flavors are up, down, strange, charm, top, and bottom. All mesons, or hadrons with integral spins, are bound states of a quark and an antiquark. All baryons, or hadrons with half-integral spins, are bound states of three quarks. No other combinations of the quarks are allowed. These simple rules generate all observed hadrons and only those observed.

The flavor-$SU(3)$ violates the Pauli exclusion principle stating no two fermions can occupy the same state within the same system. Quarks are fermions, but the flavor-$SU(3)$ scheme often assigns two identical quarks to be in the same bound state to produce a baryon. The problem of statistics was resolved by the postulation of a new quantum number for the quarks, which came to be called *color*. Each flavor of quarks comes in three colors, red, green, and blue, which form the fundamental representation of the color-$SU(3)$ group. A single rule of color combination explains why only the observed hadrons are allowed. The color rule is not phenomenological like the flavor rule or the eightfold way. The colors of the quarks are the source of the strong interaction. The color rule is an integral part of the interacting quantum chromodynamical system.

29. Wigner (1939). A more physical and less mathematical discussion is found in Schweber (1961), §§ 1d and 2c. More accurately, the transformations constitute the *proper* Poincaré group, because it exclude discrete transformations. It is also called the inhomogeneous Lorentz group. The criterion of free elementary particles as irreducible representations of the Poincaré group is not sufficient; it is satisfied by hydrogen atoms, mesons, and other entities that have decomposable structures.

30. See Sachs (1987) and the references cited therein.

31. Einstein, letter to Besso, January 3, 1916, quoted in Stachel (1989), p. 86.

32. Anderson (1967), § 10.3; Trautman (1973), p. 182ff.

33. Anderson (1967), Ch. 4. The notion of absolute objects is elaborated in Friedman (1983), § 2.1 and Earman (1989), pp. 38–40.

34. Anderson (1967), § 4.6; Weinberg (1972), p. 92.

35. Einstein (1916), p. 118; Pauli (1921), p. 145. For recent reviews on the equivalence principle, see Friedman (1983), § V.4; Norton (1985).

Chapter 4

36. For example, consider the electron gyromagnetic ratio or g factor. Write $g = 2(1 + \varepsilon)$. The QED prediction of ε is $1,159,652,140(\pm 5.3)(\pm 4.1)(\pm 27.1) \times 10^{-12}$. The best experimental result of ε, obtained for a single electron in a Penning trap, is $1,159,652,188.4(\pm 4.3) \times 10^{-12}$.

37. Yang (1981), p. 8.

38. Crease and Mann (1986) give an interesting account of the historical development of quantum field theories.

39. See, for example, Mandl and Shaw (1984), Ch. 2; Ryder (1985), Chs. 2 and 4.

40. An example of a Lagrangian for *nonlocal* fields is

$$\mathscr{L}(x) = \int dy\, f\left(\psi(x),\, \psi(y),\, \partial_\mu \psi(x),\, \partial_\mu \psi(y)\right)$$

41. The introduction of particle creation and annihilation operators is often called "second quantization," as distinct from the first quantization of replacing position and momentum by operators. The name can be confusing. As Ryder said: "There is only one quantum theory, not two; what we are doing is quantizing a *field*, rather than the motion of a *single particle*, as we do in quantum mechanics" (1985, p. 129).

42. More correctly, the energy of the field is $\Sigma_k \hbar \omega_k \left[n(\mathbf{k}) + \frac{1}{2}\right]$. The constant term $\frac{1}{2}\sum \hbar \omega_k$ is called the zero-point energy, and it is infinite because there are infinitely many modes. It is customarily removed by subtraction. The same problem occurs in general relativity, where a divergent vacuum stress–energy–momentum tensor $T^{\mu\nu}$ would give rise to an infinite cosmological constant. However, observations of the motion of distant galaxies put an upper limit on the cosmological constant.

43. The occupation number states are *not* eigenstates of the creation and annihilation operators, whose operations change the states. The electric field E is linear in creation and annihilation operators. The expectation value of E vanishes in the occupation number representation, no matter how large $n(\mathbf{k})$ is, or how many photons are detected. Therefore, the occupation number representation cannot pass to the limit of a classical field. The creation and annihilation operators have their own eigenstates, which are called *coherent states* $|c\rangle$. The coherent states are not orthogonal to each other; they form an overcomplete set. $\langle c|E|c \rangle$ represents a classical electric field in the limit of infinite number of photons, where its fluctuation vanishes. The eigenvalues of the coherent states are complex numbers for bosons and elements of a Grassmann algebra for fermions.

44. The vacuum and spontaneous symmetry breaking is treated in all textbooks on gauge field theory. For a detail discussion of the nature of the vacuum, see Aitchison (1985).

45. For a discussion of the local characteristic of classical electromagnetism, see t'Hooft (1980).

46. The derivation of the QED Lagrangian via local symmetry can be found in most textbooks on gauge field theory or quantum field theory. See, for example, Aitchison and Hey (1989), Ch. 2; Mandl and Shaw (1984), § 12.1. The electron field is represented by a spinor operator $\psi(x)$. The free electron Lagrangian is

$$\mathscr{L}(x)_{e0} = \bar{\psi}(x)(i\gamma^\mu \partial_\mu - m)\psi(x)$$

where the adjoint field $\bar{\psi} = \psi^\dagger \gamma^0$, γ^0, and γ^μ are Dirac matrices. The parameter m is to be replaced by the experimentally determined electron mass in the renormalization program. Independent variations of ψ and $\bar{\psi}$ in the action integral of the

Lagrangian formalism lead to the Dirac equation. The Lagrangian $\mathscr{L}(x)_{e0}$ is invariant under the global transformation (4.1) but not invariant under the local transformation (4.2). In terms of the transformed electron field, $\psi'(x) = e^{-i\theta(x)}\psi(x)$, it becomes

$$\mathscr{L}_{e0}(x) \rightarrow \mathscr{L}'_{e0}(x) = \bar{\psi}(x)(i\gamma^\mu\partial_\mu - m)\psi(x) + \bar{\psi}(x)\gamma^\mu\partial_\mu\theta(x)\psi(x)$$

The extraneous term that spoils its invariance comes from the partial derivative of the phase, $\partial_\mu\theta(x)$. To cancel the extraneous term we introduce (4.3) with $A_\mu(x)$ transforming according to (4.4). The electromagnetic field $F_{\mu\nu} = \partial_\mu A_\nu - \partial_\nu A_\mu$ is the derivative of the electromagnetic potential. Adding the coupling term and the kinetic term of the electromagnetic field to $\mathscr{L}(x)_{e0}$, we obtain the full quantum electrodynamic Lagrangian

$$\mathscr{L}_{\text{QED}} = \bar{\psi}(i\gamma^\mu\partial_\mu - m)\psi - \tfrac{1}{4}F_{\mu\nu}F^{\mu\nu} - e\bar{\psi}\gamma^\mu A_\mu\psi$$

The first term represents the electron field, the second the electromagnetic field, and the third the interaction between them.

47. See, for example, Trautman (1980b); Ivanenko and Sardanashvily (1983).
48. See, for example, Wu and Yang (1975); Daniel and Viallet (1980).
49. See, for example, Abraham and Marsden (1978); Thirring (1978).
50. t'Hooft (1980), p. 104.

Chapter 5

51. The working interpretation of quantum mechanics is distinct from the Copenhagen interpretation. The Copenhagen interpretation is famous but lacks a clear formulation. Its major components include Heisenberg's uncertainty principle, Born's probability interpretation, Bohr's complementarity and his doctrine on the impossibility of separating atomic objects from our instruments of observation, and the claim that quantum mechanics is the absolutely final theory. The tenets of Bohr and Heisenberg were developed by their followers, notably into various forms of observer-created reality. Such elaborations are often mixed in the bag labeled "Cophenhagen interpretation." The working interpretation incorporates the uncertainty principle and the Born postulate. Thus it is sometimes confused with the Copenhagen interpretation. It should not be; it does not include many of the Copenhagen doctrines that become the center of interpretive controversies. (Born was in Göttingen, where Heisenberg did a lot of his work. Thus "Cophenhagen" is not even descriptive.)

Gell-Mann said physicists are "brainwashed" by Bohr to think that quantum mechanics has answered everything. I think most physicists are not so much brainwashed as indifferent. It is significant that many physicists, including de Broglie, Landé, and Bohm, who were once followers of the Copenhagen school, and who have seriously thought about the problem, finally turned against it. Working physicists are mostly pragmatists, but they are not all seduced by the Cophenhagen "tranquilizing pillow," as Einstein called it. Dirac said: "I think it might turn out that Einstein will prove to be right, because the present form of quantum mechanics should not be considered as the final form . . . . It is the best that one can do up till now. But, one should not suppose that it will survive indefinitely into the future" (1975, p. 10). Feynman said that he could not pinpoint the problem with quantum mechanics, but he was not convinced there is none. He said that the origin of probability in quantum mechanics is an interesting question (1965, p. 129; 1982, p. 485). The mechanism underlying the probabilistic nature of quantum mechanics is what the Copenhagen school taboos. The general attitude of physicists can be seen in the outpouring of letters to *Physics Today* (April 1989) in reponse to a reference frame article on the quantum question (October 1988). About half of the responses do not think quantum mechanics has answered everything, others express disgust about the wanton speculations of some quantum interpreters.

52. Feynman said this in a dialogue with Bohr that occurred in the Twelfth Solvay Congress in 1961. Bohr said: "Every one is aware of the fact that the process of observation implies the use of irreversible amplifying devices." Feynman replied: "The details we are worrying about now, like what happens when we hit 2 neutrons together and produce various pions, etc. is not twisted up with the difficulties of the measurement process in the usual sense. We don't have to go all the way back; it is only a guess about what to forget about because obviously we can't always say that we have to study everything at once." (Recorded in Stoops, 1961, pp. 254, 256.)

53. Kant (1781), Bxii. I follow the standard way of referring to the *Critique of Pure Reason*, numbers preceded by A referring to pages in the first edition of 1781, B to the second edition of 1787. Both paginations are given in the margin of Kemp Smith's translation.

54. Heidegger (1926), § 16.

55. Hughes (1989), p. 302, see also Ch. 6.

56. Shimony (1986).

57. Heisenberg suggested reviving the Aristotelian "potentia" and introduced the notion of "potentiality" to interpret the state vector (1958b, pp. 53, 158). For a recent discussion of similar ideas, see Shimony (1986). "Latency" was suggested by Henry Margenau in 1954, and adopted by Hughes in his "quantum event interpretation" (1989). The interpretation of quantum mechanics in terms of "propensity" was advanced by Popper (1956). "Eventuality" was introduced in Stein (1972). The last two theories are more detailed compared to the others.

58. Stein (1972), p. 373.

59. van Fraassen (1991), p. 108. See also Healey (1979).

60. For arguments on the necessity of respects in statements about similarities of things, see Armstrong (1978), Ch. 6. Attributes are universals, whose ontological status has engendered much philosophical debate. Many references can be found in Armstrong's book.

61. Kant (1781), B207–B210. We are considering monadic properties, which include quality and quantity. Quantity is a property type that applies to composite objects and is discussed in Appendix A.

62. Penrose (1987), p. 106.

63. The classical state space is the bare differentiable manifold. We can introduce a metric in a manifold if we please, but it is optional.

64. For a discussion of spectral representations of observables in general cases, see Jauch and Misra (1965).

65. I call $a_i c_i$ the constituent value instead of component value to avoid confusion; there are also various spatial components, for example, the z-spin or y-position component, which are represented by different observables. The spatial components should not be confused with the constituent elements of each observable.

66. Healey (1979); Redhead (1987), p. 45.

67. Perhaps the most authoritative statement of the criterion of observability in the positivist philosophy was given by Carnap. Carnap said "observable" applies only to properties such as blue, hard, and hot, which are directly perceived by the senses. Thus the truth of a sentence containing an observation term can be ascertained in an act of observation that takes a short time and involves at most simple instruments. For references to Carnap and criticism of the positivist position, see Putnam (1975). For a recent empiricist defense of an observability criterion based on human physiology, see Fraassen (1980), Ch. 1, § 2.

68. The quotes are from the papers on the observation of the J/ψ particle, published in the 2 December 1974 issue of *Physical Review Letters*.

69. Kant said: "The grossness of our senses does not in any way decide the form of possible experience in general. Our knowledge of the existence of things reaches, then, only so far as perception and its advance according to empirical laws can extend" (1781, A226/B274). He argued in the "Transcendental Aesthetic" of the *Critique of Pure Reason* that the forms of intuitions are space and time ("aesthetic" is used in the original Greek sense of perception). In the general conceptual structure of experiences, the forms of intuition are incorporated as the schematization that introduces particulars. In the schematized categories, the forms are most closely associated with the categories of quantity, which are entitled "Axioms of Intuition." Quantity presupposes a diversity of individuals. For more on the position of the "Aesthetic" in the structure of the *Critique*, see note 120.

70. Dirac (1930), p. 34f.

71. Krantz et al. (1971).

72. The system of complex numbers $\mathbb{C}$ is the set of ordered pairs (x, y) of real numbers obeying the usual rules of vector addition and scalar multiplication by a real number, and the operation of complex multiplication

$$(x_1, y_1)(x_2, y_2) = (x_1 x_2 - y_1 y_2, x_1 y_2 + y_1 x_2)$$

73. Stueckelberg (1960); Mackey (1963). Bohm (1951), § 4.5 gives a simple discussion on the complex nature of wavefunctions.

74. Dirac (1930), p. 10.

75. Landé, quoted in Popper (1956), p. 46. A similar idea is expounded by Ian Hacking, who argued that we must acknowledge the reality of unobservable entities when we use them as tools (1983).

76. For a review on a wide class of neutron interferometry experiments, see Rauch (1987).

77. For various methods of measurements, see the review article by Yamamoto and Haus (1986). Such measurements involve extra noise because of the interference. The noise is tolerated because the purpose is to extract maximum information.

78. See, for example, Kaempffer (1965), pp. 11–16.

79. Corbett and Hurst (1978); Busch and Lahti (1989). Mittelstaedt et al. (1987) reports a single-photon interferometry experiment that simultaneously measures the incompatible observables.

80. Yamamoto and Haus (1986) also discusses applications to communication problems. More theoretical considerations on quantum communication and detection can be found in Helstrom (1976).

81. Kochen and Specker (1967). For a detailed discussion of the Kochen–Specker paradox and various responses to it, see Redhead (1987), Ch. 5. Mermin (1990) gives a simple unified proof for the Kochen–Specker and Bell theorems.

82. Mermin (1990).

83. Schrödinger, "Examples of Probability Predictions"; also letter to Einstein, July 13, 1935, quoted in Fine (1986), p. 75. The Spanish boot metaphor is found in Goethe's *Faust*, 1912–1915:

> For thus your mind is trained and braced,
> In Spanish boots it will be laced,
> That on the road of thought may be
> It henceforth creep more thoughtfully.

84. Dummett (1991), p. 326.
85. Bohr (1949), p. 240.
86. That is why I steer clear of the current philosophical debate on scientific realism and antirealism, which is heavily based on the notion of truth. It seems to me to be a linguistic version of the traditional argument behind the veil of perception, which, if history teaches a lesson, leads only to a dead end.
87. The stipulation of superselection rules was introduced by Wan (1980). Suppose a system is subjected to superselection rules that forbid the superposition of its states, then it is always clearly in one of its states and will manifest the corresponding spectral value with probability 1. The system can then be called classical. Unfortunately, the Hamiltonian governing its temporal evolution can no longer be an observable. For if it is, it would be subjected to superselection and cannot connect different states, so the system will be forever stuck in the initial state. The meaning of a quantum mechanics wherein the Hamiltonian is not an observable is unclear. For a discussion of superselection rules and their application to the measurement problem, see Beltrametti and Cassinelli (1981), Chs. 5 and 8.

In decoherence theories, it is argued that classical systems are never isolated, and the interaction with the environment washes out the phase relations among various states. However, granted the washing out, the system is still left as an incoherent superposition of the various states represented by the mixture. But the measurement on the single system returns only a single spectral, not a combination of them. Thus there is still an uncontrolled jump. A review on the environment-induced decoherence of quantum states can be found in Zurek (1991). Decoherence is incorporated into an interpretation of quantum mechanics in Omnés (1992). Bub (1988) considers systems with infinite degrees of freedom.

There are other criteria for classicality. The most popular one is Bohr's "irreversible act of amplification," which many of his followers take as a panacea because it sounds like a physical process. However, Bohr made no attempt to explicate the amplification more substantially. In device physics, there are theories of irreversible amplifications, such as avalanche breakdown and impact ionization. However, these theories have already invoked quantum mechanics and the collapse of the wavefunctions. Is Bohr's amplification a physical process? If so, why is it inexplicable? Devoid of substantiation, the "irreversible act of amplification" is an empty phrase that merely covers up the problem.
88. The major contending positions are well recorded in Jammer (1974). Recent interpretations tend to be quite technical. A recent realist version is Krips (1987); an antirealist version is van Fraassen (1991). Redhead (1987) gives a balanced account of the technical concepts involved in the philosophical debate. All include extended bibliographies.
89. Leggett (1986), p. 28.
90. see, e.g., Kolmogorov and Fomin (1968).
91. For example, Feller (1950); Pfeiffer (1965).

92. Popper (1957).
93. There is another way of seeing observables as schematizations of the Hilbert space. In many scientific theories, we somehow schematize an object system by projecting into it some partial structures of the real line. For instance, in Appendix A we find that such schematization underlies the theories of measure and probability. The resort to real numbers has the additional advantage of securing empirical access, for real number representability is our form of observation.

   The idea that an observable schematizes the Hilbert space by correlating some of its structures with the real line is apparent in the spectral theorem of Hilbert space, which underlies von Neumann's formulation of quantum mechanics. The spectral measure $P_A(\Delta)$ associated with an operator A assigns to every Borel set $\Delta \in \mathbb{R}$ a projection operator such that $P_A(\emptyset) = 0$, $P_A(\mathbb{R}) = I$, and $P_A(\cup_i \Delta_i) = \Sigma_i P_A(\Delta_i)$ for any disjoint sequence $\{\Delta_i\}$ of Borel sets. The spectral theorem says that to each self-adjoint operator A there corresponds a spectral measure $P_A(\Delta)$ that resolves A into a family of projection operators projecting onto a family of orthogonal subspaces. Thus the spectral measure "projects" the Borel structure of the real line into the Hilbert space and decomposes it into orthogonal subspaces, which are "indexed" by the spectral values of the operator.

   Similar interpretation has been suggested by several authors. In the lattice-theoretic formulation of quantum mechanics, an observable picks a specific Boolean subalgebra, which H. Primas regarded as a particular frame of reference (1981, p. 325f). We share the common idea that the structure of the real line is related to the state space to effect some kind of schematization. However, the specific features of the schematizations lead to divergent interpretations. In one approach, an analogy is drawn with the probability calculus and the consideration is mainly set-theoretic. The schema partitions the Hilbert space into disjoint subspaces, and the relation between a state $|\phi\rangle$ and an eigenspace M_a is that of membership. According to the usual definition of properties as sets, the subspace M_a to which $|\phi\rangle$ belongs is its property. By association, $|\phi\rangle$ is said to have the eigenvalue a_i if $|\phi\rangle \in M_a$. In a second approach, an analogy is drawn with classical mechanics, and the schema uses the full vector-space structure of the Hilbert space. It introduces eigenstates as coordinate axes, and more generally eigenspaces as "coordinate planes." The relation between a state and an eigenstate is vector component. The coordinatization is associated with real numbers through the eigenvalues, but the connection between the numbers and the amplitudes is indirect and there is no one-to-one correspondence between them. The second approach is more holistic; its schema includes more features so that its concepts are more intrinsically correlated. I have adopted the second approach because it better agrees with quantum physics.

   Primas' method of schematization is analogous to the probability calculus. It is close to quantum logic and followed in one way or another by several interpreters. They attribute to the quantum system an additional state, variously called the value state or microstate, which purports to list the observables, if any, for which it has definite eigenvalues (Hardegree, 1980; van Fraassen, 1991).
94. Wigner (1983), § 7. Bohm criticized the suggestion to generalize the meaning of "observables" to cover all quantities that can be represented by self-adjoint operators. Mathematically, it is often convenient to consider all self-adjoint operators, for we can treat them as a structure with specific features such as certain algebras. Such structures are better understood as theoretical idealizations providing powerful analytic tools. Interpreters sometimes claim that the abstract structure of operators is more fundamental because it has better operational or instrumental

meaning. However, to extrapolate the operational meaning of a handful of operators to all operators, most of which we cannot even dream of, is unwarranted.

95. Good analyses and reviews of the Bell theorem and related experiments can be found in the papers respectively by J. Cushing, A. Shimony, D. Mermin, and J. P. Jarrett in Cushing and McMullin (1989).

96. Heisenberg's original paper uses terms such as inexact, indeterminate, or lack of precise value, which are ontic descriptions. Later, the indeterminacy relation became known as the uncertainty principle, which carries a strong epistemic connotation.

97. The electric field vector of a monochromatic electromagnetic field can be written as $\mathbf{E} = \mathbf{E}_0[A \cos(\omega t) + B \sin(\omega t)]$, where the two terms represent the two quadratures of the field. The quadrature field operators A and B are analogous to the position and momentum operators of a harmonic oscillator so that $(AB - BA)|\phi\rangle = i\hbar|\phi\rangle/2$. In the *coherent state*, which characterizes the field of a laser operating far above threshold, the variance of the two quadratures are equal, and their product is $\hbar/4$, which means the coherent state is a minimum-uncertainty state. Physicists have produced another kind of minimum-uncertainty states called *two-photon coherent states* or *squeezed states*, in which the noise in one quadrature is reduced, so that $\langle\delta A^2\rangle_\phi < \hbar/2$. The noise in the other quadrature increases proportionally. $\langle\delta A^2\rangle_\phi\langle\delta B^2\rangle_\phi$ keeps to $\hbar/4$ as $\langle\delta A^2\rangle_\phi$ and $\langle\delta B^2\rangle_\phi$ change. Levi (1987) gives a brief review with references.

98. Busch and Lahti (1983). Let f be a smooth function on the real line and let g be its Fourier transform. The *support* of f is $supp(f) = cl\{x \in \mathbb{R} ; f(x) \neq 0\}$. The *support property*: If $supp(f)$ is bounded, then $supp(g)$ is the whole real line, and conversely. The *dispersion property*: $\Delta|f|^2 \cdot \Delta|g|^2 \geq 1/2$.

99. The meaning of incompatibility can be stated more precisely in general cases with continuous spectra. Instead of eigenstates $|\alpha_i\rangle$ corresponding to eigenvalues a_i, we have eigenspaces M corresponding to Borel sets Δ in the spectrum. An observable A picks a unique subspace $M_A(\Delta)$ of the state space for each Borel set Δ in its spectrum $\Lambda(A)$. An observable B does likewise. A and B are *incompatible* if for every bounded $\Delta \subset \Lambda(A)$ and $\Delta'\subset\Lambda(B)$, $M_A(\Delta)$ and $M_B(\Delta')$ are set-theoretically disjoint.

100. In the gamma-ray microscopic argument, Heisenberg proposed a kind of disturbance model, which argues that the measurement of one observable unavoidably disturbs the system and precludes the possibility of accurately measuring the conjugate observable. He later retreated from the position.

101. A generalized observable A does not restrict the values of its spectral measure $P_A(\Delta)$ to projection operators. It admits all positive operators and relaxes the orthogonality conditions on the eigenspaces. Intuitively, a positive operator is a kind of weighted average over projection operators, and it does not correspond to a sharp eigenvalue as the projection operator does. It offers an interpretation of the indeterminacy principle for single systems by building the unsharpness of measured values into the concept of observables. Generalized observables and positive operator-valued measures have been widely used in measurement problems, see Busch et al. (1991). Busch (1985) discusses their relation to the indeterminacy principle.

102. Bohr (1954), p. 210, his italics; (1934), p. 54. Heisenberg (1958), p. 100.

103. Mermin (1981), p. 405, his italics; Clauser and Shimony (1978), p. 1921. These are scholarly and well-argued papers. Once when Einstein was discussing the interpretation of quantum mechanics with A. Pais, he stopped in his walk and asked:

Do you really believe that the moon exists only when you look at it? Pais wrote in 1982: "The twentieth century physicist does not, of course, claim to have the definitive answer to this question" (Pais, 1982, p. 5).

104. W. Duch and D. Aerts, in *Physics Today*, June 1986, p. 15.

105. Feynman et al. (1963), III-2-8.

106. Einstein (1949), p. 669.

107. Bohr and Heisenberg were not positivists. However, their doctrine that the complete specification of experimental arrangements is required for a discussion of the behavior of quantum systems is positivistic. When pressed on important points, Bohr almost always fell back on positivist ideas. Positivists in turn used the Copenhagen interpretation of quantum mechanics to promote their phenomenalism.

108. For a discussion of the empiricist doctrines and the difference between idealism and phenomenalism, see Bennett (1971), Ch. 3.

109. Kant (1781), B142, his italics.

110. Dirac (1930), p. vii.

111. For the foundation of measurements, see Krantz (1971), Ch. 1. A few more remarks on utility theory is found in note 214.

112. Group theory finds its major root in the theory of solvability of algebraic equations by radicals. Up until the late eighteenth century, mathematicians were chiefly engaged in computing the roots of particular equations. Then Joseph-Louis Lagrange introduced a new way of thinking. Instead of looking at individual equations, he examined the system of all algebraic equations. He wanted to find a criterion to decide whether an equation is solvable by radicals without actually solving it. Lagrange's goal was fully met in the early 1830s by Evariste Galois, who introduced the name "group." Galois' theory enables one to read off the structure of the roots of any algebraic equation from the structure of a group of permutations associated with the equation. In the late 1860s, Felix Klein and Sophus Lie undertook to investigate geometric and analytic systems by group methods. In all cases, the surveying glance of the mathematicians turned from features of individual figures and equations to the structures of the systems of all geometries and all algebraic equations. Klein classified geometries according to groups of transformations in his Erlangen program. Lie found that most differential equations are invariant under certain easily describable sets of transformations. Lie introduced the concept of transformation groups and developed the theory of continuous groups. Lie groups play an important role in today's physics.

113. Pauli (1921), Preface to the 1956 edition, p. 3.

114. Weyl (1952), p. 126; Weinberg (1980), p. 515.

115. Shakespeare, *Macbeth*, II, i, 34–49.

116. Wigner (1967); Wald (1984), p. 56.

117. See, for example, Hanson (1958).

118. Putnam (1981), p. 49. Putnam rejected metaphysical realism and argued for an "internalist perspective."

119. Kuhn (1962).

120. Kant argued that experiences are not unconditional; even immediate observations such as seeing a tree have a host of presuppositions and involves much theory. The *Critique of Pure Reason* surveys the general structures and the fundamental bounds of theoretical reason and investigates the conditions underlying the possibility of empirical knowledge. Loosely speaking, the logic of experiences harbors a definite scheme that enables us to have intelligible observations of objects instead of

rhapsodies of sensual impressions. Kant tried to delineate those elements of the scheme without which there can be no scheme, and hence no object discernible under it, for the scheme comprises the concepts through which any object whatsoever of experience can be thought. The scheme is a web of general concepts or categories. They are intrinsic to all objective experiences as their presuppositions; hence they are *a priori*, in contradistinction to *a posteriori* concepts that pertain to the contents of experiences and must answer to them. The categories must extend beyond themselves to cover the objects of experiences, hence they are *synthetic* and can be negated without involving contradictions, as opposed to *analytic* concepts, whose negation leads to contradictions. They are totally formal and empty of contents, which must be supplied by specific experiences. They can only be applied to objects of possible experiences, otherwise they lead to illusions.

The kind of argument Kant developed and used extensively is called the transcendental argument, which has the general form:

X (e.g., we have experiences).

For X to be possible, the conditions p, must be satisfied.

Therefore p.

"How is X possible?" initiates the inquiry and is called a transcendental question. The topic X is typically some fundamental empirical knowledge in Kantian philosophy. Hence the conditions p must be internally manifested, for outside knowledge we know nothing and can say nothing.

The argument of the *Critique of Pure Reason* proceeds from the general to the specific, or we may say from the deep structure to the surface structure of empirical knowledge. The major portion of the *Critique,* the "Element," is divided into two parts: "Transcendental Aesthetic" and "Transcendental Logic," which separately investigates the sensual and conceptual aspects of knowledge and objective experience. The "Aesthetic" is a tiny part, containing Kant's official theory of space and time as the forms of intuition (see note 69). The "Logic" consists the "Analytic" and the "Dialectic." The "Analytic" contains the bulk of his positive results regarding theoretical reason. There he argued that all intelligible experiences involve the general concept of the objective world, which can be analyzed into various categories. The general concept and the categories are not derived from experiences but are spontaneous products of human understanding. Kant made great effort to justify their objective validity. The justification falls into two parts, called "Analytic of Concepts" and "Analytic of Principles." The two is separated by a section called the "Schematism," which argues that the specific categories are applicable only when coupled with the forms of intuitions. The most important part of the "Analytic of Concepts" is the "Transcendental Deduction of the Categories," which argues in broad terms for the general concept of objects with almost no mention of the specific categories. The "Analytic of Principles" breaks down the general concept of objects into the categories and individually argues for them. Kant's table contains twelve categories, divided into four groups. The most interesting arguments are those for the group of relations, called "Analogies of Experiences," where he discussed the permanence of substance and the necessity of causality. The "Analogies" is followed by the "Refutation of Idealism," which can be regarded as the culmination of the long chain of arguments starting from the "Deduction." The "Dialectic" investigates the logic of illusion. Here Kant exposes various metaphysical illusions generated by the abuse of reason.

The *Critique of Pure Reason* is the first of the triad in Kant's philosophy. It is followed by the *Critique of Practical Reason*, which treats ethics, and the *Critique of Judgment*, which treats aesthetics. The first two *Critiques* are each followed by a book of doctrines for determinate judgments. The doctrine of pure reason is the *Metaphysical Foundations of Natural Science*, which can be seen as an attempt to provide a realistic interpretation of Newtonian concepts such as force and matter. The attempt is inconclusive in Kant's own opinion.

Some people say the first *Critique* is mainly concerned with justifying Newtonian mechanics. It is much broader and deeper. It is interesting that formal science is discussed only toward the end of the *Critique*. After the long criticism of reason in the "Transcendental Dialectic," Kant inserted an Appendix to discuss reason's positive role is science. The regulative employment of reason guides scientists towards systematic unification. Kant picked up the guiding role of reason and scientific creativity in the two introductions to the *Critique of Judgment*.

Most ideas discussed in this section are derived from reading the "Transcendental Deduction" and commentaries on it, especially Bennett (1966), Heidegger (1967), and Strawson (1966). Kant's argument draws heavily on the notion of time and is much more involved. My case is much easier because my starting point has assumed much more. However, I think I have not distorted his conclusion about the basic conceptual structure of experience.

121. Russell (1950), Ch. 7.
122. Kant (1781), B137, B135.
123. Kant (1781), B137, A197/B242.
124. Kant (1781), A158/B197, his italics.
125. Bell, "Subject and Object" in (1987), p. 40, his italics.
126. In a conversation with Heisenberg, Einstein distinguished "what nature does" from "what we know about nature," and warned against conflating the two. He explained that observation is a very complicated concept, and only theory, that is knowledge of natural laws, enables us to deduce the underlying phenomena from what we perceive. Heisenberg recorded the conversation and recalled that it made a deep impression upon him and influenced his thinking in formulating the indeterminacy principle (Heisenberg, 1971, Ch. 5, pp. 77f; 1983, pp. 107–22).
127. Bohr (1934), p. 16; (1949), p. 235, his italics. Einstein, letter to Schrödinger, Dec. 22, 1950, in Przibram (1967), p. 39. Einstein made the same point in his argument with Pauli; see Born (1968), p. 226.

In the late 1920s, Einstein scrutinized the internal consistency of quantum mechanics. The arguments are more scientific than philosophic; one is never sure of the consistency unless it is thoroughly examined. They should not be confused with the later philosophical debate, although they often are. After Einstein was convinced that quantum mechanics is consistent, he thought that it is, "within the natural limits fixed by the indeterminacy-relation, *complete*. The formal relations which are given in this theory—i.e., its entire mathematical formalism—would probably have to be contained, in the form of logical inferences, in every useful future theory" (1949, p. 667, his italics). His chief contention with the Copenhagenists is the *reality* of the microscopic world (note 4). He also opposed Bohr's claim on the *finality* of quantum mechanics. He was dissatisfied that quantum mechanics disallows certain questions about the physical world, such as the moment when an atom decays, and thought some future theory that supersedes quantum mechanics will enable us to ask such questions. Bohr said that it is impossible for any theory to supersede quantum mechanics. The debate is metaphysical.

128. Leibniz, letter to des Bosses, in Leibniz (1969), p. 609. A more detailed discussion of Leibniz's theory of relations, in conjunction with his theory of space, is given in note 159. The notion of well-founding is associated with internal relations, which is discussed in § 29.

129. von Neumann (1932), p. 420.

130. The chain of quantum measurements is analyzed in von Neumann (1932), Ch. 6. London and Bauer (1939) first invoked consciousness to terminate the chain. Wigner said: "One could attribute a wave function to the joint system: friend plus object, and this joint system would have a wave function also after the interaction" (1967, pp. 172, 176). Bohr illustrated the ambiguity of the agent thus: If we hold a stick loosely in our hand, it is an object. If we hold it firmly and use it as a probe, it is part of ourselves (1934, pp. 55, 99). Bohr quoted a passage from a Danish novel: "I even think that I think of it, and divide myself into an infinite retrogressive sequence of 'I's who consider each other. I do not know at which 'I' which stops at it," and said: "it is certainly not easy to give a more pertinent account of essential aspects of the situation with which we are all faced" (1963, p. 13).

Nowadays a work in quantum interpretation is incomplete without a mention of the Schrödinger Cat, which is fast approaching the popularity of Lewis Carroll's Cheshire Cat that phases in and out. I have alluded to the Cat in the discussion of eigenvalues in § 13. Here is the original. Consider an experimental system containing a cat, a bit of radioactive material, and a diabolical device. The cat is alive when the experiment begins. The probability is one-half that one of the radioactive atoms will decay in an hour. If the decay happens, it will trigger the device, releasing poisonous gas, killing the cat. Schrödinger said: "The ψ-function of the entire system would express this by having in it the living and the dead cat (pardon the expression) mixed or smeared out in equal parts" (1935, p. 157). Interpreters love to draw cartoons of the Cat, perhaps in conjunction with Wigner's friend. Figure N.1 follows the tradition.

Figure N.1 The triple paradoxes of Schrödinger's Cat, Wigner's Friend, and Bohr's Egos illustrate that von Neumann's chain of quantum measurements never ends; for it has not touched the genuine meaning of observation.

131. See, for example, Gamut (1991), Chs. 1–3.
132. Kant (1781), B157, A341/B399.
133. Kant (1781), B421f, his italics.
134. Hume (1739), p. 252.
135. The yes–no experiment is discussed in several treatises on mathematical founda-
     tions, for example, Beltrametti and Cassinelli (1981). It is taken over in numerous
     philosophical interpretations. See, for example, Stein (1972), Healey (1989), and
     Hughes (1989). The last gives an ample bibliography.
136. Most original papers and important developments in quantum logic can be found
     in Hooker (1975).
137. Jauch (1968) and Beltrametti and Cassinelli (1981) give good accounts of the
     lattice-theoretic approach. They also discussed the yes–no experiments.
138. Margenau and Cohen (1967).
139. The "our knowledge" interpretation is most strongly defended by Rudolf Peierls,
     who said it is the only viable interpretation (1979, § 1.6).
140. See, for example, Feynman et al. (1965), Ch. III-5. Most textbook discussions of
     such experiments involve an ensemble of particles. More elaborate theories show
     that the same difference between analysis and selection holds for single systems,
     for example, for single-photon states. They show also that as long as selection is
     not made, the separated constituents can suffer a lot of disruption without losing
     their capability to recover the original state (Tan et al., 1991; Scully et al., 1991).
141. Redhead (1987), p. 58.
142. Kripke (1976), p. 416

Chapter 6

143. See, for example, Wilding (1982).
144. Parmenides, Fragment 7.
145. Strawson (1959) presents a detailed analysis of the notion of individuals. I agree
     with most of Strawson's arguments. My major objection is his notion that things
     are "*occupiers* of space" (p. 57, his italics).
146. In the philosophy of space–time, relationists argue that space–time is a set of
     relations among non-spatio-temporal entities, while substantivalists argue that
     space–time is some arena that material entities occupy. Both take the entities as
     independently given and neither discusses how they are individuated.
147. The particularity of beings is argued for in the earlier *Category*. An extended
     discussion of primary beings and substances is found in Books Zeta and Eta of
     Metaphysics. Owen (1979) gives a good account of Aristotle's analysis of beings.
148. Russell (1948), pp. 293–300. Ayer, "Individuals," in Ayer (1954), pp. 1–25.
149. Quine (1960), Chs. 5, 6; "On What There Is," in 1933, pp. 1–19; Strawson (1950).
150. In physics, there are three types of statistics, Boltzmann statistics for classical
     particles, Bose–Einstein statistics for bosons, and Fermi–Dirac statistics for fer-
     mions. Of the three, only the Fermi–Dirac statistics forbids more than one particle
     to occupy each state. The other two expressly allow many particles to be in the
     same state. We should be careful to interpret the statistics. In quantum field
     theory, the statistics are consequences of the commutation relations of field opera-
     tors and the condition of microcausality, which have certain spatio-temporal con-
     notations. In quantum mechanics, they are posited independently. However, as
     discussed in § 23, rigorously the particles in a quantum multiparticle system are not
     individuated because their phases are all entangled. In many calculations, we make

the independent-particle approximation that neglects the phase correlation. In quantum statistical mechanics, where the approximation is made, we are less concerned with the behavior of a particular particle than with the behavior of the aggregate.

151. The notion of identity criterion was explicitly introduced in Frege (1884), § 67. Among specific criteria of identity, the usual criterion for material things is their spatio-temporal continuity. Locke suggested that personal identity is determined by memory. Donald Davidson suggested that events (not in our technical sense) are identified by having the same causes and effects.

152. Wittgenstein (1918), 5.5301–3.

153. Heidegger (1957), p. 26.

154. "Sortal name" was first used by Locke to stand for abstract ideas signifying the essence of genus and species (1670, Bk. III, Ch. iii, § 15). The sortal dependence of individuation is recently analyzed by Strawson, Peter Geach, and David Wiggins.

155. There are also characterizing terms that do not usually accept supplementary individuation. We do not usually say "a bunch of intelligence" or "a cluster of whiteness." The individuation of "a beautiful thing" lies in the thing and not in beauty. However, as seen in § A1, with proper mathematical tools we can quantify certain abstract nouns, hence blurring their difference with mass terms.

156. Wittgenstein (1945), §§ 66–69. For discussions on the problem of kinds, see, for example, Putnam (1970).

157. Newton: "All things are placed in time as to order of succession, and in space as to order of situation" (Newton 1687, pp. 9ff). Leibniz: "It [space] does not depend upon such or such a situation of bodies, but it is that order, which renders bodies capable of being situated, and by which they have a situation among themselves when they exist together." "Space is . . . an order of situations, or (an order) according to which, situations are disposed" (1716, fourth letter, ¶41; fifth letter, ¶104).

158. Leibniz, fourth letter to Clarke ¶41; fifth letter ¶104; third letter ¶4.

159. Many philosophers found Leibniz paradoxical for dismissing relations and yet advancing a "relational theory of space." Early misperception is understandable because Leibniz's writings were mostly unpublished until the end of the nineteenth century. As a result of the publication and the critical studies, many Leibnizian scholars agree that the "relational theory of space" is totally out of place in Leibniz's philosophical system. For example, Broad said: "This is not the theory Leibniz really held" (1981, pp. 171–73). However, in the space–time literature, the misperception persists. For example, Sklar said: "The full doctrine of Leibnizian doctrine of space and time, as it appears in the context of his total metaphysics, is extraordinarily complex. . . . We won't pay any attention to this 'deeper' Leibnizian metaphysics" (1974, pp. 169ff). Friedman said: "In general, the Leibniz I speak of here is not the actual historical Leibniz, but, rather, the Leibniz of the positivists" (1983, p. 219). Most authors did not bother to apologize.

Leibnizian scholars generally agree that Leibniz was committed to an accident-in-substance ontology and a subject–predicate logic. He generally maintained that propositions of the form *Fa* suffice for the description of the world, so that relations in the form *Rab* are not required. However, his world is thoroughly interconnected. Thus his predicates *F* must contain relational properties. Many interpreters argue that some kind of relation is essential because relations underlie the notion of compossibility, without which his system collapses. The knot is the meanings of relations. Leibniz had a theory of relations, which is seldom mentioned in discussions of his "relational theory of space." In his chapter on Leibniz's relations and denominations, B. Mates said: "The term 'relation' is used in several senses by Leibniz, as well as by the tradition within which he writes and by the people with whom he corresponds" (1986, pp. 207f). Leibniz had tried to explicate relations in his argument with Clarke. I first concentrate on the Clarke correspondence to examine the meanings of relations and the theories of space, then venture to take a peep in his broader philosophy.

Leibniz said: "I hold space to be something merely relative, as time is; that I hold it to be an order of coexistence, as time is an order of succession." "Space is . . . an order of situations, or (an order) according to which, situations are disposed." "It [space] is that order, which renders bodies capable of being situated, and by which they have a situation among themselves when they exist together." He repeated similar formulations several times, in the correspondence and elsewhere (third letter ¶4; fourth ¶41; fifth, ¶104). "Relation" does not appear in the formulation. "Relative" is much weaker and more vague than "relational" or "consisting of relations"; it can mean no more than "not absolute." The meaning of "order" and "situation," which appear consistently, is not elaborated.

Leibniz invoked relations explicitly in ¶47 of his fifth letter. This long paragraph is the definitive source of the relational theory of space. It consists two parts; the first gives a definition of "place" in terms of relations, the second explains the meaning of relations. Let us first look at relations.

Leibniz explicated relations by two examples. One of them concerns the lengths of two line segments, L and M. Leibniz said there are three ways to express the relation "L is longer than M." The first way expresses the relation in terms of a ratio:

1. r is the ratio of the lengths of L to M and $r > 1$.

The other two ways of expression do not involve the ratio explicitly,

2. L has a length that is r times the length of M and $r > 1$;
3. M has a length that is $1/r$ times the length of L and $r > 1$.

Leibniz specifically picked on (1), because in it neither L nor M is the subject. The ratio "is indeed out of the subjects; but being neither a substance, nor an accident, it must be a mere ideal thing." However, he condoned (2) and (3). In (3), M is the subject with the property of having a certain length. The property is relational because it depends on L having a certain property. Leibniz said: "M the lesser is the subject of that accident, which philosophers call relations." He had made his point clear. First, there are *two kinds of relations*, those expressed in (1), and those expressed in (2) and (3). Both kinds are well founded, that is, based on the properties of the relata. Second, relations can be a kind of individual properties, in such case they are expressed as relational properties. Third, only relations of the kind (2) and (3) are individual properties, and they are the only one approved. The ratio in (1) is not an individual property; it can be ascribed to neither L nor M, for it has "one leg in one and the other in the other." These points are all consistent with his general insistence on subject predicate logic.

Leibniz's second example of the family tree illustrates the same idea of different kinds of relations among members of the family. It is expressed more succinctly in another passage: "You will not, I believe, admit an accident which is in two subjects at once. My judgment about relations is that paternity in David is one thing, sonship in Solomon another, but that the relation common to both is a merely mental thing whose basis is the modification of the individuals" (letter to des Bosses, in Leibniz, 1969, p. 609). The more complicated family tree shows an order comprising paternity, sonship, brotherhood, and so on. But the genealogical lines and place, as relations common to two entries, "would only be ideal thing."

Leibniz said he offered the examples on relations immediately after the definitions of place and space "to show how the mind uses, upon occasion of accidents which are in the subject, to fancy to itself something answerable to those accidents, out of the subject." The wording is curious; it sounds like a critique of some illusion rather than the announcement of a theory. Even more curious is the chief definiendum, *place*. "Place" seldom appears in other formulations of Leibniz's theory of space. It is rather the important concept in Newton's theory; Newton said in the *Principia*: "Place is a part of space which a body fills up. . . . I say, a part of space, not the situation" (1686, p. 6). Leibniz's own theory is usually formulated in terms of *situations*. And at a critical point in the paragraph he made the distinction: "And here it may not be amiss to consider the difference between place and the relation of situation, which is in the body that fills up the place." Thus Newton and Leibniz agreed that place and situation are distinct concepts. For Leibniz, the relation of situation is of the type (2), for it is "in the body"; it is the right kind of relation. The same cannot be said of place. That is the crucial difference.

"I will here show, how people come to form to themselves the notion of space," Leibniz began ¶47. People see that the world of things exhibits an order of coexistence or order of situation. The situation of a body is its "individual accident." People notice that a few bodies, say *A* and *B*, move, while many others remain fixed. A moving body distinguishes itself by containing the cause of change in its intrinsic properties: "we call a motion in that body, wherein is the immediate cause of change." People compare the orders of situations at different times, and find that the situation of *A* at time t_1 *agrees* with the situation of *B* at time t_2. So far the exposition agrees with Leibniz's formulations of his own theory; everything is stated in terms of *individual properties*, which include situations, order, and relations of the type (2).

"But the mind not contended with an agreement, looks for an identity, for something that should be the same; and conceive it as being extrinsic to the subject: and this is what we call *place* and *space*. But this can only be an ideal thing; containing a certain order, wherein the mind conceives the application of relations. In like manner, as the mind can fancy to itself an order made up of genealogical lines." At this point, as the mind fancies, *the meaning of relations changes from type (2) to type (1)*. People turn the relational properties in the subjects into an explicit relation out of the subjects and form the notion of "place" to answer for that "something that should be the same." They define:

4. Two objects *A* and *B* have the *same place* at different times t_1 and t_2 if, at the respective times, there are fixed objects $\{F_i: i = 1, 2, 3, \ldots, N, N \gg 1\}$ to which *A* and *B* hold the same explicit relation of coexistence.

Taking the totality of places, people form the notion of space. The "place" defined in (4) is akin to the "ratio" in (1); it is extrinsic to the subject and a mere ideal thing.

Once the relation is taken out of the subject, people easily mistake the place in (4) or the ratio in (1) to be something real. They can quantify over them and write

$(\exists x)$ (x is a place and x is occupied by A at t_1 and x is occupied by B at t_2). Leibniz expressly warned against this: We can understand the notion of place "without needing to fancy any absolute reality out of the things whose situation we consider." He concluded the paragraph: "And 'tis this analogy, which makes men fancy places, traces, and spaces."

(4) is the famous relational theory of space. Scholars find it incompatible with Leibniz's philosophy. Russell suggested Leibniz had a "good" philosophy and a "bad" philosophy and "it probable that as he [Leibniz] grew older he had forgot the good philosophy he kept to himself, and remembered only the vulgarized version by which he won the admirations of Princes and (even more) of Princesses" (1900, p. vi).

In view of the incompatibility, perhaps it is more sensitive not to count the arch rationalist himself as one of those "people" who are so prone to fancy—Leibniz was never shy to say "I reason thus and thus"—Perhaps those people were supposed to be Newtonians, with whom Leibniz was debating. Perhaps Leibniz was saying: Let me show how you people have gone wrong. This interpretation is made more plausible by the distinction Leibniz drew between "place" and "situation." *I suggest Leibniz advanced not a positive theory of space but a diagnosis of a passage to the illusory Newtonian substantival space.* Its starting point is Leibniz's own theory of space in terms of situations, and its end point is the substantival space. The "relational theory" (4) is the way station, which is itself confused if not illusory. The exit point, "but the mind not contended . . . ," illustrates a logic similar to what Kant later elaborated in his Dialectics: Illusion results when reason gets carried away.

The exact meaning of "orders" and "situations" is not clear. Leibniz said reality consists only of extensionless simple substances, monads, and their properties. The prime model for the monad is the mind, and its properties are called perceptions. The properties of each monad mirror every other and mirror God, so that each monad is like a miniature universe in itself. Among the infinite number of internal properties programmed into a monad by the preestablished harmony, there is a peculiar one called its "point of view." The name is metaphoric, for the monads do not actually look at each other; they have no windows. The points of view distinguish one monad from others and make sure that their internal properties are in relative agreement. Implicit relations obtain among the monads by virtue of their points of view, though there is no explicit communication.

Leibniz also associated the monad's the point of view with its situation, position, or locality, and expressed it externally by a mathematical point. He said: "They [the monads] may be called metaphysical points; they have something of the nature of life and they have a kind of perception, and mathematical points are their points of view for expressing the universe." "Position is, without doubt, nothing but a mode of a thing, like priority or posteriority. A mathematical point itself is nothing but a mode, namely an extremity" (Leibniz, quoted in Russell, 1900, pp. 254, 250). The important idea behind the point of view and position seems to be that a certain process of abstraction is involved to generate the concept of situation. The point of view is extracted from other monadic properties and suffers a change of status in the process. As an internal property, it is an attribute of the monad and hence real. However, when extracted and expressed as a position or a situation or a mathematical point, it becomes only a modality. As an abstraction, position is ideal.

Since extended bodies cannot be formed of aggregate of extensionless monads, they are only phenomena. "Matter is not composed of these constitutive unities but results from them, since matter or extended mass is nothing but a phenomenon grounded in things, like the rainbow or the mock-sun, and all reality belongs to unities. . . . Substantial unities are not parts but foundations of phenomena" (letter to de Volder, 1969, p. 536). Extended phenomena result from foundational unities by some kind of "diffusion" and integration. "A simple substance, though it has no extension, yet has position, which is the foundation of extension, since extension is the simultaneous continuous repetition of position" (in Russell, 1900, p. 255). Thus the phenomenon of extended bodies is well founded because it is based on positions, which are in turn based on the points of view of the monads.

It is a central tenet of Leibniz that the complete concept of a simple substance contains all its properties, including relational properties. Extensionless does not imply nonspatial, for there are other spatial concepts such as positions and situation, which are properties of simple substances. After an extensive review, Mates said: "When all is said and done, what is most clear about Leibniz's view on space and time is that he considered the spatial and temporal relations of things in the actual world to be completely determined in each case by the individual accidents of the relata" (1986, p. 231). This is a far cry from the vulgar theory of the "Leibniz of the positivists." Perhaps it is more appropriate to say it is Leibniz's prediction of how positivists fall into the error of defining space as conventional relations among nonspatial entities.

160. In the coordinate bundles approach of Steenrod, we start with "spatio-temporal coordinate patches" and the symmetry group, and construct the total space representing the objective world by taking their product (Appendix B).

161. See R. M. Dancy, "Aristotle and Existence," in Knuuttila and Hintikka (1986), p. 50.

162. Kant (1781), A264/B320, A272/B328. Kant's precritical view on space is somewhat different. Strawson gave a "low-key" interpretation of space and time as forms of intuition: "the duality of intuition and concept is merely the epistemological aspect of the duality of particular instance and general type. . . . Clearly the thought, at its most general, is of some peculiarly intimate connexion between space and time, on the one hand, and the idea of particular item, the particular instance of the general concept, on the other" (1966, p. 48). He also noted how little Kant used the result of the "Aesthetic" in the "Analytic." The "Analytic" is pure conceptual analysis; space and time are conceptualized even in the "Schematism." In the "Analytic," Kant distinguished between space and time, which are not conceptual, and the concepts of space and time, which presuppose the synthesis of apperception (footnote in B160). Thus the "Analytic" embodies a theory of the concepts of space and time, which can be understood with little reference to the "Aesthetic."

163. Weyl (1949), p. 131.

164. Newton (1962), p. 136. Earlier in the paper, Newton has denied that extension is a substance or accident or nothing at all, for "it has its own manner of existence, which fits neither substance nor accidents" (p. 131f). Earman delineated five senses of absoluteness in Newton's Scholium (1989, p. 11). The first is substantivalism; space is a substratum underlying physical events. Two others, the independent absoluteness of space and time, are specific features of the Newtonian space, which are abandoned by absolutists after the advent of relativity. The remaining two senses of absoluteness, first space–time has intrinsic structures, second the

structures are immutable in the actual world and across physically possible worlds, are satisfied by the primitive spatio-temporal structure *M*.

165. The *orientability* of a manifold involves two notions: first, local orientation distinctions; second, whether the local distinctions vary continuously over the manifold (Fig N.2). It does not invoke a space into which the manifold is embedded. Kant had used orientability in his incongruous counterparts argument against relationism, but the argument did not entice him into a substantival space.

166. Einstein, "On the Generalized Theory of Gravitation," in Einstein (1954), p.349.

167. Being continuous has two meanings: There are no gaps, and there are no jumps. When we say "time flows," we think of both its seamlessness and its evenness, and

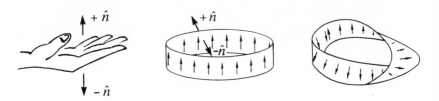

Figure N.2 At each point in a manifold we define a local coordinate system, depicted by the little arrows. In *orientable* manifolds such as a hand and a cylindrical system, the local coordinate strip returns to itself after being brought about continuously around the manifold. Thus we can define global directions $+n$ and $-n$, or palm and back, or two sides of the manifold. In *nonorientable* manifolds such as a Möbius strip, the local coordinate system is "flipped" when brought around the strip. Thus we cannot establish a parity definition.

often do not bother to differentiate the two. The two notions are distinct. "No gaps" is captured by the mathematical concept of completeness, which underlies a *continuum*. "No jumps" is captured by the mathematical concept of *continuous functions*.

The most intuitive idea of a continuum is a line drawn without lifting the pen from the paper, a line has no gaps. Continuity in this sense should not be confused with infinite divisibility or being dense. An ordered system is *dense* if between any two members there is always a third. The rational numbers constitute a system that is infinitely divisible and everywhere dense, but it is not a continuum. It is full of gaps or lacks the property of completeness, as is demonstrated by the *Dedekind cut*. Suppose we divide all rational numbers into two exclusive classes, *A* and *B*, such that any number in *A* is smaller than any one in *B*. There are cases in which *A* has no least upper bound and B has no greatest upper bound. For example, if we stipulate $A = \{a: a^2 < 2\}$, $B = \{b: b^2 > 2\}$, then the "cut" belongs to neither class, for $\sqrt{2}$ is not a rational number. The example exhibits the gaps in the rational line. In the real line or the real number system , these gaps are completely filled by the irrational numbers. Dedekind's stipulation of the *completeness* of $\mathbb{R}$ is: Any cut of $\mathbb{R}$ in *A* and *B* as above is either a least upper bound for *A* or a greatest lower bound for *B*. The real line is complete and hence a continuum. A continuum is not amorphous; it comprises discrete points, as the real line comprises real numbers.

Continuous functions are familiar in calculus, where it is usually defined in terms of real numbers. There is a more general definition in topology that is independent of real numbers. Consider two topological spaces *X* and *Y*, and a map or function $f: X \to Y; x \mapsto y = f(x)$. f is *continuous* if for all points $x \in X$ and any neighborhood O_y of $f(x) \in Y$, there is a neighborhood O_x of $x \in X$ such that x' $\in O_x$ implies $f(x') \in O_y$. Intuitively speaking, f is continuous at a point x means $f(x')$ is as close to $f(x)$ as we demand if x' is sufficiently close to x. The notion of continuous functions places no restriction on the nature of the domain X; we can define continuous functions on discrete sets. For example, the mappings $x \to x +$ a and $x \to ax$ of the rational line into itself are both continuous, although the rational line is not a continuum.

Most classical quantities are functions with space–time as the domain. People sometimes infer the continuity of space–time from the continuity of motion, thereby arriving at the notion of a substantival backdrop. Leibniz protested; he realized there are two notions of continuity. He rejected the continuum of absolute space, arguing that "in actual bodies there is only a discrete quantity." His world consists only of discrete monads. Yet he placed great emphasis in his Principle of Continuity: "Nothing takes place suddenly, and it is one of my great and best confirmed maxim that *nature never makes leaps.*" The principle originates in mathematics, but it is as effective in physics; for example, rest can be considered as an infinitely small velocity or infinite slowness. In measurements, if we vary the physical system slightly, we expect the measured value to change as slightly. If the value swings violently with a minor perturbation of the system, we say something is wrong. Thus if a function is to be physically meaningful, its value should change sufficiently little if the argument stays sufficiently close to the original value. This is Leibniz's Principle of General Order, which is reminiscent of Cauchy's ϵ-δ definition of continuous functions (Leibniz, 1969, pp. 351–53, 519, 539, 544). Some commentators said the notion of continuity is an unresolved contradiction in Leibniz's system. The clear distinction of the two concepts of continuity shows that it is not a contradiction. Leibniz's monadology, fantastic as it seems, is surprising consistent. His frustration lied in the lack of a good formulation of the continuum, which was not developed until 1872 by Richard Dedekind.

168. Stein (1967), p. 194; Broad (1981), p. 171.
169. More discussions of internal and external relations are found in § 29.
170. Hesiod, Theogony, 116, 176–82. Hesiod repeated the separation theme in anthropic terms: Kronos (Time) severed the coupling of Ouranos (Heaven) and Gaia (Earth). The repetition shows the symmetry of space and time. The meaning of *chaos* is found in Kirk, et al. (1983), Ch. I, Sec. 5. The Pythagorean cosmology is found in Aristotle's *Physics*, 213b; for more discussion, see Guthrie (1962), pp. 277–282. Aristotle's own definition of space is found in *Physics*, 212b, 211a.
171. Locke (1690), second ed., II, xxvii, 1; Popper (1953), p. 107.
172. Aristotle, *Physics*, 214a.
173. Leibniz (1765), p. 230. Sanford (1970) also argues to the same effect.
174. Strawson (1959), Ch. 2.
175. For a catalogue of the myriad arguments on rotation, see Earman (1989), Ch. 4.
176. In 1925, George E. Uhlenbeck and Samuel Goudsmit announced that the observed hydrogen spectra can be explained accurately if the electron has, besides its orbital angular momentum, an additional intrinsic angular momentum of half Planck's constant, $\hbar/2$. In 1928, Dirac developed the relativistic wave equation for electrons and showed that the spin can only be understood as a consequence of relativistic invariance.
177. Older textbooks sometimes introduced a "relativistic mass" as distinct from "rest mass." The relativistic mass is supposed to change with the velocity of the particle,

that is, coordinate dependent. The distinction is a confusion. Physics defines only one mass m, which is an invariant. See Weinberg (1972), p. 33; Okun (1989).

178. Goodman (1954), p. 74.

179. In his 1916 paper laying the foundation of general relativity, Einstein argued that epistemologically the most basic spatio-temporal concept is the coincidence of material bodies. Leaving aside the epistemological consideration, coincidence is crucial to modern physics, for it underlies the concept of point interaction.

180. See, for example, Feynman et al., Vol. 3, Chs. 3 and 4.

181. Friedman (1983), pp. 62f, 217.

182. Einstein specifically challenged this point. The major question is, as Einstein put it, "Is geometry—looked from the physical point of view—verifiable (viz., falsifiable) or not? . . . Poincaré says no and consequently is condemned by Reichenbach" (1949, pp. 677f). Einstein did not subscribe to Poincaré's conventionalism, and neither did Eddington. Yet both pointed out the significance of Poincaré's idea, which is, in Eddington's words, "he brings out interdependence between geometrical law and physical law" (1920, Prologue). Weyl also argued that "geometry, mechanics, and physics form an inseparable whole," (1921, pp. 67, 93).

   In positivist theories, a privileged status is given to the distance function $d(a, b)$, which is seldom used in physical theories but which answers to the familiar notion of the "metrical structure" of the world. In modern geometry, the distance function stipulates an abstract rule to define distance, but it does not stipulate any standard of extensive magnitude. If $d(a, b)$ and $d(a', b')$ turn out to be equal, we say the distances are the same. But this equality comes only *after* integration and does not constitute the basis of a universal standard. In the positivist instrumental definition of metrical properties, some lengths $d(a, b)$ and $d(a', b')$, where (a, b) and $(a' b')$ denote the ends of a rigid rod at different locations, are stipulated to be equal *in advance*. This stipulation of a universal standard is *a priori* and arbitrary, and is the source of the conventionalist conclusions.

183. According to this reasoning, "$(\exists x)(x$ is the present king of France and x is bald)" is false because "x is the present king of France" has no instantiation. The various senses of being were first discriminated by Frege and have become the cornerstone for the interpretation of predicate logic. Recently, many classical scholars argued that the copula had been traditionally *the* meaning of being, and the Frege–Russell classification of the senses of being played little or no role in philosophy until their time. See the collection of essays in Knuuttila and Hintikka (1986).

184. Kant (1781), A598/B626, A600f/B628f.

185. Newton said: "That gravity should be innate, inherent, and essential to matter, so that one body may act upon another at a distance through a *vacuum*, without the mediation of anything else, by and through which their action and force may be conveyed from one to another, is to me so great an absurdity that I believe no man has in philosophical matters a competent faculty of thinking can ever fall into it" (letter to Richard Bentely, February 25, 1692/3, his italics).

   Leibniz said: "There is no vacuum at all in the tube or in the receiver; since glass has small pores, which the beams of light, the effluvia of the load stone, and other very thin fluids may go through" (fifth letter, ¶34). He is essentially right about the vacuum. Even if the bottle is made of lead, it would not stop neutrinos from zipping through. Some people blamed Leibniz and others who argued against the vacuum for dogmatically ignoring the hard facts provided by the vacuum pump. The critics are even more dogmatic in their arbitrary exclusion of less apparent but no less relevant evidences in their narrow definition of "hard facts."

186. For instance, Friedman's absoluteness is essentially a negative concept; absolute is not relational, not conventional, and not dynamical (1983, pp. 62–70).

187. Friedman (1983), pp. 217ff. He defined: "space–time: the set of all places-at-a-time or all actual and possible events. Our theories postulate various types of geometry structures on this set and picture the material universe—the set of all *actual* events—as embedded within it" (p. 32).

188. Friedman (1983), p. 223.

189. Anderson (1964).

190. Einstein (1954), p. 246.

191. Weinberg (1972), p. vii.

192. "The hypothetical employment of reason has, therefore, as its aim the systematic unity of the knowledge of understanding, and this unity is the *criterion of the truth* of its rules. The systematic unity (as a mere idea) is, however, only a *projected* unity, to be regarded not as given in itself, but as a problem only" (A647/B675). Kant distinguished between the definition of truth, which is "the agreement of knowledge with its objects," and the criterion of truth, and argued that a general and sufficient criterion cannot possibly be given (A58/B82). The coherence criterion is thus hypothetical.

193. Field (1980), p. 35; Earman (1989), p. 155.

194. Einstein (1954), p. 268.

195. Teller (1987); Earman (1989), pp. 14, 115.

196. Earman (1989), pp. 159–163, 14.

197. Aitchison (1985), p. 390.

198. Many papers on the debate on reference, modality, and intensional logic are collected in Linsky (1971) and Schwartz (1977).

199. Frege (1892); Russell (1905).

200. Mill (1843), Bk. 1, Ch. 2; Kripke (1972).

201. It may be argued that the reference is descriptive, for since the identity x is unique to the event $\psi(x)$, it can be counted as a property. "Being itself," "being Pegasus," "being the bearer of the name 'Pegasus,'" or anything that can be said about the object can be called descriptions because words are intrinsically meaningful. However, to lump everything under the predicate letter indiscriminately is to muddle the logical structure of our thoughts when a distinction is commonly used and clearly articulated. The identity is a formal concept, it differs from manifest properties. Descriptions involve determinate properties or explicit relations between the referent and some other objects, for example, "the teacher of Alexander" or "the king of France." Neither is involved in the references of field theories. No definite relation has been introduced yet.

202. The original idea of possible worlds came from Leibniz. It has become a major concept in semantics. Kripke suggested "possible state (or history) of the world" or "counterfactual situation" as less misleading terms. For possible worlds are not given sets of entities with dubious identities, but stipulations we make in circumscribing a set of identified entities in discourse, (1972).

A systematic assignment of qualities is not trivial, for there may be many possibilities open to each individual. In field theories, this is shown by the monstrous phase space each event possesses; it is something like a circle for the simple $U(1)$ symmetry and a sphere for $SU(2)$. A rule would pick out one point in the sphere for each event. Mathematically, a set of compossible qualities can be represented by the notion of continuous assignment. There are at least two concepts that satisfy this criterion, a *section* or a *horizontal lift* in the fiber bundle formulation

(Appendix B). The latter is more substantial, for it is involved in the notion of causality in field theories.

203. Russell (1948), pp. 74–80. Russell was criticizing Carnap's proposal to eliminate proper names and replace them by coordinates. For example, instead of "the object a is blue" one says "the position (3) is blue" (Carnap, 1936, pp. 11–13). My interpretation is totally different from that of Carnap, who substituted the concept of finite objects by the spatio-temporal regions they occupy. Russell pointed out that Carnap's space is substantival. I do not postulate any substantival regions of space.

204. Mill said, for example, Dartmouth was originally so named because it was located on the mouth of the River Dart. However, once the name was bestowed, it was emancipated from its original meaning. The Dart may be silted so that Dartmouth is no longer at its mouth, yet "Dartmouth" sticks regardless of the fact.

205. Itzykson and Zuber (1980), p. 110.

206. Weinberg (1977); Haag (1992), p. 105.

207. Field-theoretic methods are powerful in treating many-particle systems. Symmetrized states $|\alpha\alpha \ldots \alpha'\alpha' \ldots \rangle_{\pm}$ can be conveniently cast in the occupation number representation $|n_{\alpha}n_{\alpha'} \ldots \rangle_{\pm}$. Fock space formulation that accommodates variable number of particles can be introduced. Creation and annihilation operators analogous to that of field quanta can be defined. So can field operators, which are a superposition of creation and annihilation operators. The procedure is usually called second quantization. It must be emphasized that the formalism does not change the physical content. Using field-theoretic methods in many-body problems does *not* make the many-particle system into a field. Second quantization does not bring a second quantum theory but only extends the original theory to an indefinite number of particles to take advantage of powerful analytic tools. For a clear discussion of second quantization in the context of quantum many-particle systems, see Negele and Orland (1988), Ch. 1.

208. For example, see Dieks (1990), van Fraassen (1991), Chs. 11, 12, and references cited therein.

209. See, for example, Gottfried, § 41; Sakurai (1985), Ch. 6.

210. Leibniz said: "Two terms are the *same* if one can be substituted for the other without altering the truth of any statement. If we have A and B and A enters into some true proposition, and the substitution of B for A wherever it appears, results in a new proposition which is likewise true, and if this can be done for every such proposition, then A and B are said to be the *same*; and conversely, if A and B are the same, they can be substituted for one another as I have said." See Mates (1986), Ch. 7, for more discussion.

211. Dieks (1990). He also argued that the closest we can come to a single-particle state is to take the partial trace of the symmetrized state. The result is not a pure state but a mixture. The meaning of a "single particle" with a mixture state is unclear. Furthermore, the mixture states of all "particles" are the same, so that we cannot distinguish the "particles" by their different states.

212. See, for example, Gottfried (1966), § 20.2. The approximation that individuates the two beams in the Stern–Gerlach experiment occurs between Eqs. (11) and (12). Note that the argument invokes the center-of-mass position as the criterion of individuation.

Chapter 7

213. I say that the Cartesian coordinate systems enable us to describe directions, which are qualities. I do not mention positions, which pertain to identity. The global coordinate systems roll together two functions that are made distinct in theories with local symmetries. Local theories have two symmetry groups, the spatio-temporal group and the local group, which split the work of the single symmetry group in global theories. In general relativity the Poincaré group, which is the group for special relativity, is localized to each spatio-temporal point and released from the duty of identifying events. Identification becomes the job of the spatio-temporal group of diffeomorphism (§ 7).

214. Globally applicable predicates are so prevalent in our thinking we seldom notice their conflict with the idea of individual entities. It helps to consider an analogous case in economics, where the notion of individuals strikes a more responsive note. Utility or value is an important predicate in economics. In classical economics, the value of a commodity is determined by the amount of labor and/or capital required for its production. It is a universal measure accepted by all individuals. This is unsatisfactory because different individuals may value the same commodity differently. Paul Samuelson said: "If one were looking for a single criterion by which to distinguish modern economic theory from its classical precursors, he would probably decide that this is to be found in the introduction of the so-called subjective theory of value into economic theory" (1947, p. 90). In modern economics theories, utility is the measure of commodities according to individual preferences, and there is no universal standard for utilities among individuals. Each consumer orders his preferences for bundles of commodities. The preferential order can be numerically represented by a utility function, which assigns a number to each commodity bundle according to its preferential status. Thus a consumer can say to himself of a commodity bundle: its value is so much. The utility function of a consumer is not unique; any monotone transformation of it preserves the order of preferences and serves equally well. Thus there is an arbitrary element involved in choosing a particular utility function. The arbitrariness prevents the comparison of utilities for different consumers. The end of interpersonal comparison of utility has great repercussion in general welfare considerations, for it makes sayings such as "the greatest utility for the greatest number" meaningless.

It is important to note that the psychological nature of utility is inconsequential in the argument—economists ignore psychological factors and the results of psychological experiments showing that their model of consumers is unrealistic—logically, what matters is individuality. The preference orders of various consumers are logically not different from things with various properties. Each consumer observes a preference order, which he describes by a utility function. Being individuals, each chooses a utility function as he pleases, as long as the choice falls within a group of functions related by monotone transformations. In physicists' terminology, the utility function has a monotone symmetry, and the source of incomparability stems from the individuality or "locality" of the symmetry.

Economic theories are not interested in a collection of solipsistic consumers. Presumably the consumers can talk to each other and agree to conform to some global convention. However, talk and conformity are relations, which are inessential and perhaps even obscuring to the only kind of relations that concerns economics, trade. Consumers can reconcile their differences through trade and exchange, where each seeks to maximize his utility under his budget constraint. Trade results in equilibrium market prices for the commodities. The full respect for the individuality of consumers simultaneously brings out the full force of the market mechanism.

Similar ideas are found in gauge field theories, where the sharp formulation of individual events brings out the significance of interaction dynamics. In both cases, some universal standard supporting dubious relations that tacitly bind individuals is discarded and replaced by a substantive mechanism that explicitly relates the individuals. There are also big differences. The consumers in competitive markets are passive price takers whose characteristics are not changed by the institution of the market. In contrast, the world described by gauge field theories is fully inter-active and dynamical. The institution of interactions changes the characteristics of the individual events, so that the events themselves become the source of interaction.

215. Kant's argument centers on the distinction between subjective and objective temporal sequences. Experiential representations occur in a sequence, but the order of events in the sequence is accidental, as stream-of-consciousness novels vividly portray. By themselves the representations do not yield a coherent picture of the objective world. We do experience the world coherently because we have already used certain concepts and rules to synthesize the representations and to determine the objective temporal order. Since time is itself not perceivable, the objective time order can only be obtained through the temporal relations among objects. The three modes of time are duration, succession, and coexistence, and the three corresponding temporal relations are the three analogies, permanence, causality, and reciprocity. The "Analogies of Experience" argues for the necessity of these concepts in intelligible experiences. It is followed by the "Refutation of Idealism." Idealism asserts the primacy of inner consciousness and argues that the existence of external objects are either doubtful or impossible to demonstrate. Kant countered that inner experiences depend on outer experiences: "The mere, but empirically determined, consciousness of my own existence proves the existence of objects in space outside me" (1781, B275).

216. For instance, Kant said: "If you represent time by a straight line produced to infinity, and simultaneous things at any point of time by lines drawn perpendicular to it, the plane thus generated will represent the phenomenal world, both as to its substances and as to its accidents" (Inaugural Dissertation, § 14).

217. Most people visualize time linearly as an ordered sequence of events. Some regard time as a circle, in the eternal recurrence of events. Lines and circles are both geometric. In thinking of time as a curve, we have already spatialized time in some ways. The concept of place can be employed both spatially and temporally. The temporal analog of extension is duration, of here is now, of geography is history. Change usually means temporal change. However, it is just as meaningful to define change as variation in geography instead of in history. In the first instance, the object gains and loses properties across its spatial extension, as a rod is hot on one side and cold on the other. Two objects can move closer and apart in space, their distance measured by the instantaneous separation of their positions. Analogously, two objects can move closer and apart in time, their temporal separation measured by the time separating their being in the same place. For more details, see Taylor (1955).

218. Aristotle, *Physics*, 219b.

219. Parmenides' notion of timelessness seems to mean a tenseless present (Owen, 1966). The distinction between a static time and a flowing time was adumbrated by Plato, who called our time "a moving image of the eternity which remains for ever at one" (*Timaeus*, 37). It was clearly brought out by the Neoplatonist Iamblichus of Syria, who died around A.D. 325. Iamblichus differentiated a generative time or a higher "now," which is beyond the sensible world, and a generated time or many lower "nows" residing in participating events and things. He said: "And where should we think that the flow (*rhoe*) and shifting (*ekstasis*) of time occur? We shall say in the things which participate in time. For these are always coming into being and cannot take on the stable nature of time without changing, but touch that nature with ever different parts of themselves" (quoted in Sorabji, 1983, p. 38).

220. Kant (1781), B224. The statement of the First Analogy in the first edition of the *Critique* is: "All appearances contain the permanent (substance) as the object itself, and the transitory as its mere determination, that is, as a way in which the object exists" (A182). In the second edition it is replaced by: "In all change of appearance substance is permanent; its quantum in nature is neither increased nor diminished" (B244). I think the second edition's formulation involves a more general concept. "Permanent" in the first edition's statement clearly means endurance, but in the second edition's statement it shifts closer to being atemporal, which is presupposed by the notion of endurance. Kant did not explicitly distinguish two meanings of permanence, but he did distinguish between a singular "substance" and plural "substances," as Strawson noted. It is interesting to make a list and see that the two meanings and usages can be separated. The overall movement of the "First Analogy" is from the general to the specific. The conservation of matter is not the general principle central to the "First Analogy," but a particular case to illustrate the principle of permanence. Finally, "substance" is a concept and not a substratum that bears properties. Kant had rejected the substantival space. He also attacked Locke's notion of substance as distinct from properties: "But the accidents are not particular things that exist, rather only particular ways of considering existence; they therefore do not need to be borne, but rather signify only the manifold determination of one and the same thing" (quoted in Guyer, 1987, p. 212).

221. The possible world argument can be found in Leibniz's fifth letter to Clarke. The kind of possible worlds we need here is more radical than the kind discussed in § 21, where the set of events is fixed and only their qualities can vary across the possible worlds.

222. See, for example, Sklar (1974).

223. Hume's considered definition is: "We may define a CAUSE to be 'An object precedent and contiguous to another, and where all the objects resembling the former are plac'd in like relations of precedence and contiguity to those objects, that resemble the latter'" (Hume, 1739, p. 170; see also pp. 69ff, 79). In the *Treatise* Hume did not make a clear separation between objects and appearances of objects, and used "idea" and "object" interchangeably. Here I only consider the meaning of relations between objects.

It is commonly agreed that cause is a relation not between objects but between events or the changes of objective states. For example, we say striking a match causes it to burst into flame; we do not say a match causes fire. However, Hume's arguments center not on events but on successive objects, impacts of objects and their subsequent motion. This suits my purpose because I am considering a four-dimensional world, and it is clearer to talk about objects than their "changes," which will need additional definition. Thus I use "cause" to mean relations among objects in general. ("Event" in this note is understood in the ordinary sense, not to be confused with "event" in the main text, which is a technical term for four-dimensionally extensionless objects.)

224. Hume said: "In single instances of the operation of bodies, we never can, by our utmost scrutiny, discover anything but one event following another; without being able to comprehend any force or power by which the cause operates, or any connexion between it and its supposed effect. . . . All events seem entirely loose and separate. One event follows another; but we never can observe any tie between them. They seem conjoined, but never connected" (Hume, 1748, pp. 73ff).

225. Wittgenstein (1953), §§ 189–240.

226. Hume (1739), p. 70.

227. See, e.g., Piaget (1954), Ch. 1.

228. Hume (1739), Bk. I, Part IV, Sec. 2.

229. Hume (1739), p. 225.

230. Hume (1739), p. 89.

231. Kant (1788), p. 54.

232. Weinberg (1972), p. 92.

233. Rorty (1967) gives a good review on the philosophical debate on internal and external relations.

234. Glashow (1985), p. 584.

235. Einstein (1949), p. 680. "*Das Wirkliche ist uns nicht gegeben, sondern aufgegeben (nach Art eines Rätsels).*" Schilpp's translation is "The real is not given to us, but put to us (by way of a riddle)." Einstein attributed the sentence to Kant, but emphasized that he was not raised in the Kantian tradition. *Gegeben* and *aufgegeben* are Kant's words (1781, A498/B526), and I follow Kemp Smith's translation of *aufgegeben* as "set as a task." However, I do not know the exact source of Einstein's sentence. The same words were used by Weyl, who said that objective reality is *nicht gegeben, sondern aufgegeben*, (1949, p. 117).

Appendixes

236. The mathematical theory of probability emerged in the decade around 1660, during which Pierre Fermat, Blaise Pascal, and Christiaan Huygens pioneered probability, and John Graunt published the first extensive statistical inference based on the London mortality record. Probability theory took its classical form in Pierre Simon Laplace's 1812 treatise. It was found to be too restrictive as high-level abstract construction in mathematics developed in the nineteenth century. In 1933, A. N. Kolmogorov advanced the present axiomatization. There are several dissents to the Kolmogorov axiomatization. Richard Mises had a different system based on the notion of a collective; his limiting relative frequency is not countably additive. Jeffrey and Hartigan relaxed Kolmogorov's axiom of unitary total probability and allowed infinite probabilities. Also, there is quantum probability, whose probability space goes beyond a σ algebra. Quantum probability is pure mathematics inspired by quantum mechanics; it has little to do with the physics.

237. T. Fine (1973) and Weatherford (1982) are two recent reviews on various theories of probability. Fine emphasizes the logical side and Weatherford the philosophical side. The frequency view was pioneered by J. Veen in 1866. It was developed by R. Mises and H. Reichenbach into the limiting frequency theory. B. C. van Fraassen

made limiting frequency countably additive in his modal interpretation. The idea of propensity was adumbrated by C. S. Peirce and formally advanced by K. Popper in 1957. Kyburg (1974) gives a good review of various proponents.

Classical statistical inference was developed in the 1920s by R. A. Fisher, E. S. Pearson, and J. Neyman, who are commonly called frequentists. However, they are less extreme than those of limiting frequentists, whose views are conditioned by the positivist philosophy. The pioneering works in subjective probability are done by F. P. Ramsey and B. de Finetti in the 1930s. The probability axioms appear as criteria of rationality that broadly constrains the person's judgment. de Finetti introduced the concept of exchangeability and showed that given enough evidence, persons with diverse initial opinions can rapidly learn and reach intersubjective agreement. In 1954, L. J. Savage combined their ideas with classical statistical inference and the concepts of decision and utility to lay the foundation of modern Bayesian decision theory. For a recent review on subjective probability, see Fishburn (1986) and the attending comments by various authors.

238. Loéve (1977), p. 151.

239. Aristotle said the only things that are fundamentally quantitative are numbers, speeches, lines, surfaces, bodies, time, and place. Everything else is quantitative in a secondary sense (*Category*, 4b20, 5a35).

240. Kant (1781), A163/B203. Kant distinguished between extensive and intensive magnitudes, the latter corresponding to qualities.

241. Russell (1903), Ch. XI, argues that numbers describe sets. Quine (1960), p. 118, opts for the dissolution of number words. Benacerraf (1965) considers various interpretations of number words.

242. The classic text in measure theory is Halmos (1950) (see especially §§ 4, 7, 8). This material is also covered in many textbooks on probability and analysis, e.g. Loéve (1977), §§ 1.1, 1.2, 3.1.

243. For the proof, see, for example, Herbaek and Jech (1984), Ch. 11.

244. A σ algebra S is a collection of subsets E_i of Ω such that: (1) If E_i and E_j belong to S, then their difference E_i - E_j belongs to S; (2) if $E_i, i = 1, 2, \ldots$, belong to S, then their countable union $E_1 \cup E_2 \cup \ldots$ belongs to S; (3) the set Ω belongs to S.

245. More exactly it is the extended real line, which is set of all real numbers x plus the positive and negative infinities, $\mathbb{R} = \{x : -\infty \leq x \leq \infty\}$.

246. Pfeiffer (1965), § 3.1, p. 119; Loéve (1977), § 5.3, p. 168.

247. Consider a bounded function f on a set Ω with elements ω. Partition the range of f into intervals Δx_i, each with length less than ε. The inverse function partitions Ω into subsets $E_i = \{\omega: \omega \in f^{-1}(\Delta x_i)\}$. The *indication function* of the set E_i is defined by

$$\chi_E(\omega) = \begin{cases} 1, & \text{if } \omega \in E_i \\ 0, & \text{if } \omega \notin E_i \end{cases}$$

Now pick $x_i \in \Delta x_i$. In the subset E_i, the values of $f(\omega)$ all fall within ϵ of x_i. A bounded measurable function f can be uniformly approximated by $f(\omega) \approx \Sigma x_i \chi_E$. Now suppose E_i represents a property, say, a specific color, red. The indicator function χ_E singles out this color. The function f is a rule that systematically represents each specific color E_i by a value x_i. I call f the property type and x_i its values.

248. Pfeiffer (1967), pp. 120ff; Loéve (1977), p. 168.

249. As an example, consider the indicator function $\chi_E(\omega)$. It has only two values, 1 and 0, depending on whether ω falls within E or not. Now consider any Borel set Δ. It can contain 1 or 0, or both, or neither; there are only four possibilities. Hence the σ algebra S_f generated by $\chi_E(\omega)$ contains only four elements $(E, \Omega\text{-}E, \Omega, \phi)$. It is a very simple and measurable system, which essentially carves out the set E.

250. The distribution function $F_f(x)$ gives the proportion of those individuals whose values of f are less than x. The derivative of the distribution function is called the *density function*. It gives the relative magnitude of the set of ω whose values of f lie in the infinitesimal range dx. The familiar bell-shaped curves are density functions.

251. The concept of measure was originally introduced by Henri Lebesgue in the context of integration, where its meaning is most apparent. The familiar kind of integration is the Riemann integral, in which we divide the *domain* of integration into tiny parts, sum the contributions of the parts, and evaluate the integral by taking the limit in which the parts shrink to zero (Fig. N.3a). The domain of the Riemann integral must be analytic; the integral makes no sense if the domain is some abstract space.

Lebesgue's innovation is to change the independent variable from the function's domain to its range. The *range* of the function is divided into tiny parts, the partition is projected by the inverse function into the domain, the contributions of the parts are summed, and the integral is evaluated in the proper limit. The Lebesgue integral relaxes the restriction on the domain of f, so that integration can be generalized to functions defined on abstract spaces.

In applications, the domain of the function usually represents objective systems. The real number system has rich structures. Instead of ascribing all the structures to the object systems as in the Riemann integral, the Lebesgue integral projects only those structures that are necessary to support quantitative measurements. Thus it demands a weaker presupposition about the system. The range includes contributions from both objective attributes and the general form of our measurements, hence it is in some way under our control.

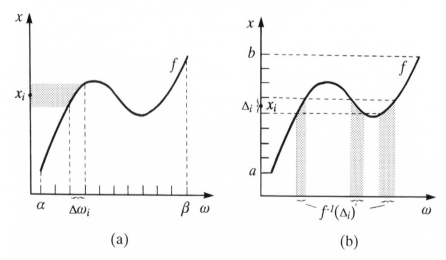

(a)   (b)

Figure N.3. (a) In a Riemann integral, the domain of the function f is partitioned. (b) In a Lebesgue integral, the range of f is partitioned, and the inverse function f^{-1} induces a corresponding partition of the domain.

252. Kyburg (1976).
253. Nagel (1986).
254. Pfeiffer (1965), pp. 58, 157, 290f; Loéve (1977), pp. 235f.
255. Kolmogorov (1933), p. 9.
256. Reichenbach said: "We define probability as limiting frequency within a infinite sequence" (1935, p. 68). H. Cramér expressed the general attitude of statistical frequentists. He said the probability of E with respect to experiment ε is P means "in a long series of repetitions of ε, it is practically certain that the frequency of E will be approximately equal to P" (1946, p. 149). The difference between the two was noted by Mises, who complained that Cramér omitted giving a clear definition of probability, (1957, pp. 81f).
257. Reichenbach allowed a "fictitious meaning" for singular cases (1935, § 72). It is unclear how this fictitious meaning fits into his limiting frequency definition. The early limiting frequentists were hostile to modal concepts. Recently, van Fraassen refined some technical points of limiting frequency theories and broadened its conceptual base, mainly to hypothetical sequences. However, his modal interpretation still reduces probability to statistics of "long runs" (1978). Summaries of criticisms of the limiting frequentist view, together with references, can be found in Fine (1973), Ch. IV, and Weatherford (1982), Ch. IV, § 10. The statistical frequency conception is not often discussed in philosophical reviews, for it is not an independently articulated theory but is embodied in the working of classical statistical inference. Its critics are other working statisticians, mainly the Bayesians.
258. The most commonly quoted source of epistemic probability is P. S. Laplace's *A Philosophical Essay on Probabilities*. Laplace believed that the world is deterministic, and argued that we have to use probability calculus because we lack complete knowledge. He said: "The curve described by a simple molecule of air or vapor is regulated in a manner just as certain as the planetary orbits; the only difference between them is that due to our ignorance. Probability is relative, in part to this ignorance, in part to our knowledge. . . . The theory of chance consists in reducing all the events of the same kind to a certain number of cases equally possible, that is to say, to such as we may be equally undecided about in regard to their existence, and in determining the number of cases favorable to the event whose probability is sought. The ratio of this number to that of all the cases possible is the measure of this probability, which is thus simply a fraction whose numerator is the number of favorable cases and whose denominator is the number of all the cases possible" (1820, pp. 6f). This is the classical definition of probability based on the principle of insufficient reason. Some commentators see it as epistemic probability or the measure of our ignorance. The readers will decide for themselves whether Laplace has regarded probability as a measure of ignorance, confusing the calculation made in partial ignorance with the calculation of ignorance.
259. Trautman (1980a, 1980b) compares the formalisms of general relativity and gauge field theories. Daniel and Viallet (1980) review the fiber bundle formulation of gauge field theories. Abraham and Marsden (1978) give the modern formulation of classical mechanics.
260. Steenrod (1951) §§ 1–3; Kobayashi and Nomizu (1963), pp. 52f.
261. Kobayashi and Nomizu (1963), § 1.5; Daniel and Viallet (1980).
262. Kobayashi and Nomizu (1963), pp. 63f, 68–71.
263. Mayer (1977), pp. 61–63, 74–76; Daniel and Viallet (1980).

264. For the tetrad formulation of general relativity, see Held, et al. (1976). For the fiber bundle formulation, see Trautman (1980a, 1980b). Ramond (1989), § 6.4, presents gravity as a gauge theory with a parallel development for the Yang-Mill theories.

265. For the evolution of the early universe, see Weinberg (1977); Silk (1980).

266. Gottfried and Weisskoff (1984), pp. 155, 169.

267. See Krauss (1986).

268. Guth and Steinhardt (1984).

269. For an account of the grand unifying theories, see Georgi and Glashow (1980).

270. Weinberg (1977), p. 144; A. Camus, "The Myth of Sisyphus"; J. P. Sartre, *The Flies*, Act III; B. Russell, "A Free Man's Worship."

271. Kant, *Critique of Judgment*, § 28. He concluded the *Critique of Practical Reason*: "Two things fill the mind with ever new and increasing admiration and awe, the oftener and more steadily we reflect on them: the starry heavens above me and the moral laws within me."

BIBLIOGRAPHY

Abraham, R., and J. E. Marsden, 1978, *Foundations of Mechanics*, second ed., Addison–Wesley.

Aitchison, I. J. R., 1985, "Nothing's Plenty, the Vacuum in Modern Quantum Field Theory," *Contemp. Phys.* **26**, 333–91.

Aitchison, I. J. R., and A. J. G. Hey, 1989, *Gauge Theories in Particle Physics*, 2nd ed., Adam Hilger, Philadelphia.

Albert, D. Z., 1992, *Quantum Mechanics and Experience*, Harvard University Press.

Alexander, H. G., ed., 1956, *The Leibniz–Clarke Correspondence*, Barnes & Noble.

Anderson, J. L., 1964, "Relativity Principles and the Role of Coordinates in Physics," in *Gravitation and Relativity*, ed. by H. Y. Chiu and W. F. Hoffmann, Benjamin, pp. 175–94.

Anderson, J. L., 1967, *Principles of Relativity Physics*, Academic Press.

Armstrong, D. M., 1978, *Nominalism and Realism*, Cambridge University Press.

Asquith, P. D., and R. N. Giere, ed., 1980, *PSA 1980*, Phil. of Sci. Assoc., East Lansing, MI.

Ayer, A. J., 1954, *Philosophical Essays*, Macmillan.

Ballentine, L. E., 1990, "Limitations of the Projection Postulate," *Foundations of Phys.* **20**, 1329 –43.

Belinfante, F. J., 1973, *A Survey of Hidden Variable Theories*, Pergamon.

Bell, J. S., 1987, *Speakable and Unspeakable in Quantum Mechanics*, Cambridge University Press.

Benacerraf, P. 1965, "What Numbers Could Not Be," *Phil. Rev.* **74**, 47–73.

Beltrametti, E. G., and G. Cassinelli, 1979, "Properties of States in Quantum Logic," in *Problems in the Foundation of Physics*, ed. by G. T. de Francia, North Holland, pp. 29–70.

Beltrametti, E. G., and G. Cassinelli, 1981, *The Logic of Quantum Mechanics*, Addison–Wesley.

Bennett, J., 1966, *Kant's Analytic*, Cambridge University Press.

Bennett, J., 1971, *Locke, Berkeley, Hume*, Oxford University Press.

Berzi, V., and V. Gorini, 1969, "Reciprocity Principle and the Lorentz Transformations," *J. Math. Phys.* **10**, 1518–24.

Bohm, D., 1951, *Quantum Theory*, Dover.

Bohr, N., 1934, *Atomic Theory and the Description of Nature*, Cambridge University Press.

Bohr, N., 1949, "Discussion with Einstein on Epistemological Problems in Atomic Physics," in Schilpp (1949), pp. 199–242.

Bohr, N., 1963, *Essays 1958–1962*, Wiley.

Born, M., 1968, *The Born–Einstein Letters*, tr. by I. Born, Walker & Co., NY.

Broad, C. D., 1981, "Leibniz's Last Controversy with the Newtonians," *Leibniz: Metaphysics and Philosophy of Science*, ed. by R. S. Woolhouse, Oxford University Press, pp. 157–74.

Brown H. R., and R. Harré, ed, 1988, *Philosophical Foundations of Quantum Field Theory*, Oxford University Press.

Bub, J., 1988, "How to Solve the Measurement Problem of Quantum Mechanics," *Found. of Phys.* **18**, 701–22.

Busch, P., 1985, "Indeterminacy Relations and Simultaneous Measurements in Quantum Theory," *Int. J. Th. Phys.* **24**, 63–92.

Busch, P., and P. J. Lahti, 1983, "A Note on Quantum Theory, Complementarity and Uncertainty," *Phil. Sci.* **52**, 64–77.

Busch, P., and P. J. Lahti, 1989, "The Determination of the Past and the Future of a Physical System," *Found. Phys.* **19**, 633–78.

Busch P., P. J. Lahti, and P. Mittelstaedt, 1991, *The Quantum Theory of Measurement*, Springer–Verlag.

Carnap, R., 1936, *The Logical Syntax of Language*, tr. by A. Sweaton, Routledge.

Clauser, J. F., and A. Shimony, 1978, "Bell's Theorem, Experimental Tests and Implications," *Rep. Prog. Phys.* **41**, 1881–927.

Corbett, J. V., and C. A. Hurst, 1978, "Are Wave Functions Uniquely Determined by Their Position and Momentum Distributions?" *J. Austral. Math. Soc.* **B20**, 182–201.

Cramér, H., 1946, *Mathematical Methods of Statistics*, Princeton University Press.

Crease, R. P., and C. C. Mann, 1986, *The Second Creation*, Macmillan.

Cushing, J. T., and E. McMullin, ed., 1989, *Philosophical Consequences of Quantum Theory*, University of Notre Dame Press.

Daniel, M., and C. M. Viallet, 1980, "The Geometrical Setting of Gauge Theories of the Yang-Mills Type," *Rev. Mod. Phys.* **52**, 175–97.

Davies, E. B., 1976, *Quantum Theory of Open Systems*, Academic Press.

Dieks, D., 1990, "Quantum Statistics, Identical Particles, and Correlations," *Synthesis* **82**, 127–55.

Dirac, P. A. M., 1930, *The Principles of Quantum Mechanics*, 4th ed., 1958, Oxford University Press.

Dirac, P. A. M., 1975, "Directions in Physics," ed. by H. Hora and J. R. Shepanski, Wiley.

Dummett, M., 1991, *The Logical Basis of Metaphysics*, Harvard University Press.

Earman, J., 1989, *World Enough and Spacetime*, MIT Press.

Einstein, A., 1916, "The Foundation of the General Theory of Relativity," reprinted in *The Principle of Relativity*, Dover, 1952, pp. 111–64.

Einstein, A., 1949, "Remarks Concerning the Essays Brought Together in this Co-operative Volume," in Schilpp (1949), pp. 665–88.

Einstein, A., 1954, *Ideas and Opinions*, Crown Publishers, NY.

Elliott, J. P., and P. G. Dawber, 1979, *Symmetries in Physics*, 2 vols., Oxford University Press.

Fano, G, 1971, *Mathematical Methods of Quantum Mechanics*, McGraw-Hill.

Feller,W., 1950, *An Introduction to Probability and Its Applications*, 3rd ed., Wiley, 1968.

Feynman, R. P., 1965, *The Character of Physical Laws*, MIT Press.

Feynman, R. P., 1982, "Simulating Physics with Computers," *Int. J. Theo. Phys.* **21**, 467–88.

Feynman, R. P., R. B. Leighton, and M. Sands, 1963, *The Feynman Lectures on Physics*, 3 vols., Addison–Wesley.

Field, H., 1980, *Science without Numbers*, Princeton University Press.

Fine, A., 1986, *Shaky Game, Einstein Realism and the Quantum Theory*, University of Chicago Press.

Fine, T. L., 1973, *Theories of Probability*, Academic Press.

Fishburn, P. C., 1986, "Axioms of Subjective Probability," *Stat. Sci.* **1**, 335–58.

Fraassen, B. C. van, 1980, *The Scientific Image*, Oxford University Press.

Fraassen, B. C. van, 1991, *Quantum Mechanics, an Empiricist View*, Oxford University Press.

Frege, G., 1884, *The Foundations of Arithmetic*, 2nd ed., tr. by J. L. Austin, Northwestern University Press, 1980.

Frege, G., 1892, "On Sense and Meaning," in *Philosophical Writings of Gottlob Frege*, ed. by P. Geach and M. Black, Basil Blackwell, 1952, pp. 56–78.

Friedman, M., 1983, *Foundations of Space–Time Theories*, Princeton University Press.

Galelei, G., 1629, *Dialogue Concerning the Two Chief World Systems*, University of California Press, 1967.

Gamut, L. T. F., 1991, *Logic, Language, and Meaning, Vol. 2, Intensional Logic and Logical Grammar*, University of Chicago Press.

Georgi, H., and S. L. Glashow, 1980, "Unified Theory of Elementary–Particle Forces," *Physics Today*, September, pp. 30–39.

Geroch, R., and G. T. Horowitz, 1979, "Global Structure of Spacetimes", in *The Large Scale Structure of Space–Time*, ed. by S. W. Hawking and G. F. R. Ellis, Cambridge University Press, pp. 212–93.

Glashow, S. L., 1985, "Topics in Elementary Particle Physics," in *Perspectives in Particles and Fields*, ed. by M. Lévy et al., Plenum, pp. 583–96.

Glymour, C., 1977, "The Epistemology of Geometry," *Nous* **11**, 227–57.

Goodman, N., 1954, *Fact, Fiction, and Forecast*, Harvard University Press, 4th ed., (1983).

Gottfried, K., 1966, *Quantum Mechanics*, Vol. 1., Benjamin.

Gottfried, K., and V. F. Weisskopf, 1984, *Concepts of Particle Physics*, Vol. 1., Oxford University Press.

Greenberger, D. M., M. A. Horne, A. Shimony, and A. Zeilinger, 1990, "Bell's Theorem without Inequalities," *Am. J. Phys.* **58**, 1131–1143.

Gudder, S. P., 1979, "A Survey of Axiomatic Quantum Mechanics," in Hooker (1979), Vol 2., pp. 323–63.

Guth, A. H., and P. J. Steinhardt, 1984, "The Inflationary Universe," *Scientific American* **255**, October, pp. 48–59.

Guyer, P., 1987, *Kant and the Claims of Knowledge*, Cambridge University Press.

Haag, R., 1992, *Local Quantum Theory*, Springer–Verlag.

Hacking, I., 1983, *Representing and Intervening*, Cambridge University Press.

Halmos, P. R., 1944, "The Foundation of Probability," *Am. Math. Monthly* **51**, 493–510.

Halmos, P. R., 1950, *Measure Theory*, Van Nostrand.

Halmos, P. R., 1951, *Introduction to Hilbert Space*, Chelsea Pub. Co., NY.

Hanson, N. R., 1958, *Patterns of Discovery*, Cambridge University Press.

Hardegree, G. M., 1980, "Micro-States in the Interpretation of Quantum Theory," in Asquith and Giere (1980), pp. 43–54.

Hawking, S. W., and G. F. R. Ellis, 1973, *The Large Scale Structure of Space–Time*, Cambridge University Press.

Healey, R. A., 1979, "Quantum Realism, Naivete is No Excuse," *Synthesis* **42**, 121–144.

Healey, R. A., 1989, *The Philosophy of Quantum Mechanics*, Cambridge University Press.

Heidegger, M., 1926, *Being and Time*, tr. by J. Macquarrie and E. Robinson, Harper & Row, 1962.

Heidegger, M., 1957, *Identity and Difference*, tr. by J. Stambaugh, Harper.

Heidegger, M., 1967, *What Is a Thing?*, tr. by W. B. Barton, Jr., Regnery/Gateway, South Bend, IN.

Heisenberg, W., 1958, "Quantum Theory and Its Interpretations," *Daedalus* **87**(*3*), 95–108.

Heisenberg, W., 1971, *Physics and Beyond*, Harper.

Heisenberg, W., 1983, *Tradition in Science*, Seabury Press, NY.

Held, A., ed., 1980, *General Relativity and Gravitation*, Vol. 1, Plenum.

Held, A., P. Heyde, and G. D. Kerlick, 1976, "General Relativity with Spin and Torsion, Foundations and Prospects," *Rev. Mod. Phys.* **48**, 393–416.

Helgason, S., 1978, *Differential Geometry, Lie Groups, and Symmetric Space*, Academic Press.

Helstrom, C. W., 1976, *Quantum Detection and Estimation Theory*, Academic Press.

Hiley, B. J., and F. D. Peat, eds., 1987, *Quantum Implications*, Routledge & Kegan Paul.

Hong, C. K., and L. Mandel, 1986, "Experimental Realization of a Localized One-Photon State," *Phys. Rev, Lett.* **56**, 58–60.

Hooker, C. A., ed., 1975, *The Logico-Algebraic Approach to Quantum Mechanics*, Reidel.

Hrbacek, K., and T. Jech, 1984, *Introduction to Set Theory*, 2nd ed., Marcel Dekker, NY.

Huang, K., 1987, *Statistical Mechanics*, 2nd ed., Wiley.

Hughes, R. I. G., 1989, *The Structure and Interpretations of Quantum Mechanics*, Harvard University Press.

Hume, D., 1739, *A Treatise of Human Nature*, Oxford University Press.

Hume, D., 1748, *Essays Concerning Human Understanding*.

Itzykson, C., and J. Zuber, 1980, *Quantum Field Theory*, McGraw–Hill.

Ivanenko, D., and G. Sardanashvily, 1983, "The Gauge Treatment of Gravity," *Phys. Repts.* **94**, 1–45.

Jammer, M., 1974, *The Philosophy of Quantum Mechanics*, Wiley.

Jauch, J. M., 1968, *Foundations of Quantum Mechanics*, Addison–Wesley.

Jauch, J. M., and B. Misra, 1965, "The Spectral Representation," *Helv. Phys. Acta* **38**, 30–52.

Jordan, T. F., 1969, *Linear Operators for Quantum Mechanics*, Wiley.

Kaempffer, F. A., 1965, *Concepts in Quantum Mechanics*, Academic Press.

Kant, I., 1781, *Critique of Pure Reason*, tr. by N. Kemp Smith, St. Martin's Press, NY, 1965.

Kant, I., 1788, *Critique of Practical Reason*, tr. by L. W. Beck, Macmillan.

Knuuttila, S. and J. Hintikka, eds., 1986, *The Logic of Being*, Reidel.

Kobayashi, S., and K. Nomizu, 1963, *Foundations of Differential Geometry*, Vol. 1, John Wiley & Sons.

Kochen, S., and E. Specker, 1967, "The Problem of Hidden Variables in Quantum Mechanics," *J. Math. and Mech.* **17**, 59–87.

Kolmogorov, A., 1933, *Foundations of the Theory of Probability*, Chelsea, 1951.

Kolmogorov, A., and S. V. Fomin, 1968, *Introduction to Real Analysis*, Dover.

Krantz, D. H., R. D. Luce, P. Suppes, and A. Tversky, 1971, *Foundations of Measurement*, Vol. 1, Academic Press.

Krauss, L. M., 1986, "Dark Matter in the Universe," *Scientific American*, December, pp. 58–68.

Kripke, S. A., 1972, *Naming and Necessity*, Harvard University Press.

Kripke, S. A., 1976, "Is There a Problem about Substitutional Quantification?" in *Truth and Meaning*, ed. by G. Evans and J. Mcdowell, Oxford University Press.

Krips, H., 1987, *The Metaphysics of Quantum Theory*, Oxford University Press.

Kuhn, T. S., 1962, *The Structure of Scientific Revolutions*, University of Chicago Press.

Kyburg, H. E., 1976, "Chance," *J. Phil. Logic* **5**, 355–93.

Lahti, P., and P. Mittelstaedt, eds., 1987, *Symposium on the Foundations of Modern Physics*, World Scientific, NJ.

Laplace, P. S., 1820, *A Philosophical Essay on Probabilities*, Wiley, 1917.

Leggett, A. J., 1986, "The Superposition Principle in Macroscopic Systems," in Penrose and Isham (1986), pp. 28–40.

Leibniz, G. W., 1969, *Philosophical Papers and Letters*, tr. and ed. by L. E. Loemker, 2nd edition, Reidel.

Levi, B. G., 1987, "Still More Squeezing of Optical Noise," *Phys. Today*, March, pp. 20–22.

Linsky, L., ed. 1971, *Reference and Modality*, Oxford University Press.

Locke, J., 1690, *Essays Concerning Human Understanding*, Dover.

Loéve, M., 1977, *Probability Theory*, 4th ed, Springer–Verlag.

London, F., and E. Bauer, 1939, "The Theory of Observation in Quantum Mechanics," in Wheeler and Zurek (1983), pp. 218–59.

Mackey, G. W., 1963, *The Mathematical Foundation of Quantum Mechanics*, Benjamin.

Mandl, F., and G. Shaw, 1984, *Quantum Field Theory*, Wiley.

Margenau, H., and L. Cohen, 1967, "Probability in Quantum Mechanics," in *Quantum Theory and Reality*, ed. by M. Bunge, Springer–Verlag, pp. 71–89.

Mates, B., 1986, *The Philosophy of Leibniz*, Oxford University Press.

Mayer, M. E., 1977, "Intoduction to the Fiber-Bundle Approach to Gauge Theories," in *Fiber Bundle Techniques in Gauge Theories*, Springer–Verlag.

Mermin, N. D., 1981, "Quantum Mysteries for Anyone," *J. Philosophy* **78**, 397–408.

Mermin, N. D., 1990, "Simple Unified Form for the Major No-Hidden-Variables Theorems," *Phys. Rev. Lett.* **65**, 3373–76.

Mill, J. S., 1843, *A System of Logic*.

Mises, R. von, 1957, *Probability, Statistics, and Truth*, Dover.

Mittelstaedt, P., A. Prieur, and R. Schieder, 1987, "Unsharp Joint Measurement of Complementary Observables in a Photon Split Beam Experiment," in Lahti and Mittelstaedt (1987), pp. 405–18.

Nagourney, W., J. Sandberg, and H. Dehmelt, 1986, "Shelved Optical Electron Amplifier, Observation of Quantum Jumps," *Phys. Rev. Lett.* **56**, 2797–99.

Namiki, M., ed., 1987, *Foundations of Quantum Mechanics in the Light of New Technologies*, Phys. Soc. Japan.

Ne'eman, Y., 1981, ed., *To Fulfill a Vision: Jerusalem Einstein Centennial Symposium on Gauge Theories and Unification of Physical Forces*, Addison–Wesley.

Nagel, T., 1986, *The View from Nowhere*, Oxford University Press.

Negele, J. W., and H. Orland, 1988, *Quantum Many-Particle Systems*, Addison–Wesley.

Neumann, J. von, 1932, *Mathematical Foundations of Quantum Mechanics*, tr. by R. T. Beyer, Princeton University Press, 1955.

Newton, I., 1686, *Mathematical Principles of Natural Philosophy*, tr. by A. Notte, Encyclopedia Britannica, Inc.

Newton, I., 1962, "On the Gravity and Equilibrium of Fluids," in *Unpublished Scientific Papers of Isaac Newton*, selected and translated by A. R. Hall and M. B. Hall, Cambridge University Press.

Norton, J., 1985, "Einstein's Principle of Equivalence," *Stud. Hist. Phil. Sci.* **16**, 203–46.

Nussbaum, M. C., 1986, *The Fragility of Goodness*, Cambridge University Press.

Okun, L. B., 1989, "The Concept of Mass," *Physics Today*, June 1989, pp. 31–36.

Omnes, R., 1992, "Consistent Interpretations of Quantum Mechanics," *Rev. Mod. Phys.* **64**, 339–82.

Owen, G. E. L., 1966, "Plato and Parmenides on the Timeless Present," *The Monist* **50**, 317–40.

Owen, G. E. L., 1979, "Particular and General," *Proc. Aristotelian Soc.* **79**, 1–21.

Pais, A., 1982, *Subtle is the Lord, The Science and Life of Albert Einstein*, Oxford University Press.

Pauli, W., 1921, *Theory of Relativity*, Dover, 1958.

Peierls, R., 1979, *Surprises in Theoretical Physics*, Princeton University Press.

Penrose, R., 1987, "Quantum Physics and Conscious Thought," in Hiley and Peat (1987), pp. 105–20.

Penrose, R., and C. J. Isham, ed., 1986, *Quantum Concepts in Space and Time*, Oxford University Press.

Petersen, A, 1963, "The Philosophy of Niels Bohr," *Bull. of Atomic Scientists* **19**(7), 8–14.

Pfeiffer, P. E., 1965, *Concepts of Probability Theory*, Dover, 1978.

Piaget, J., 1954, *The Construction of Reality in the Child*, Ballantine Books, NY.

Popper, K. R., 1956, *Quantum Theory and the Schism in Physics*, Rowman and Littlefield, Totowa, NJ.

Popper, K. R., 1957, "The Propensity Interpretation of Probability and the Quantum Theory," in *Observation and Interpretation in the Philosophy of Physics*, ed. by S. Körner, pp. 24–52, Dover.

Primas, H., 1981, *Chemistry, Quantum Mechanics, and Reductionism*, Springer–Verlag.

Przibram, K., ed., 1967, *Letters on Wave Mechanics*, Philosophical Library, NY.

Putnam, H., 1970, "Is Semantics Possible?," in *Language, Belief, and Metaphysics*, ed. by H. E. Kiefer and M. K. Munitz, pp. 50–63, State University of New York Press.

Putnam, H., 1975, "What Theories are Not," collected in *Mathematics, Matter, and Method*, Cambridge University Press, pp. 215–27.

Putnam, H., 1981, *Reason, Truth and History*, Cambridge University Press.

Quine, W. V. O., 1960, *Word and Object*, MIT Press.

Quine, W. V. O., 1962, *Ontological Relativity*, Columbia University Press.

Rauch, H., 1987, "Quantum Measurements in Neutron Interferometry," in Namiki (1987), pp. 3–17.

Raymond, P., 1989, *Field Theory, A Modern Primer*, 2nd ed., Addison–Wesley.

Redhead, M., 1987, *Incompleteness, Nonlocality, and Realism*, Oxford University Press.

Reichenbach, 1935, *The Theory of Probability*, University of California Press, 1949.

Riemann, G. F. B., 1854, "On the Hypotheses Which Lie at the Foundations of Geometry," in *A Source Book in Mathematics*, ed. by D. E. Smith, Dover, 1959, pp. 411–25.

Rorty, R. M., 1967, "Relations, Internal and External," in *The Encyclopedia of Philosophy*, ed. by P. Edwards, Macmillan, Vol. 7, pp. 125–33.

Russell, B., 1900, *A Critical Exposition of the Philosophy of Leibniz*, George Allen & Unwin, London.

Russell, B., 1903, *Principles of Mathematics*, Norton.

Russell, B., 1905, "On Denoting," in *Logic and Knowledge*, Capricorn Books, NY, 1956, pp. 39–56.

Russell, B., 1948, *Human Knowledge*, Touchstone Book.

Russell, B., 1950, *An Inquiry into Meaning and Truth*, Unwin Paperbacks, London.

Ryder, L. H., 1985, *Quantum Field Theory*, Cambridge University Press.

Sachs, R. G., 1987, *The Physics of Time Reversal*, University of Chicago Press.

Sakurai, J. J., 1985, *Modern Quantum Mechanics*, Benjamin.

Samuelson, P. A., 1947, *Foundations of Economic Analysis*, Harvard University Press, 1965.

Schilpp, P. A., ed., 1949, *Albert Einstein, Philosopher, Scientist*, Open Court.

Schrödinger, E., 1935, "The Present Situation in Quantum Mechanics," in Wheeler and Zurek (1983), pp. 152–67.

Schrödinger, E., 1961, *My View of the World*, tr. by C. Hastings, Ox Bow Press, 1983.

Schwartz, S. P., ed., 1977, *Naming, Necessity, and Natural Kinds*, Cornell University Press.

Schweber, S. S., 1961, *An Introduction to Relativistic Quantum Field Theory*, Row, Peterson and Co., Evanston, IL.

Scully, M. O., B. Englert, and H. Walther, 1991, "Quantum Optical Tests of Complementarity," *Nature* **351**, 111–16 (May).

Shimony, A., 1986, "Events and Processes in the Quantum World," in Penrose and Isham (1986), pp. 182–293.

Silk, J., 1980, *The Big Bang, The Creation and Evolution of the Universe*, Freeman.

Sklar, L., 1974, *Space, Time, and Spacetime*, University of California, Berkeley.

Sorabji, R., 1983, *Time, Creation, and the Continuum*, Cornell University Press.

Srinivas, M. D., 1975, "Foundations of a Quantum Probability Theory," *J. Math. Phys.* **16**, 1672.

Stachel, J., and D. Howard, ed., 1989, *Einstein and the History of General Relativity*, Birkhäuser, Boston.

Steenrod, N., 1951, *The Topology of Fiber Bundles*, Princeton University Press.

Stein, H., 1967, "Newtonian Space-time," *Texas Quarterly*, Vol. X, no. 3, pp. 174–200.

Stein, H., 1970, "On the Notion of Field in Newton, Maxwell and Beyond," in *Historical and Philosophical Perspectives of Science*, ed. by R. H. Stuewer, pp. 264–87, University of Minnisota Press.

Stein, H., 1972, "On the Conceptual Structure of Quantum Mechanics," in *Paradigms and Paradoxes*, ed. by R. G. Colodny, University of Pittsburgh Press, pp. 367–438.

Stoops, R., ed., 1961, *The Quantum Theory of Fields*, Interscience.

Strawson, P. F., 1950, "On Referring," *Mind* **54**, 320–44.

Strawson, P. F., 1959, *Individuals*, Methuen, London.

Strawson, P. F., 1966, *The Bounds of Sense*, Methuen, London.

Stueckelberg, E. C. G., 1960, "Quantum Theory in Real Hilbert Space," *Helv. Phys. Acta.* **33**, 727–52.

Tan, S. M., D. F. Walls, and M. J. Collett, 1991, "Nonlocality of a Single Photon," *Phys. Rev. Lett.* **66**, 252–55.

Taylor, R., 1955, "Spatial and Temporal Analogies and the Concept of Identity," reprinted in *Problems of Space and Time*, ed. by J. J. C. Smart, Collier Books, 1964, pp. 381–96.

Teller, P., 1987, "Space–Time as a Physical Quantity," in *Kelvin's Baltimore Lectures and Modern Theoretical Physics*, ed. by R. Kargon and P. Achinstein, MIT Press.

Thirring, W., 1978, *A Course in Mathematical Physics, I. Classical Dynamical Systems*, tr. by E. M. Harrell, Springer–Verlag.

't Hooft, G., 1980, "Gauge Theory of the Forces Between Elementary Particles," *Scientific American* **242**, 104–36 (June).

Trautman, A., 1973, "Theory of Gravitation," in *Physicist's Conception of Nature*, ed. by J. Mehra, Reidel, pp. 179–98.

Trautman, A., 1980a, "Yang-Mills Theory and Gravitation: A Comparison," in *Geometric Techniques in Gauge Theories*, ed. by R. Martini and E. M. Jager, Springer–Verlag, pp. 179–88.

Trautman, A., 1980b, "Fiber Bundles, Gauge Fields, and Gravitation," in Held (1980), pp. 287–305.

Wald, R. M., 1984, *General Relativity*, University of Chicago Press.

Wald, R. M., 1986, "Correlations and Causality in Quantum Field Theory," in Penrose and Isham (1986), pp. 193–301.

Wan, K., 1980, "Superselection Rules, Quantum Measurement, and Schrödinger Cat," *Canadian J. Phys.* **58**, 976–82.

Weatherford, R., 1982, *Philosophical Foundations of Probability Theory*, Routledge & Kegan Paul.

Weinberg, S., 1972, *Gravitation and Cosmology*, Wiley.

Weinberg, S., 1977, "The Search for Unity, Notes for a History of Quantum Field Theory," *Daedalus* **106**, No. 4, pp. 17–35.

Weinberg, S., 1980 "Conceptual Foundations of the Unified Theory of Weak and Electromagnetic Interactions," *Rev. Mod. Phys.* **52**, 515–523.

Weyl, H., 1921, *Space, Time, Matter*, Dover, 1952.

Weyl, H., 1949, *Philosophy of Mathematics and Natural Science*, Princeton University Press.

Weyl, H., 1952, *Symmetry*, Princeton University Press.

Wheeler, A., and W. H. Zurek, ed., 1983, *Quantum Theory and Measurement*, Princeton University Press.

Wightman, A. S., 1976, "Hilbert's Sixth Problem, Mathematical Treatment of the Axioms of Physics," in *Proc. of the Sym. in Pure Math.* **28**, pp. 147–240.

Wigner, E. P., 1939, "On Unitary Representations of the Inhomogeneous Lorentz Group," *Ann. Math.* **40**, 149.

Wigner, E., 1961, "Remarks on the Mind–Body Problem," in Wheeler and Zurek (1983), pp. 168–81.

Wigner, E., 1967, *Symmetries and Reflections*, MIT Press.

Wigner, E., 1971, "The Subject of Our Discussions," in *Foundations of Quantum Mechanics*, pp. 1–19, ed. by B. d'Espagnat, Academic Press.

Wigner, E., 1983, "Interpretation of Quantum Mechanics," in Wheeler and Zurek (1983), pp. 260–314.

Wilding, J. M., 1982, *Perception, From Sense to Object*, Hutchinson, London.

Wittgenstein, L., 1918, *Tractatus Logico-Philosophicus*, tr. by C. K. Ogden, Routledge & Kegan Paul, 1922.

Wittgenstein, L., 1953, *Philosophical Investigations*, tr. by G. E. M. Anscombe, Macmillan.

Wu, T. T., and C. N. Yang, 1975, "Concept of Nonintegrable Phase Factors and Global Formulation of Gauge Fields," *Phys. Rev. D* **12**, 3845–57.

Yamamoto, Y., and H. A. Haus, 1986, "Preparation, Measurement, and Information Capacity of Optical Quantum States," *Rev. Mod. Phys.* **58**, 1001–20.

Yang, C. N., 1981, "Geometry and Physics," in Ne'eman (1981), pp. 3–11.

Yang, C. N., and R. L. Mills, 1954, "Conservation of Isotopic Spin and Isotopic Gauge Invariance," Phys. Rev. **96**, 191–95.

Zurek, W. H., 1991, "Decoherence and the Transition from Quantum to Classical," *Phys. Today*, October, pp. 36–44.

INDEX

a priori, 62, 93, 156, 205, 243
A-amplitude, 69–70, 85
Abelian (commutative), 44, 59
absolute rotation, 140
absoluteness, 135, 137, 255
action-at-a-distance, 24, 119, 146, 193
activity of the mind, 10, 103, 195, 205
actual state, 67, 132
affine structure, 28–9
Aharanov-Bohm effect, 55, 186
algebraic geometry, 26
algebraic structure, 26
amplitude, 18, 68–71, 74, 76–8, 86, 95, 111–2
analytic and synthetic, 243
Anderson, James, 39, 147
antiparticle, 50, 227
Aquinas, Thomas, 139
argument from illusion, 90–1
Aristotle, 120, 127, 237
  on being *qua* being, 123
  on quantity, 198, 261
  on space, 139, 140
  on time, 173
asymptotic freedom, 45
Atomist, 120, 139
Augustine, St, 6
Ayer, A. J. 124

Ballentine, Leslie, 23
basis, 20, 68, 161–2
Bayesian, 197, 209, 211, 261
being *qua* being, 123, 135
Belinfante, F. J. 22
Bell experiments, 24, 87, 165, 193
Bell, John, 24, 104
Bell's inequality, 24
Beltrametti, Enrico, 22
Berkeley, George, 9, 89
Bernoulli, James, 209
Big Bang, 4, 174, 223–4
Birkhoff, Garrett, 114, 230
Bohm, David, 231, 236, 240
Bohr, Niels, 73, 108, 236, 239, 242, 244
  on deep truth, 81
  on finality of quantum mechanics, 105

on inseparability of object and instrument, 89, 105, 108, 237, 245
  on quantum property, 63, 86
  on reality, 229
Borel line, 200–1, 204–5
Borel set, 201, 205–6, 211, 240–1
Born, Max, 16, 229, 236
Born postulate, 20–21, 79, 83–5, 117, 165
boson, 45, 53, 159, 163, 246
Broad, C. D. 137, 247
Broglie, Louis de, 236
Busch, Paul, 22, 77, 88

canonical quantization, 50
Carnap, Rudolph, 238, 256
Cartan, Elie, 214
Cartesian space, 30, 72
Cartesian subject, 106, 112
Cassinelli, Gianni, 22
categorical framework, 12–3, 91, 93–4, 106, 109, 140
  of conventions, 9, 98–9
  expansion of, 10–1, 91, 106, 129, 168, 183
  impoverished, 14, 98
  of objects, 8, 10, 95, 97, 99–101
  phenomenal, 9, 97–8
  of things, 8, 97–8
  versus substantive conceptual scheme, 94
causal mechanism, 185
causal relation, 168, 189–90
causal theory of reference, 98, 156–7
causality, 12, 175, 182, 259–60
change, 170–72
charge conjugation, 38
Clarke, Samuel, 248
classical mechanics, 60, 73, 95, 104, 119, 161, 186, 188, 221, 240. *See also* Newtonian mechanics
cloud chamber, 22
cluster of qualities, 124–5, 133, 159
co-reference, 110–1, 153–4
Cohn, L, 115
coincidence, 142–3, 254
collapse of the wavefunction, 22, 82
color charge, 44, 234

274   **Index**

commutation relation, 20, 49, 51, 53, 73, 246
comparison of events, 132, 177–80, 187–8
comparison of vectors, 29, 175–6
complex number, 73, 231, 238
complex quantity, 68–9, 72–4
compossible, 155, 247
connection, 29, 41, 147, 168, 183, 187–8, 214, 218–22
consciousness, 105–6, 108, 112–3
conservation law, 35, 49, 170
constant predication, 166–7, 177–80
content of experience and judgment, 109–13, 209
continuity, two senses of, 136, 252–3
continuous function, 253–4
continuum, 136–7, 231, 252
convention, 9, 98–9
   global, 167, 174, 176–7, 183
convention of predication, 14
conventionism, 10, 94, 101, 146
coordinate-free, 33, 86, 95, 97, 99–100, 132–3, 152
coordinate representation, 33, 67, 86, 95
coordinate system, 33, 41, 67–8, 92, 95, 152–4, 167
   distinct from tetrad, 41
coordinate transformation, 33, 153–4, 220–1
Copenhagen interpretation, 4, 63, 89, 105, 108, 236–7, 242, 244
Corbett, J. V. 77
counterfactual assertion, 21
coupling term, 57, 59, 185
CPT theorem, 38
curvature, 145, 147, 221

Davidson, Donald, 247
Davies, E. B., 89, 230
de Finetti, Bruno, 210, 261
decoherence effect, 82, 239
defective mode of knowledge, 13, 93–4, 100–1, 156
definite description, 65–6, 95, 127, 158, 166
definite value principle, 63, 70
demonstrative, 61, 102, 111, 152, 156
Descartes, René, 92, 120, 200
determinism, 15, 206, 212, 229
Dickens, Charles, 106
Dieks, Dennis, 165
differentiable manifold, 30, 39, 65, 67, 95, 132, 148–9, 169, 214, 216, 232
   defined, 27–8
Dirac, Paul Arien Maurice, 3, 16, 231, 253
   on observable, 72, 74
   on quantum mechanics, 236
   on transformation theory, 61, 91, 101

Dirac equation, 49–50
discrete individuals, 120
dispositional attribute, 63, 212
distance, 27, 30–1, 143–4, 254
distribution, 203–4
Dummett, Michael, 80
dynamical coupling, 184
dynamical system, 59–60
dynamical variable, 16, 35
   not observable, 53, 72
Dyson, Freeman, 43

Earman, John, 146, 148–151, 251
Eddington, Arthur, 144, 254
Edwards, C. M. 230
egocentric particular, 103
eigenstate, 20, 68, 78, 86, 88, 116
eigenvalue, 19, 20, 64, 69–70, 73, 86, 116
   not quantum property, 78–80
   observed, 81, 104–5, 107, 111, 117, 165
Einstein, Albert, 3, 24, 25, 60, 92, 101, 236
   on equivalence principle, 42
   on gauge invariance, 43
   on general relativity, 26, 39
   on geometry, 144, 254
   on observability, 104, 244
   on quantum interpretation, 5, 89, 105
   on quantum mechanics, 244
   on reality, 11, 195, 229, 231, 241–2, 260
   on relativity, 147
   on space-time, 119, 136, 150
Einstein-Podolsky-Rosen paradox, 165
Eleatics, 120
empirical access, 96–7
empirical ramification, 75
empiricism, 90–1, 176
empty space, 135–6, 139, 146, 189. *See also* vacuum
endurance, 172–5
enduring object, 175, 179–80, 189
ensemble, 21
entity, 122–4, 134
epistemology, 115
equivalence principle, 32, 40–41
ether, 43, 148–50
event, 122–3, 129–32, 152–4, 159, 167, 169, 187–9, 216
event (in probability calculus), 198, 202, 207
exclusion, 139–40
exist, 135
existence change, 173–4
expansion, 142
expectation value, 20–1
experimental metaphysics, 24
extensional, 110, 154
extensive and intensive variables, 198